Innovative Pedagogical Practices for Higher Education 4.0

Innovative and creative teaching methods tailored to meet the demands of the current era of Industrial Revolution 4.0 are becoming increasingly prevalent in higher education institutions. Educators must cultivate and implement these strategies in their classrooms to incorporate the essential skills of Education 4.0 and digitalization in education. This book aims to explore and showcase various aspects of innovative pedagogies for Higher Education 4.0, drawing from best practices worldwide to provide a comprehensive understanding of this evolving field.

Innovative Pedagogical Practices for Higher Education 4.0 highlights the cutting-edge technologies essential for Education 4.0 and equips educators with the skills required for 21st-century teaching. Through international case studies, this book explores the integration of educational technology, illuminating the obstacles encountered by educational institutions aiming to embrace Education 4.0 in higher education settings. Furthermore, it presents inventive tools and strategies for harnessing technology in teaching, offering a new outlook on innovative pedagogies.

This book meets the need for solutions that tackle the challenges of Education 4.0 and demonstrates how these challenges can be overcome through innovative pedagogies for student learning in higher education institutions. It is a valuable resource for educators, specialists, academic institutions, and policymakers seeking practical solutions in the ever-evolving landscape of education.

Innovative Pedagogical Practices for Higher Education 4.0

Solutions and Demands of the Modern Classroom

Edited by
Muhammad Mujtaba Asad,
Prathamesh Padmakar Churi,
Fahad Sherwani, and Razali Bin Hassan

CRC Press
Taylor & Francis Group
Boca Raton London New York

CRC Press is an imprint of the
Taylor & Francis Group, an **informa** business

First edition published 2025
by CRC Press
2385 NW Executive Center Drive, Suite 320, Boca Raton FL 33431

and by CRC Press
4 Park Square, Milton Park, Abingdon, Oxon, OX14 4RN

CRC Press is an imprint of Taylor & Francis Group, LLC

Library of Congress Cataloging-in-Publication Data
Names: Asad, Muhammad Mujtaba, editor. | Churi, Prathamesh, editor. |
Sherwani, Fahad, editor. | Hassan, Razali Bin, editor.
Title: Innovative pedagogical practices for higher education 4.0 :
solutions and demands of the modern classroom / edited by Muhammad
Mujtaba Asad, Prathamesh Padmakar Churi, Fahad Sherwani and Razali Bin Hassan.
Description: First edition. | Boca Raton, FL : CRC Press, 2025. |
Includes bibliographical references and index.
Identifiers: LCCN 2024013213 (print) | LCCN 2024013214 (ebook) |
ISBN 9781032508597 (hbk) | ISBN 9781032510125 (pbk) | ISBN 9781003400691 (ebk)
Subjects: LCSH: College teaching—Methodology. | Education, Higher—Effect
of technological innovations on. | Blended learning. | College
teachers—Training of. | College teachers—In-service training.
Classification: LCC LB2331 .I55145 2025 (print) | LCC LB2331 (ebook) |
DDC 378.1/25—dc23/eng/20240703
LC record available at https://lccn.loc.gov/2024013213
LC ebook record available at https://lccn.loc.gov/2024013214

ISBN: 978-1-032-50859-7 (hbk)
ISBN: 978-1-032-51012-5 (pbk)
ISBN: 978-1-003-40069-1 (ebk)

DOI: 10.1201/9781003400691

Typeset in Times
by codeMantra

This book is dedicated to Professor Nisar Ahmed Siddiqui (Late) and to those who embrace change, foster creativity, and work tirelessly to meet the evolving demands of education.

Contents

Preface

In the ever-evolving landscape of education, the buzzwords of our time are undoubtedly innovation and creativity. As we navigate the complexities of the Fourth Industrial Revolution (IR 4.0), educators find themselves at the forefront, tasked with crafting lessons that are not just informative but engaging, innovative, and collaborative. This book, *Innovative Pedagogical Practices for Higher Education 4.0: Solutions and Demands of the Modern Classroom*, emerges as a guiding light in the realm of teaching and learning for this contemporary era.

The discourse surrounding education in the era of IR 4.0 underscores the imperative for educators to embrace pedagogical approaches that align with the dynamic needs and demands of Education 4.0. This collection of chapters delves into the essence of innovative and creative teaching methodologies, positioning them as essential components for higher education institutions.

As we embark on this exploration, we recognize the pivotal role teachers play in not only developing but adapting teaching strategies that seamlessly integrate the skills demanded by Education 4.0 and the ongoing digitalization of education. The book aims to illuminate all facets of innovative pedagogies for Higher Education 4.0, drawing from the best practices observed across the globe.

Amidst the plethora of books available on general pedagogies for 21st-century teachers, this book stands out by focusing specifically on the innovative pedagogies vital for higher education. It endeavors to comprehensively cover all aspects, addressing the challenges faced by educators and offering innovative solutions tailored to meet the requirements of modern classrooms.

This book is not just an addition to the existing literature; it is a pioneering endeavor that addresses the challenges and outlines how these challenges can be effectively navigated through innovative pedagogies. The chapters herein serve as a valuable resource for educators seeking practical insights into enhancing students' learning experiences in higher education institutions, aligning seamlessly with the needs of IR 4.0.

In essence, *Innovative Pedagogical Practices for Higher Education 4.0* is not just a book; it is a collaborative effort to shape the future of education, one innovative pedagogy at a time.

Muhammad Mujtaba Asad

Sukkur IBA University
Sukkur, Pakistan
and
Honorary Adjunct Faculty
Sohar University
Sohar, Oman

Prathamesh Churi

Senior Member, IEEE and Independent Researcher

Fahad Sherwani

National University of Computer and Emerging Sciences
Karachi, Pakistan

Razali Bin Hassan

Universiti Tun Hussein Onn Malaysia
Parit Raja, Batu Pahat

Editors

Muhammad Mujtaba Asad is an Assistant Professor and Lead Researcher of Educational Technologies and TVET Research at Sukkur IBA University, Sukkur, Pakistan, and Honorary Adjunct Faculty at Sohar University, Sohar, Oman. Dr Mujtaba has national and international experience of more than 10 years in the industrial and educational sectors in educational technologies, work- and product-based education, and technical and vocational education. He has published more than 85 research articles (Web of Science and Scopus Indexed) in international journals and conferences worldwide. Dr Mujtaba has authored three books with Taylor & Francis and CS Publishers. He has received three gold, one silver, and one bronze medal in international research and development completions, with seven best research awards in Malaysia, Korea, Saudi Arabia, Pakistan, and UAE on commercialized projects. Moreover, he is an active member of seven national and international research grants. Dr Mujtaba is an internationally certified educational technologist (Malaysian Board of Technologist) and holds certifications and qualifications in industrial safety and health education from NIOSH and the OSH Department of Ontario, Canada. He is an Editorial Board Member to 12 reputed and indexed international journals. Dr Mujtaba holds several international copyrights and has filed one product patent under the title of HAZFO Expert 1.0 for Drilling Operation. He was awarded an honorary lifetime International Ambassadorship by University Tun Hussein Onn Malaysia based on his academic achievements and community services. Dr Mujtaba also was awarded the title of approved PhD and MPhil/MS supervisor by the Higher Education Commission of Pakistan.

Prathamesh Padmakar Churi (SMIEEE, LMCSI, LMISTE) is a Senior member, IEEE and Independent Education Technology Researcher. He was a faculty member in the Computer Engineering Department in the School of Technology, Management, and Engineering, NMIMS University, Mumbai, India. He is an Associate Editor of *International Journal of Advances in Intelligent Informatics*, *International Journal of Innovative Teaching and Learning in Higher Education*, and *International Journal of Security and Privacy*. He is involved in the peer-review process of reputed IEEE and Springer journals, such as *IEEE Transactions on Education*, *Springer Education and Information Technologies*, and 20 other journals. He has published 80+ research papers in national/international conferences and journals (Scopus and SCI Indexed), 6 Patents (including 3 Australian patents) and 5 international books (CRC Press/Taylor & Francis publications) in the field of data privacy and education technology. He has been a keynote speaker, chair, and convener at international conferences, including flagship conferences such as IEEE TALE 2017–2020, Springer ICACDS, etc. He received the Best

Young Researcher Award by GISR Foundation for his research contribution in the field of data privacy and security and education technology. He has also received appreciation and best faculty awards from NMIMS University. He received a Best Reviewer Award at IEEE TALE 2020, Springer ETRD, JCHE, and Elsevier IJIM Data Insights. He is an active leader, coach, mentor, and volunteer in many nonprofit organizations. He is a Board of Study member at many universities for curriculum development and educational transformations.

Fahad Sherwani is an Assistant Professor at the National University of Computer and Emerging Sciences, Karachi, Pakistan. He has published several research articles as well as three books with Taylor & Francis and CS Publishers. Dr Sherwani was awarded the title of approved PhD and MPhil/MS supervisor by the Higher Education Commission of Pakistan.

Razali Bin Hassan is the Director of the Malaysia Research Institute for Vocational Education and Training (MyRIVET). He earned a PhD in education at the University of Warwick, UK, an MS in TVET at the University Putra Malaysia, a BE of technology with education (electrical technology) at the Malaysia University of Technology, and a Certificate of Teaching at the Technical Teachers Training College. Dr Hassan has experience as a teacher, a lecturer, a consultant, a researcher, and a motivator in various ASEAN, European, Middle Eastern, and African countries. He has published several research articles and books in educational technologies and TVET.

Contributors

Muhammad Kamran Abbasi
Department of Distance, Continuing and Computer Education
Faculty of Education
University of Sindh
Elsa Kazi Campus
Hyderabad, Pakistan

Hasnida Ab-Samat
School of Mechanical Engineering
Universiti Sains Malaysia
Penang, Malaysia

H. M. Affendi
Faculty of Build Environment
Universiti Kebangsaan Malaysia
Bangi, Malaysia

A. Ahmad
School of Education
Faculty of Social Sciences and Humanities
Universiti Teknologi Malaysia
Skudai, Malaysia

Noor Ahmed
Department of Education
Sukkur IBA University
Sukkur, Pakistan

Ahmad Al Yakin
Pancasila Civic education department
Universitas Al Asyariah Mandar Sulawesi Barat Indonesia
Polewali, Indonesia

Muhammad Mooneeb Ali
Higher Education Department Punjab
Punjab, Pakistan

R.H. Raja Mohd Ali
Department of Audit and Information Systems
Universiti Utara Malaysia
Changlun, Malaysia

Norah Almusharraf
Linguistics and Translation Department
Prince Sultan University
Riyadh, Saudi Arabia

Unaeza Alvi
Department of Education
Sukkur IBA University
Sukkur, Pakistan

N. Amrous
Department of Computer Science
Mohammadia School of Engineers (EMI)
Mohammed V University
Rabat, Morocco

Muhammad Mujataba Asad
Department of Education
Sukkur IBA University
Sukkur, Pakistan

Anis Masturina Azhan
School of Mechanical Engineering
Universiti Sains Malaysia
Penang, Malaysia

Noor-ul-Ain Bhellar
Department of Management Sciences and Technology
Begum Nusrat Bhutto Women University, Sukkur
Rohri, Pakistan

Muskan Butt
Department of Education
Begum Nusrat Bhutto Women University, Sukkur
Rohri, Pakistan

Rameen Butt
School of Education
Lahore University of Management Sciences
Lahore, Pakistan

Wong Ling Chai
Faculty of Hospitality and Tourism Management
Asia Pacific University of Technology and Innovation
Kuala Lumpur, Malaysia

Hala Dalbani
Linguistics and Translation Department
Prince Sultan University
Riyadh, Saudi Arabia

Nadira Dayo
Faculty of Psychology and Educational Sciences
KU Leuven
Leuven, Belgium

N. El Faddouli
Department of Computer Science
Mohammadia School of Engineers (EMI)
Mohammed V University
Rabat, Morocco

K. Erarslan
The Faculty of Engineering
Kütahya Dumlupınar University
Kütahya, Türkiye

G. Faouzi
Department of Computer Science
Mohammadia School of Engineers (EMI)
Mohammed V University
Rabat, Morocco

Raja Nor Safinas Raja Harun
English Language and Literature Department
Faculty of Languages and Communication
Universiti Pendidikan Sultan Idris (UPSI)
Tanjung Malim, Malaysia

Pham Quang Huy
Training Department
School of Accounting
University of Economics Ho Chi Minh City (UEH)
Ho Chi Minh City, Vietnam

Muhammad Imran
Linguistics and Translation Department
Prince Sultan University
Riyadh, Saudi Arabia

D. Kalay
The School of Foreign Languages
Kütahya Dumlupınar University
Kütahya, Türkiye

Devinder Kaur
School of Marketing and Management
Asia Pacific University of Technology and Innovation
Kuala Lumpur, Malaysia

M. Khabouze
Department of Computer Science
Faculty of Economics and Management,
Hassan 1st University
Settat, Morocco

Nurul Salwana Khair
Faculty of Education, Languages, Psychology and Music
SEGi University
Petaling Jaya, Malaysia

Sana Khalid
School of Education
Lahore University of Management Sciences
Lahore, Pakistan

Sahib Khatoon
Center of English Language and Linguistics
Mehran University of Engineering and Technology
Jamshoro, Pakistan

Tee Poh Kiong
School of Marketing and Management
Asia Pacific University of Technology and Innovation
Kuala Lumpur, Malaysia

J. Koh
School of Engineering and Technology (SET)
Central Queensland University
Australia

Song Bee Lian
School of Marketing and Management
Asia Pacific University of Technology and Innovation
Kuala Lumpur, Malaysia

N. Mafarja
Department of Professional Education and Postgraduate
Faculty of Technical and Vocational Education
Universiti Tun Hussein Onn Malaysia
Batu Pahat, Malaysia

Fareed Hussain Mangi
Energy Systems Engineering Department
Sukkur IBA University
Sukkur, Pakistan

M.M. Mohamad
Department of Professional Education and Postgraduate
Faculty of Technical and Vocational Education
Universiti Tun Hussein Onn Malaysia
Batu Pahat, Malaysia

Nur Amalina Muhammad
School of Mechanical Engineering
Universiti Sains Malaysia
Penang, Malaysia

Muthmainnah Muthmainnah
Indonesian Language Department
Universitas Al Asyariah Mandar Sulawesi Barat Indonesia
Polewali, Indonesia

Habibullah Pathan
Sohar University
Sohar, Oman

Anna-Marie Pelser
Faculty of Economic and Financial Sciences
North-West University
Potchefstroom, South Africa

Vu Kien Phuc
School of Accounting
University of Economics Ho Chi Minh City (UEH)
Ho Chi Minh City, Vietnam

Aneela Pirzado
Department of Education
Sukkur IBA University
Sukkur, Pakistan

Zafarullah Sahito
Department of Education
Sukkur IBA University
Sukkur, Pakistan

A. Sanjeev
Faculty of Commerce and Management
Shree Guru Gobind Singh Tricentenary University
Gurugram, India

R. Sharma
Amity College of Commerce and Finance
Amity College of Commerce
Amity University
Gurugram, India

Amreet Kaur Jageer Singh
English Language Unit
Centre for Language Competencies
Faculty of Languages and Communication
Universiti Pendidikan Sultan Idris (UPSI)
Tanjung Malim, Malaysia

Sukanya Singh
Department of Education
Central University of Rajasthan
Ajmer, India

Sheiladevi Sukumaran
Faculty of Education, Languages, Psychology and Music
SEGi University
Petaling Jaya, Malaysia

Muhammad Usman Thaheem
Department of English
National University of Modern Languages
Hyderabad, Pakistan

Liew Chee Yoong
Department of Accounting and Finance
UCSI University
Kuala Lumpur, Malaysia

Jacqueline Żammit
Department of Languages and Humanities in Education
University of Malta
Msida, Malta

Mubeen Zehra
Department of Education
Al-Falah University
Faridabad, Haryana, India

Muhamad Nazrin Muhd Zukri
School of Mechanical Engineering
Universiti Sains Malaysia
Penang, Malaysia

1 The Achievement of Quality 4.0 for Green Entrepreneurship Education in the Lens of Education 4.0

Digital Education Ecosystem as the Most Favorable Environment for Digital Pedagogy Implementation

Pham Quang Huy and Vu Kien Phuc

1.1 INTRODUCTION

Building on the perspectives of Yi (2021), creating enterprises with the intention of resolving environmental issues and advancing sustainability is referred to as "green entrepreneurship." Green entrepreneurship is a flourishingly common method of performing business that puts an emphasis on environmental sustainability as more and more organizations become aware of the requirement to lessen their environmental impact and contribute to the planet (Ali et al., 2023). In order to reach a balance between economic growth, social advancement, and environmental sustainability, nations all over the world are pushing for the development of green entrepreneurship initiatives (Anghel & Anghel, 2022). The foundation of green entrepreneurship is green education, which is seen as the area where capabilities for long-term development are formed and developed (Makhloufi et al., 2021). According to Cai et al. (2022), "green entrepreneurship education" refers to the development of businesspeople who possess both environmental and economic ideals. Due to their ability to start firms and their higher level of education, students in this demographic play a vital role in the development of green entrepreneurship projects (Alwakid et al., 2021). Higher education institutions (HEIs) play a significant role in promoting green entrepreneurship by combining the idea of environmental values with academics.

DOI: 10.1201/9781003400691-1

This helps students create positive perceptions of green companies. Notably, the mobile and digital connection of the Fourth Industrial Revolution (Industry 4.0) is significantly accelerating the green shift. Industry 4.0 gives rise to Quality 4.0, which focuses on integrating quality management principles with newly developed Industry 4.0 capabilities to enable businesses achieve excellence (Saihi et al., 2023). Industry 4.0's requirement for a highly skilled workforce gave rise to the concept of Education 4.0. To prepare students and the next generation of learners for the next industrial revolution, which necessitates new technologies and skill sets, Education 4.0, a new paradigm in teaching and learning, is intended. Simply defined, Quality 4.0 in green entrepreneurship education (QGEE) involves upholding quality in the modern era of Education 4.0. Understanding Quality 4.0, which exemplifies the role of the quality function in the Education 4.0 scenario, is important in order to properly comprehend the underlying principles of the new standard and apply interventions to adopt it. The requirement of current study programs in green entrepreneurship education determines how current these concerns are. The quintuple helix model is accommodated by green entrepreneurship, according to Romanowski and Gnusowski (2019), with the help of intermediaries for sustainable innovation. Due to these perspectives, green entrepreneurship entails coordination with all relevant parties. To secure the skilled workforce needed for Industry 4.0, it is critical to encourage data and information flow among the players in the ecosystem of Education 4.0 and to facilitate technology and knowledge transfer. Thus, it is essential to guarantee interoperability across the board in the Industry 4.0 ecosystem (Chituc, 2022). In order to orchestrate and facilitate the entire educational process for students, teachers, and administrators, the digital transformation of education goes beyond the simple development of online courses. This is done by building comprehensive and holistic digital education ecosystems (DEEs). In light of the society's transition to a new technological setup, the networkization of the global educational space, and the active implementation of green initiatives, the concept of a DEE has been established as a contemporary approach of setting up the educational environment. Despite the importance of Education 4.0, the domains of DEEs and Education 4.0 are developing without a clear understanding of how DEEs assist in accomplishing the objectives of Quality 4.0 via the lens of Education 4.0.

Remarkably, Chituc (2022) describes a framework for implementing revolutionary teaching–learning systems for the Industry 4.0 vision, which is based on three dimensions: technological, pedagogical, and organizational. Aiming to promote the growth of active learners through the construction and restructuring of cognitive structures as well as enhanced connection, digital pedagogy (DP) also transforms learning. The usage of technology in the classroom may have an impact on the teacher's pedagogical approaches. DP is thought of as a technique to integrate digital resources to improve the learning process. It can provide not only the core competencies required for business operations but also horizontal competencies that are environmentally conscious and aid in the development of a green entrepreneurial mindset in business students. Taken together, this prompts the need to reconsider and develop a paradigm for how DEE shows an impact on the attainment of QGEE that is statistically consistent and trustworthy. The examination of the possible role of DP in the relationship between DEE and QGEE served as the other impetus for this research and highlighted opportunities for theoretical and practical contributions, starting from these thoughts on the underutilization of DP. Also, these research gaps spark the following fascinating research questions:

RQ1. *What impact does the DEE have on QGEE?*
RQ2. *Does DP act as a mediator in the interconnection between the DEE and QGEE?*

The present study provides contributions for the academic and practitioner communities by filling several gaps based on the examination of the overall findings and key insights. According to Burzyńska et al. (2018), green entrepreneurship is a relatively young and recent phenomenon. Because sustainable development is essential, green entrepreneurship is growing in popularity. As a result, education programs in this field have been steadily adjusting to the transition brought on by circularity and sustainability. Researchers are still striving to identify important and pertinent elements of quality management with regard to Industry 4.0, and empirical research on the subject is still in its early stages. As a result, the research's culmination is to expand the frontier of knowledge for quality management and measurement in green entrepreneurial education through the lens of Education 4.0. Even though Education 4.0 is a young field, its significance is well recognized (Da Motta Reis et al., 2020). Without a clear vision for how to satisfy the requirements for the specified competencies and skills of Industry 4.0, research and development effort focusing on Education 4.0 and DEEs is dispersed (Chituc, 2022). According to the best knowledge of the researchers, this may be the first academic paper to present a new and important perspective on the highly unique impacts of DEE on QGEE. Adding to this, the novelty of this research is to identify and evaluate the function of DP that leads to Quality 4.0 because the adoption of DP is thought to be both a social and a pedagogical necessity.

In the practical significance, drawing on the theoretical foundation for QGEE in this study, HEIs can gain a detailed understanding of what Quality 4.0 means for their organizations and how to implement it successfully in order to improve and enhance sustainability, with a focus on education for sustainable development. Students and society at large will gain value from the simultaneous development of a Quality 4.0 community and provision of ongoing assistance through critical digital pedagogy implementation (CDPI). HEIs are at the center of education in the age of Education 4.0, which has rekindled efforts to revolutionize education. As a result, they must deal with previously unimaginable changes and difficulties in order to prepare students to meet shifting educational needs. This study offers managers, administrators, and educators of HEIs with management and operational pointers as well as a plan for implementing Quality 4.0 with an overall focus on green entrepreneurship, its pedagogical strategy, and DEE. The remaining sections are configured as follows. The main goal of Section 1.2 is to lay forth the theoretical underpinnings of the suggested conceptual framework graphic, which is followed by the subsequent formulation of hypotheses in Section 1.3. Section 1.4 provides the research methodology. The statistical results are reported in Section 1.5. In Section 1.6, there are conclusions with management implications and suggestions for potential future research.

1.2 LITERATURE REVIEW AND THEORETICAL BASIS

1.2.1 Theoretical Lenses Adoption

Contingency theory. In this work, the theoretical lenses used to approach the empirical environment were based on contingency theory. Because it challenges the tenet of classic management theory—that there is just one optimal way to do things—it has gained more and more traction (Csaszar & Ostler, 2020). Contingency theory is one of the top theories offering a strong platform for study on organizational design (Donaldson, 2007). This type of theory is developed by researchers from Ohio State University in 1950, and it is used in studies that are centered on organizational issues (Sauser et al., 2009) because of its deeply established ideas in organizational literature. Other fields, such as information systems and information technology management literature, have made extensive use of contingency theory (Abedin, 2022). The fundamental objective of the organizational design is to account for three important contingencies: unpredictability, strategy, and size. Although it was first used to examine systems design and implementation in the literature on information systems, it was subsequently utilized to investigate how the environment and other factors affect task performance fitness (Reinking, 2011). According to Wang (2010), organizations in this condition must act more resolutely in order to adjust to the difficult circumstances brought on by the ambiguity of unit responsibilities. The contingency theory states that different circumstances call for different approaches. One significant contribution is the identification of variables that have a significant impact on an organization's overall design and the forecasting of the implications resulting from variations in these variables (De Morais & Barbieri, 2022). According to Munyon et al. (2019), the organization's productivity is significantly influenced by the compatibility or alignment of the many sources of environmental instability, technology, organizational scale, organizational structure quirks, and organizational information systems. According to contingency theory, in order to effectively build and use information systems, users and other stakeholders must be engaged and contingent elements must be taken into account. This means that there is no one ideal way to use an information system in every context (Shao et al., 2016). As technologies are seen to contribute significantly to the variations in these organizational features, formal structures are often consolidated or well-accorded with the usage of diverse technologies according to the contingency aspect.

1.2.2 Conceptual Respects

Quality 4.0 for green entrepreneurship education in the lens of Education 4.0. Academics have focused much more attention on the idea of "green entrepreneurship" because they want "going green" to be recognized as a strategy to improve environmental sustainability. This idea combines the phrases "environment" and "entrepreneurship," which were first used in the 1990s. Given this perspective, academics (Yi, 2021; Demirel et al., 2017) advocate that the solution to ecological and social issues is green entrepreneurship. According to Cai et al. (2022), "green entrepreneurship education" refers to the development of businesspeople who have an awareness of both the environment and economic principles. Green entrepreneurship education entails the strategic training and enlightenment of people, communities, and institutions on the contribution of green entrepreneurship to sustainable development. According to Margulieux et al. (2016), educational institutions are restructuring in order to provide the workforce with the skills needed in both businesses and society. The term was first used to describe a continuum, which included mechanization, water, and steam power in Industry 1.0; mass production, assembly lines, and electricity in Industry 2.0; computers and automation in Industry 3.0; and cyber-physical systems and data-driven analytics in Industry 4.0 (Ashaari et al., 2021). As argued by Adnan et al. (2019), education undergoes a similar transition as a result of this advancement. More concretely, Education 1.0 is designed to concentrate on the demands that the agricultural community currently faces, Education 2.0 on the needs of the industrial community, and Education 3.0 on the needs of the society that has been influenced by globalization. Education 1.0, 2.0, and 3.0 are considered by Butt et al. (2020) as reflections of the Web's evolution, while according to Adnan et al. (2019), Education 4.0 is designed to meet the demands of an inventive society and helps people stay up to date with the times (Mukul & Büyüközkan, 2023). By implementing pertinent skills in accordance with the need for improvement and requalification, unlearning and relearning, Education 4.0 takes the lead in the new educational paradigm. According to Mukul and Büyüközkan (2023), Education 4.0 is a personalized approach that uses modern technology to create innovative learning formats and encourages students to interact with various materials in accordance with their academic and skill levels. With the aid of Industry 4.0's cutting-edge technologies, Education 4.0 places the student at its core and makes them the center of learners' learning (Moraes et al., 2022). Although there are an increasing number of studies that have been published, the notion of Quality 4.0 is still relatively new, and very few academics have attempted to define it (Saihi et al., 2023). Using Industry 4.0 technology, integration, and digitalization, Quality 4.0 is a contemporary quality management method (Maganga & Taifa, 2022). Building on the perspectives of Sony et al. (2020), it assists in bringing quality management into line with Industry 4.0 in order to improve performance, efficiency, and innovation inside businesses. In the meantime, as advocated by Ramezani and Jassbi (2020), it is an Industry 4.0 branch that seeks to improve quality by implementing clever fixes as well as intelligent algorithms. According to Broday (2022) and Thekkoote (2022), Quality 4.0 is the process of integrating quality management principles with Industry 4.0 technologies like artificial intelligence and digitalization to improve innovation enterprise efficiency and performance. The surveying and highlighting of QGEE, as seen through the Education 4.0 lens, should be viewed as a significant resolution that seeks to coordinate the institution's strategy with its organizational capabilities and with the specialized skills of its staff in order to enhance the program's economic and social value.

Hinged on the viewpoints of Qian et al. (2023), increased interaction made possible by digitalization enables students to actively study the course material online or through other sources.

Digital education ecosystem. As a result of the digitalization of education and the growth of digital technologies, DEEs are created (Chituc, 2022). The DEE consists of interconnected heterogeneous and geographically dispersed e-learning information and communication technology (ICT) infrastructures, information systems, and "things" used in educational activities, which are made possible by the Internet of Things and Cyber-Physical Systems (Chituc, 2021), which are the cornerstones of Industry 4.0. Smartness of the system, which affects how adaptive, predicative, and guiding the system can be for a learner or instructor, is an intriguing idea for the maturity of a DEE. In this study, DEE is taken into account as a sociocultural setting where the process of teaching learners in a tailored way is based on an analysis of their requirements and requests in the interaction of all participants in educational relationships with the use of variable, adaptive digital content.

Digital pedagogy. As ICTs have developed, DP has, in the words of Mourtzis et al. (2022), expanded to open education, exemplifying the forward-thinking educational technologies of Education 3.0 and Education 4.0. Electronic devices are employed as teaching aids in a teaching methodology known as DP (Spiros et al., 2022). According to Lähdesmäki et al. (2023) as well asVäätäjä and Ruokamo (2021), DP is the socio-constructivist approach to making use of digital technology in education. In accordance with Lewin and Lundie (2016), DP has made it possible for online, blended, and in-person learning environments. Because of this, the term "digital pedagogy" should not be limited to the application of digital instruments (Mourtzis et al., 2022). It should be viewed as the combination of technological know-how and the teacher's expertise and abilities.

1.3 HYPOTHESIS DEVELOPMENT

According to Anghel and Anghel (2022), the definition of green entrepreneurship is provided in the context of the business environment's requirement to respond to the transition to a green economy and sustainable development. As a result, green entrepreneurship education is crucial for building business support and providing a framework for entrepreneurs who want to run successful companies with a particular emphasis on their effects on the environment and society. According to Lahikainen et al. (2021), HEIs can be viewed as catalysts for entrepreneurship, innovation, technical advancement, and economic growth. Due to this, HEIs are looking for ways to develop their entrepreneurial skills by rethinking and harmonizing their governance structures, mission statements, and strategies, as well as their pedagogical practices. They are also forming public and private collaborations with local, national, and international stakeholders. With a focus on using digital technologies to enhance an organization's capacity to meet customer expectations with high quality, Quality 4.0 is the digitalization of traditional quality methodologies (Alzahrani et al., 2021). Applying Quality 4.0 principles to green entrepreneurship education has the potential to spur technical innovation across a range of domains.

The development of DEE is a result of the democratization of digital technology and the digitization of education (Chituc, 2022). The advancement of the DEE is one of two main priorities, which implies the need for the infrastructure and equipment mentioned above as well as the improvement of teachers' and students' digital competence and confidence and also the provision of high-quality educational materials, user-friendly tools, and secure platforms that respect users' privacy and ethical standards. With the support of DEE, HEIs can address numerous issues with infrastructure, size and data quality indicators, departmental cooperation, and internal transparency. Based on the preceding analysis, the hypothesis of the current research is rationally stated as follows.

Hypothesis 1 (H1). *DEE demonstrates a substantial and positive influence on QGEE.*

The necessity to upgrade the educational system with contemporary pedagogical models that take into account the requirements of Industry 4.0 makes the relevance of applying DEE concepts to Education 4.0 significant. In order for Education 4.0 to be possible, interoperability in DEEs must be attained (Chituc, 2021). Digital educational tools and platforms provide visibility and transparency to learning, teaching, and evaluation while also acknowledging significant and all-encompassing components of the learning process for all interested parties. Students may establish covert underground communication channels if they do not obtain timely and accurate information and data from the university, which also causes them to lose faith in the support of the institution. When DEE is properly created, it can address the difficulties that HEIs and students encounter when pursuing their education. With the help of the DEE, a teacher can plan, coordinate, oversee, and assess these competence development pathways and the corresponding student journeys (Wolff et al., 2021). The tools that support students' actions across the many stages of the learning process enable them to monitor their progress, continuously improve feedback, and request help in a less risky setting. Based on the preceding analysis, the hypotheses of the current research are rationally stated as follows.

Hypothesis 2 (H2). *DEE demonstrates a substantial and positive influence on Pedagogical orientation (PO).*

Hypothesis 3 (H3). *DEE demonstrates a substantial and positive influence on Pedagogical practices (PP).*

Hypothesis 4 (H4). *DEE demonstrates a substantial and positive influence on Digital pedagogical competencies (DPC).*

DP is thought of as a technique to integrate digital tools in order to improve the learning process, and that using technology in the classroom may impact the teacher's pedagogical approach. Building on the perspective of Hinostroza et al. (2015), teachers most typically combine digital tools with conventional teaching methods to assist classroom management, information delivery, and instruction. In their research, Butler et al. (2017) argue that students take more ownership of their education when digital technologies are used in teaching and learning, which enhance their participation and collaboration. The adoption of DP is seen as having both a social and pedagogical imperative. Indeed, evidence has grown to support the function and significance of DP in maintaining the standard of higher education. Maslin and Smith (2017) propose the formation of a community to transform open networked educational environments into rich critical spaces in order to assist HEI in achieving a higher level of quality. Based on the preceding analysis, the hypotheses of the current research are rationally stated as follows.

Hypothesis 5 (H5). *PO demonstrates a substantial and positive influence on QGEE.*

Hypothesis 6 (H6). *PP demonstrates a substantial and positive influence on QGEE.*

Hypothesis 7 (H7). *DPC demonstrates a substantial and positive influence on QGEE.*

1.4 METHODS AND MATERIALS

1.4.1 Operationalization of Research Variables

A key problem in the field of social sciences is the operationalization of a construct. The use of preestablished scales is advocated by Bashir et al. (2020) as they are more reliable and valid than newly developed scales. To guarantee the quality of the translated scales, four bilingual experts back-translated all of the scale items from English into Vietnamese and examined each one. As advised by Guimarães et al. (2018), great attention is taken to keeping the questionnaire's syntax and terminology as plain as possible to guarantee that respondents can comprehend the tool readily. Four academic experts are consulted for advice on how to make each measuring construct more realistic after the measuring constructs are finalized. The constructs are changed to get rid of ambiguous measurement items based on the opinions of experts. A 5-point Likert scale, with 1 denoting "not important" and 5 denoting "extremely important," is used to ask respondents for their perspectives.

Prior to data collection, a pilot test is conducted to find any survey questionnaire design faults. According to Hair et al. (2019), Cronbach's alpha is utilized to assess the validity and reliability of the study methodologies. For the internal consistency of a trustworthy scale, Cronbach's alpha range values between 0.50 and 0.70 are also acceptable. The findings indicate that every item has a Cronbach's alpha of at least 0.70. The results of the pilot testing show that the items complied with the standards for determining their reliability, and all of them are retained for further analysis.

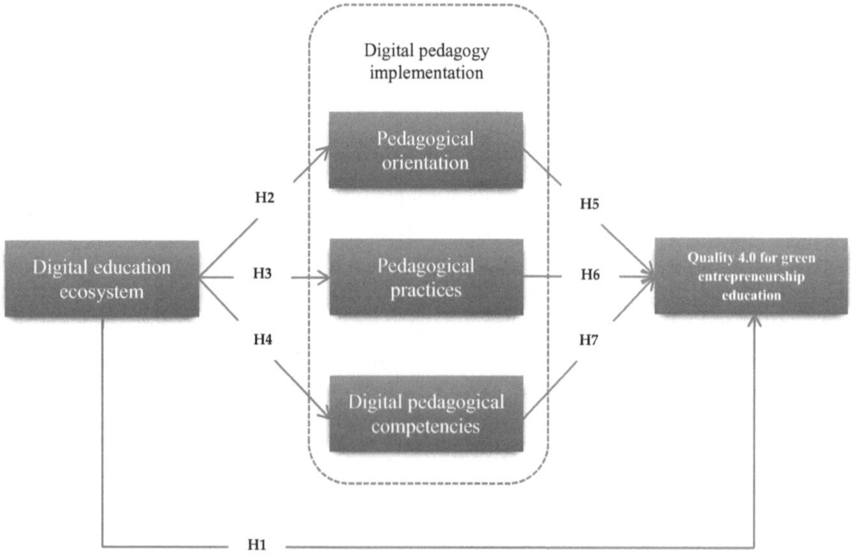

FIGURE 1.1 Hypothesized model.

Quality 4.0 for green entrepreneurship education. The first-order construct of QGEE encompasses three second-order constructs, namely Customer Centricity; Quality Management System; Innovation, which spring from the findings of Sureshchandar (2021).

Digital education ecosystem. The first-order construct of DEE encompasses five second-order constructs namely ICT infrastructure; Interoperability; Artificial intelligence tools; Implementation; Security, data protection, and privacy, which are inherited from the contribution of Chituc (2022).

Digital pedagogy. The first-order construct of DP encompasses three second-order constructs, namely PO; PP; DPC, which arise from the works of Väätäjä and Ruokamo (2021).

1.4.2 Target Population and Sampling Procedure

A paper-and-pencil questionnaire is used to collect the study's main data. The analysis focuses on HEIs, and instructors should be able to respond to the questionnaire. The strategy utilized in this study is a convenient, snowball, and purposive sampling mix. Respondents are informed of the purpose of the survey, how their information will be used, and how it will be protected before they agree to participate. According to the recommendation, the sample size should fall between 5:1 and 20:1, with 5 and 20 denoting the sample size of each item (Gupta, 2021). The period of data collection was from November 2022 to May 2023. The dataset will be purged of any responses with more than 15% of their elements missing (Hair et al., 2022). A sample size of 583 cases with a data loss rate of 11.67% remained after screening and analyzing the questionnaires. SPSS 29.0 and AMOS 28.0 are used to analyze the statistical data for this investigation.

According to the sample's sociodemographic profile, females made up the majority of participants—63.12%—while males made up 36.88%. Regarding the respondents' ages, the group "40–Under 50" made up 39.11% of the sample as a whole, and the group "30–Under 40" came in second with roughly 35.51%. The group "over 50" occupied the bottom spot among the listed groups with a meager 25.38% of the total. Almost all of the participants had at least a postgraduate degree, therefore they all had academic qualifications. These responders accumulated more than ten years of teaching experience.

1.5 RESULTS AND ELUCIDATION OF ANALYTICAL OBSERVATIONS

1.5.1 Measurement Model Evaluation

Construct validity and reliability. The reliability analysis is used to verify the reliability of the data. As the lowest Cronbach's alpha value is 0.817, higher than the referenced value of 0.7, which is used to demonstrate the trustworthiness of the scale, the results show that the questionnaire scales are reliable (Hair et al., 2019).

Confirmatory factor analysis is used to determine the degree of measurement accuracy of the measuring device. Using the results of this research as a foundation, the average extraction variance (AVE), composite reliability (CR), and loading factor values for each item are calculated. On their underlying construct, all elements statistically strongly load, with the lowest values being 0.639. In other words, according to Hair et al. (2019), all of the construct's indicators have loadings over the cutoff of 0.5. A good level of convergent validity is shown by constructs with an AVE value of 0.5 or higher (Hair et al., 2019). Additionally, all of the proposed constructions have CR values above 0.8, exceeding the cutoff point of 0.7 (Hair et al., 2019). When these criteria are statistically verified, the results in Table 1.1 show that the constructs had good reliability and validity.

Correlations and discriminant validity. According to Ma et al. (2018), discriminant validity is evaluated to see if the constructs do not overlap with one another and revealed distinctive variance in each of the individual constructs. By comparing the square root of the AVE to the values of the inter-construct correlations, discriminate validity is tested (Tsou et al., 2019). The diagonal is always greater than the correlations among the constructs. Additionally, according to Campbell and Fiske (1959), all inter-construct correlation coefficients must be less than 0.80, or considerably different from 1.0. All of the suggested model's constructs show discriminant validity for the empirical data because the correlation matrix shown in Table 1.2 satisfies these requirements.

The proposed model fitting. The fitness of the data was assessed using a variety of model fitness indices, including the Root Mean Square Error of Approximation

TABLE 1.1
Summary of Results of Convergent Validity and Construct Reliability

		Convergent Validity		Construct Reliability		
Construct	Item Acronyms	Factor Loadings Ranges	AVE	Cronbach's Alpha	Composite Reliability	Inference
		Digital Education Ecosystem				
Information and Communication Technology infrastructure	ICTI	0.639–0.779	0.510	0.900	0.903	Retained
Interoperability	INT	0.791–0.819	0.644	0.851	0.853	Retained
Artificial intelligence tools	AIT	0.814–0.909	0.721	0.891	0.894	Retained
Implementation	IMP	0.823–0.887	0.715	0.888	0.891	Retained
Security, data protection, and privacy	SDP	0.797–0.812	0.636	0.842	0.848	Retained
		Digital Pedagogy				
Pedagogical orientation	PO	0.733–0.871	0.654	0.853	0.858	Retained
Pedagogical practices	PP	0.729–0.852	0.592	0.817	0.820	Retained
Digital pedagogical competencies	DPC	0.803–0.848	0.686	0.874	0.876	Retained
		Quality 4.0 for Green Entrepreneurship Education				
Customer Centricity	CC	0.693–0.941	0.651	0.884	0.890	Retained
Quality Management System	QMS	0.667–0.928	0.587	0.848	0.856	Retained
Innovation	INN	0.783–0.856	0.677	0.867	0.871	Retained

TABLE 1.2
Summary of Results of Discriminant Validity

	ICTI	CC	QMS	AIT	IMP	DPC	INN	PO	INT	SDP	PP
ICTI	1										
CC	0.082	1									
QMS	0.165	0.347	1								
AIT	0.120	0.079	0.108	1							
IMP	0.092	−0.013	0.089	0.023	1						
DPC	0.101	0.120	0.206	0.119	0.113	1					
INN	0.121	0.013	0.094	0.152	0.151	0.083	1				
PO	0.157	0.204	0.128	0.074	0.053	0.191	0.090	1			
INT	0.065	0.046	0.223	0.008	0.243	0.218	0.143	0.129	1		
SDP	0.065	−0.033	0.045	0.001	0.252	0.079	0.132	0.072	0.209	1	
PP	0.152	0.153	0.169	0.084	0.111	0.104	0.267	0.116	0.191	0.088	1

TABLE 1.3
The Outcomes of Measurement and Structural Model Analysis

The Goodness of Fit Measures	Minimum Cutoff	Parameter Estimates of Measurement Model	Parameter Estimates of Structural Model	Recommended By
Chi-square/df	<3	1.601	1.729	Kline (2016)
TLI	≥0.9	0.969	0.963	Kline (2016)
CFI	≥0.9	0.973	0.965	Kline (2016)
GFI	≥0.9	0.939	0.931	Kline (2016)
RMSEA	<0.08	0.026	0.029	Kline (2016)

(RMSEA), Comparative Fit Index (CFI), Goodness of Fit Index (GFI), and Tucker-Lewis index (TLI). To assess the strength of the measurement model and structural model, the chi-square/df absolute index is also used. The predicted models are supported to match the collected data based on the statistical findings because all of the recorded indicators adhered to generally recognized criteria. Table 1.3 provides detailed information on the measurement model's and the structural model's fitness indices.

1.5.2 Structural Model Evaluation

Direct effect. The first set of main hypotheses in Table 1.4 postulated that DEE can have significant impacts on each dimension of DP. The effect of DEE (H2: $\beta=0.273$, $p<0.001$) illustrates a positive association with PO, while the effect of DEE (H3: $\beta=0.346$, $p<0.001$) highlights a strictly positive interconnection with PP. The relationship between DEE and DPC (H4: $\beta=0.343$, $p<0.001$) is staunchly supported.

TABLE 1.4
Structural Coefficients (β) of the Hypothesized Model

Hypothesis No.	Hypothesized Path	Standardized	S.E.	C.R.	Inference
H1	DEE → QGEE	0.245*	0.289	2.076	Buttressed
H2	DEE → PO	0.273***	0.025	3.604	Buttressed
H3	DEE → PP	0.346***	0.036	3.913	Buttressed
H4	DEE → DPC	0.343***	0.030	3.955	Buttressed
H5	PO → QGEE	0.157**	0.441	2.644	Buttressed
H6	PP → QGEE	0.182**	0.391	2.794	Buttressed
H7	DPC → QGEE	0.158*	0.444	2.521	Buttressed

Notes: *p < 0.05; **p < 0.01; ***p < 0.001.

TABLE 1.5
The Summary of the Mediation Effects

Hypothesis	Standardized	Lower	Upper	Result
DEE → DP → QGEE	0.200**	0.076	0.345	Partial mediation

Notes: *p < 0.05; **p < 0.01; ***p < 0.001.

Moreover, the effect of DEE is exposed to positively influence QGEE (H1: $\beta = 0.245$, $p < 0.05$). As such, H1–H4 are accepted.

The second set of main hypotheses in Table 1.4 predicated that each dimension of DP can lead to significant influences on QGEE. Concretely, the PO has significantly positive association with QGEE (H5: $\beta = 0.157$, $p < 0.01$). Besides, the PP is substantiated to be related to QGEE (H6: $\beta = 0.182$, $p < 0.01$) in a significantly positive manner. The investigation on the relationship between DPC and QGEE (H7) denotes that the standardized path coefficient (β) is 0.158 at $p < 0.05$. Thus, H5–H7 are supported.

Indirect effect. The meditation effect is examined in this study using the indirect effect of Gaskin and Lim's (2018) AMOS plugin study. The results in Table 1.5 show that all mediation requirements have been satisfied. A positive indirect connection between DEE (an independent variable) and QGEE (a dependent variable) is observed in the presence of DP (a mediator) ($\beta = 0.200$; $p < 0.01$). As a result, a connection of partial mediation exists. The bootstrapping analysis reveals a significant effect that is (95% confidence interval (CI)) transmitted from the DEE to the QGEE via the DP.

1.5.3 Discussion

Conforming to the expectation of the researchers, the mathematical results obtained in this study cast light on the interconnection between DEE and QGEE. Quality 4.0 is the digitization of traditional quality approaches, with an emphasis on leveraging digital technology to improve an organization's ability to satisfy customer expectations with high quality (Alzahrani et al., 2021). The integration of Quality 4.0 concepts into the teaching of green entrepreneurship has the capacity to promote technological innovation in several fields. When DEE is established appropriately, it can help with the challenges that students and HEIs have when seeking an education. These competence development routes and the accompanying student journeys can be planned, coordinated, supervised, and evaluated by a teacher with the assistance of the DEE (Wolff et al., 2021). Furthermore, HEIs can solve several challenges related to infrastructure, departmental cooperation, size and data quality indicators, and internal transparency with the use of DEE.

In consonance with the anticipations of the researchers, DP is corroborated to act as a partial mediator in the interconnection between DEE and QGEE. A teacher's pedagogical approach may be impacted by utilizing technology in the classroom. DP is perceived as a method for incorporating digital resources to enhance the learning process. As stated by Lewin and Lundie (2016), learning settings that are in-person, hybrid, and online are now achievable thanks to DP. Along this line, Maslin et al. (2017) recommend the establishment of a community to convert open, networked learning settings into rich, critical places as a means of helping HEI attain excellence.

1.6 CONCLUDING REMARKS

1.6.1 Practical Implication

As a result, all the leaders of HEIs should carefully carry out the strategic actions necessary to develop and implement DEE by taking part in government initiatives and closely working with other ecosystem players. To do this, the first step will be to focus on digital innovation. Leaders in HEIs should have a clear digital concern in every aspect of the organization, from the mindset of the staff to internal procedures to the experience of stakeholders, with the goal of constantly innovating. Second, HEI leaders should have an all-encompassing perspective and work together to administer and control HEIs in accordance with a shared goal. On the other hand, the partial mediating role of DP in the interconnection between DEE and QGEE should be taken into consideration. In this regard, all the leaders in HEIs must be open to experimenting with new technology and pedagogical approaches and be willing to embrace change and innovation. Human factors will have a significant impact on an organization's capacity for digitalization given that businesses have been urged to concentrate on the aspects that are prioritized. The leaders of HEIs should be open to new inventive and creative ideas as well as focused on enhancing teachers' competencies through ongoing education and training. It is advised that they should invest in teacher training and professional development to provide them the knowledge and skills they need to use digital devices in the classroom successfully. This can involve giving instructors access to workshops, online training courses, and networking and collaboration options. Additionally, they need to involve educators and learners in decision-making, as well as foster a climate that supports the use of technology in education.

Regarding policymakers and other governmental influencers, their roles have made a significant contribution to the process of digitalization within HEIs. With government assistance, HEIs will have the resources and fortitude to meet the challenges of digitalization. This would be accomplished by the government providing all the necessary policies, incentives, programs, and execution-level guidance for each stage of HEI digitalization. To guarantee that all students have equitable access to digital educational resources, governments must give top priority to the development of digital infrastructure, including internet and technology access. Second, the government might support HEIs by offering digital education or training. This solution's impact on the digital capacities of HEIs can be explained in part by the fact that digital training can enhance and increase HEIs' grasp of the digital world. HEIs may reduce their costs associated with hiring and training new employees if the government helped them put up a digital learning and training system for their workforce. Third, as a prerequisite for generating an ecosystem that will ensure the synchronization of scientific and technological progress in numerous sectors of the economy, as well as amplify the efficiency and effectiveness of synergistic impact from the interaction of ecosystem members, the government can underpin HEIs by formulating and operationalizing a collaborative ecosystem, that is, DEE that allows HEI to shape up their network and collaborate with other actors.

1.6.2 Limitation and Agenda for Coming-Up Studies

There are a few restrictions that should be addressed in the current research, and these can be used as the foundation for future investigations. First off, it's critical to be cautious when drawing conclusions from data because the details of the research environment may limit what may be drawn from the results. Since all of the samples came from Vietnam, additional cross-regional analysis is needed to confirm the findings before they can be applied widely. The second bottleneck, caused by convenience, snowball, and purposive selection strategies, may limit the generalizability of the study. To obtain sample data for upcoming study and ensure that the results are representative and scientific, quota sampling should be used. The relatively small sample size of the study is its third problem, which calls for additional research. Last but not the least, in order to generate a more accurate representation of the circumstance, additional constructs are recommended to be added to the model in future investigations.

ACKNOWLEDGMENT

The research of this chapter was funded by the University of Economics, Ho Chi Minh City (UEH).

REFERENCES

Abedin, B. (2022). Managing the tension between opposing effects of explainability of artificial intelligence: A contingency theory perspective. *Internet Research, 32*(2), 425–453. https://doi.org/10.1108/intr-05-2020-0300.

Adnan, A., Karim, R. A., Tahir, M. H. M., Kamal, N. N. M., & Yusof, A. L. (2019). Education 4.0 technologies, Industry 4.0 skills and the teaching of English in Malaysian tertiary education. Arab World English Journal, *10*(4), 330–343. https://doi.org/10.24093/awej/vol10no4.24.

Ali, M. A. S., Ammer, M. A., & Elshaer, I. A. (2023). Born to be green: Antecedents of green entrepreneurship intentions among higher education students. *Sustainability*, *15*(8), 6668. https://doi.org/10.3390/su15086668.

Alwakid, W., Aparicio, S., & Urbano, D. (2021). The influence of green entrepreneurship on sustainable development in Saudi Arabia: The role of formal institutions. International Journal of Environmental Research and Public Health, *18*(10), 5433. https://doi.org/10.3390/ijerph18105433.

Alzahrani, B., Bahaitham, H., Andejany, M., & Elshennawy, A. K. (2021). How ready is higher education for quality 4.0 transformation according to the LNS research framework? Sustainability, *13*(9), 5169. https://doi.org/10.3390/su13095169.

Anghel, G. A., & Anghel, M. A. (2022). Green entrepreneurship among students-social and behavioral motivation. *Sustainability*, *14*(14), 8730. https://doi.org/10.3390/su14148730.

Ashaari, M. A., Singh, K. S. D., Abbasi, G. A., Amran, A., & Liébana-Cabanillas, F. (2021). Big data analytics capability for improved performance of higher education institutions in the era of IR 4.0: A multi-analytical SEM & ANN perspective. Technological Forecasting and Social Change, *173*, 121119. https://doi.org/10.1016/j.techfore.2021.121119.

Bashir, A., Amir, A., Jawaad, M., & Hasan, T. (2020). Work conditions and job performance: An indirect conditional effect of motivation. Cogent Business & Management, *7*(1), 1801961. https://doi.org/10.1080/23311975.2020.1801961.

Broday, E. E. (2022). The evolution of quality: From inspection to quality 4.0. International Journal of Quality and Service Sciences, *14*(3), 368–382. https://doi.org/10.1108/ijqss-09-2021-0121.

Burzyńska, D., Jabłońska, M., & Dziuba, R. (2018). Opportunities and conditions for the development of green entrepreneurship in the Polish textile sector. Fibres & Textiles in Eastern Europe, *26*(2), 13–19. https://doi.org/10.5604/01.3001.0011.5733.

Butler, D., Leahy, M., Hallissy, M., & Brown, M. (2017). Different strokes for different folks: Scaling a blended model of teacher professional learning. Interactive Technology and Smart Education, *14*(3), 230–245. https://doi.org/10.1108/itse-01-2017-0011.

Butt, R., Siddiqui, H., Soomro, R. A., & Asad, M. M. (2020). Integration of industrial revolution 4.0 and IOTs in academia: A state-of-the-art review on the concept of Education 4.0 in Pakistan. Interactive Technology and Smart Education, *17*(4), 337–354. https://doi.org/10.1108/itse-02-2020-0022.

Cai, X., Hussain, S., & Zhang, Y. (2022). Factors that can promote the green entrepreneurial intention of college students: A fuzzy set qualitative comparative analysis. Frontiers in Psychology, *12*. https://doi.org/10.3389/fpsyg.2021.776886.

Campbell, D. T., & Fiske, D. W. (1959). Convergent and discriminant validation by the multitrait-multimethod matrix. *Psychological Bulletin*, *56*(2), 81–105. https://doi.org/10.1037/h0046016.

Chituc, C. (2021). A framework for Education 4.0 in digital education ecosystems. In *IFIP Advances in Information and Communication Technology* (pp. 702–709). https://doi.org/10.1007/978-3-030-85969-5_66.

Chituc, C.-M. (2022). An analysis of technical challenges for education 4.0 and digital education ecosystems. In *2022 IEEE German Education Conference* (*GeCon*) (pp. 1–5). https://doi.org/10.1109/gecon55699.2022.9942758.

Csaszar, F. A., & Ostler, J. (2020). A contingency theory of representational complexity in organizations. *Organization Science*, *31*(5), 1198–1219. https://doi.org/10.1287/orsc.2019.1346.

Da Motta Reis, J. S., Costa, A. C. F., Espuny, M., Batista, W. J., Francisco, F. E., Gonçalves, G. S., Tasinaffo, P. M., Dias, L. A . V., Da Cunha, A. M., & De Oliveira, O. J. (2020). Education 4.0: Gaps research between school formation and technological development. In *Advances in Intelligent Systems and Computing* (pp. 415–420). https://doi.org/10.1007/978-3-030-43020-7_55.

De Morais, D. O. C., & Barbieri, J. C. (2022). Supply chain social sustainability: Unveiling focal firm's archetypes under the lens of stakeholder and contingency theory. *Sustainability*, *14*(3), 1–31. https://doi.org/10.3390/su14031185.

Demirel, P., Li, Q. C., Rentocchini, F., & Tamvada, J. P. (2017). Born to be green: New insights into the economics and management of green entrepreneurship. Small Business *Economics*, *52*(4), 759–771. https://doi.org/10.1007/s11187-017-9933-z.

Donaldson, L. (2007). Ethics problems and problems with ethics: Toward a pro-management theory. *Journal of Business Ethics*, *78*(3), 299–311. https://doi.org/10.1007/s10551-006-9336-6.

Gaskin, J., & Lim, J. (2018). *Indirect Effects*. AMOS Plugin. Gaskination's StatWiki.

Guimarães, A. G., Vaz-Fernandes, P., Ramos, M. D. R., & Martinho, A. P. (2018). Co-processing of hazardous waste: The perception of workers regarding sustainability and health issues in a Brazilian cement company. Journal of Cleaner Production, *186*, 313–324. https://doi.org/10.1016/j.jclepro.2018.03.092.

Gupta, A. K. (2021). Innovation dimensions and firm performance synergy in the emerging market: A perspective from dynamic capability theory & signaling theory. Technology in Society, *64*, 101512. https://doi.org/10.1016/j.techsoc.2020.101512.

Hair, J. F., Babin, B. J., Anderson, R. E., & Black, W. C. (2019). *Multivariate Data Analysis* (8th ed.). England: Pearson Prentice.

Hair, J.F., Hult, G.T.M., Ringle, C.M., & Sarstedt, M. (2022). *A Primer on Partial Least Squares Structural Equation Modeling* (*PLS-SEM*). Thousand Oaks, CA: Sage.

Hinostroza, J. E., Ibieta, A., Claro, M., & Labbé, C. (2015). Characterisation of teachers' use of computers and Internet inside and outside the classroom: The need to focus on the quality. Education and Information Technologies, *21*(6), 1595–1610. https://doi.org/10.1007/s10639-015-9404-6.

Kline, R. B. (2016). *Principles and Practice of Structural Equation Modeling*. New York: The Guilford Press.

Lähdesmäki, S., Maunumäki, M., & Nurmi, T. (2023). Play is the base! ECEC leaders' views on the development of digital pedagogy. Early Childhood Education Journal. https://doi.org/10.1007/s10643-023-01530-7.

Lahikainen, K., Peltonen, K., Oikkonen, E., & Pihkala, T. (2021). Students' perceptions of the entrepreneurial culture in Finnish higher education institutions. Industry and Higher Education, *36*(5), 583–594. https://doi.org/10.1177/09504222211066798.

Lewin, D., & Lundie, D. (2016). Philosophies of digital pedagogy. Studies in Philosophy and Education, *35*(3), 235–240. https://doi.org/10.1007/s11217-016-9514-7.

Ma, D., Ullah, F., Khattak, M. S., & Anwar, M. (2018). Do international capabilities and resources configure firm's sustainable competitive performance? Research within Pakistani SMEs. Sustainability, *10*(11), 4298. https://doi.org/10.3390/su10114298.

Maganga, D. P., & Taifa, I. W. R. (2022). Quality 4.0 conceptualisation: An emerging quality management concept for manufacturing industries. *The TQM Journal*, *35*(2), 389–413. https://doi.org/10.1108/tqm-11-2021-0328.

Makhloufi, L., Laghouag, A., Meirun, T., & Bélaïd, F. (2021). Impact of green entrepreneurship orientation on environmental performance: The natural resource-based view and environmental policy perspective. Business Strategy and the Environment, *31*(1), 425–444. https://doi.org/10.1002/bse.2902.

Margulieux, L. E., McCracken, W., & Catrambone, R. (2016). A taxonomy to define courses that mix face-to-face and online learning. *Educational Research Review*, *19*, 104–118. https://doi.org/10.1016/j.edurev.2016.07.001

Maslin, P. R., & Smith, N. V. (2017). Practicum as Nexus: Using student voice to improve digital pedagogy within ITE. Waikato Journal of Education, *22*(3). https://doi.org/10.15663/wje.v22i3.376

Moraes, E. B., Kipper, L. M., Kellermann, A. C. H., Austria, L., Leivas, P., Moraes, J. A. R., & Witczak, M. (2022). Integration of Industry 4.0 technologies with Education 4.0: Advantages for improvements in learning. Interactive Technology and Smart Education, *20*(2), 271–287. https://doi.org/10.1108/itse-11-2021-0201.

Mourtzis, D., Panopoulos, N., & Angelopoulos, J. (2022). A hybrid teaching factory model towards personalized education 4.0. International Journal of Computer Integrated Manufacturing, 1–21. https://doi.org/10.1080/0951192x.2022.2145025.

Mukul, E., & Büyüközkan, G. (2023). Digital transformation in education: A systematic review of education 4.0. Technological Forecasting and Social Change, *194*, 1–21. https://doi.org/10.1016/j.techfore.2023.122664.

Munyon, T. P., Jenkins, M. T., Crook, T. R., Edwards, J. R., & Harvey, N. (2019). Consequential cognition: Exploring how attribution theory sheds new light on the firm-level consequences of product recalls. Journal of Organizational Behavior, *40*(5), 587–602. https://doi.org/10.1002/job.2350.

Qian, Y., Vaddiraju, S., & Khan, F. (2023). Safety education 4.0 – A critical review and a response to the process industry 4.0 need in chemical engineering curriculum. *Safety Science*, *161*, 1–19. https://doi.org/10.1016/j.ssci.2023.106069

Ramezani, J., & Jassbi, J. (2020). Quality 4.0 in action: Smart hybrid fault diagnosis system in plaster production. *Processes*, *8*(6), 634. https://doi.org/10.3390/pr8060634

Reinking, J. (2011). Contingency theory in information systems research. In *Integrated Series on Information Systems* (pp. 247–263). https://doi.org/10.1007/978-1-4419-6108-2_13.

Romanowski, R., & Gnusowski, M. (2019). Green entrepreneurship in the quintaple helix model. In *Bogucki Wydawnictwo Naukowe eBooks* (pp. 103–118). https://doi.org/10.12657/9788379862764-7.

Saihi, A., Awad, M., & Ben-Daya, M. (2023). Quality 4.0: Leveraging Industry 4.0 technologies to improve quality management practices – a systematic review. International Journal of Quality & Reliability Management, *40*(2), 628–650. https://doi.org/10.1108/ijqrm-09-2021-0305.

Sauser, B., Reilly, R. R., & Shenhar, A. J. (2009). Why projects fail? How contingency theory can provide new insights – A comparative analysis of NASA's Mars Climate Orbiter loss. International Journal of Project Management, *27*(7), 665 679. https://doi.org/10.1016/j.ijproman.2009.01.004.

Shao, Z., Feng, Y., & Hu, Q. (2016). Effectiveness of top management support in enterprise systems success: A contingency perspective of fit between leadership style and system life-cycle. European Journal of Information Systems, *25*(2), 131–153. https://doi.org/10.1057/ejis.2015.6.

Sony, M., Antony, J., & Douglas, A. (2020). Essential ingredients for the implementation of Quality 4.0. *The TQM Journal*, *32*(4), 779–793. https://doi.org/10.1108/tqm-12-2019-0275.

Spiros, M. C., Plemons, A. M., & Biggs, J. (2022). Pedagogical access and ethical considerations in forensic anthropology and bioarchaeology. *Science & Justice*, *62*(6), 708–720. https://doi.org/10.1016/j.scijus.2022.03.008.

Sureshchandar, G. (2021). Quality 4.0 – a measurement model using the confirmatory factor analysis (CFA) approach. International Journal of Quality & Reliability Management, *40*(1), 280–303. https://doi.org/10.1108/ijqrm-06-2021-0172.

Thekkoote, R. (2022). Enabler toward successful implementation of Quality 4.0 in digital transformation era: A comprehensive review and future research agenda. International Journal of Quality & Reliability Management, *39*(6), 1368–1384. https://doi.org/10.1108/ijqrm-07-2021-0206.

Tsou, H. T., Chen, J., Chou, C. Y., & Chen, T. W. (2019). Sharing economy service experience and its effects on behavioral intention. *Sustainability*, *11*(18), 5050. https://doi.org/10.3390/su11185050.

Väätäjä, J., & Ruokamo, H. (2021). Conceptualizing dimensions and a model for digital pedagogy. Journal of Pacific Rim Psychology, *15*, 1–12. https://doi.org/10.1177/1834490921995395.

Wang, P. (2010). Restructuring to repair legitimacy – a contingency perspective. Corporate Governance: An International Review, *18*(1), 64–82. https://doi.org/10.1111/j.1467-8683.2009.00782.x.

Wolff, C., Reimann, C., Mikhaylova, E., Aldaghamin, A., Pampus, S., & Hermann, E. (2021). Digital education ecosystem (DEE) for a virtual master school. In *2021 IEEE International Conference on Smart Information Systems and Technologies* (*SIST*). https://doi.org/10.1109/sist50301.2021.9465914.

Yi, G. (2020). From green entrepreneurial intentions to green entrepreneurial behaviors: The role of university entrepreneurial support and external institutional support. International Entrepreneurship and Management Journal, *17*(2), 963–979. https://doi.org/10.1007/s11365-020-00649-y.

2 From Classroom to Castle

The Effectiveness of Interactive Games for Learning Maltese Prepositions

Jacqueline Żammit

2.1 INTRODUCTION

In recent years, game-based learning has emerged as a promising pedagogical approach in higher education. The idea behind game-based learning is that students can acquire knowledge and skills more effectively through engaging and interactive activities that mimic the mechanics of games (Karagiorgas & Niemann, 2017). This approach has gained traction due to its potential to promote active learning, enhance student engagement, and foster deeper understanding (Ismaizam et al., 2022). One area where game-based learning has shown promise is in language acquisition, particularly for adult learners.

This chapter explores the impact of a specific game, *The Mdina Castle's Treasure Quest*, on prepositional learning in adult students. The game provided a captivating and immersive environment for 32 international adult students to learn and practice Maltese prepositions. By integrating prepositions into the gameplay, the students navigated through the virtual castle, solved puzzles, and overcame challenges to discover the hidden medieval gold coin treasure.

The primary objective of this study was to investigate the effectiveness of game-based learning in higher education, specifically in the context of prepositional learning. This study aimed to evaluate the effect of the game on student performance and engagement by comparing the learning results of the game-based education group with a control group that received conventional lecture-based instruction. Additionally, it aimed to explore students' attitudes and motivations toward game-based language learning.

DOI: 10.1201/9781003400691-2

2.1.1 Research Questions

This chapter seeks to address the following research questions:

1. What is the impact on adult students of learning Maltese prepositions by playing the game *The Mdina Castle's Treasure Quest*?
2. How does game-based instruction compare to traditional lecture-based instruction in terms of prepositional learning outcomes?
3. What are the students' perceptions, attitudes, and levels of engagement toward learning prepositions through game-based instruction?

2.1.2 The Aims of the Study

The aims of this study were twofold: firstly, to examine the effectiveness of game-based instruction in enhancing prepositional learning outcomes among adult students; and secondly, to explore the students' perceptions and attitudes toward game-based language learning. By investigating these aspects, this study sought to shed light on the potential benefits of game-based instruction for Maltese language acquisition in higher education and provide practical insights for educators and practitioners.

In general, this chapter contributes to the ongoing discourse on innovative pedagogical practices for higher education 4.0, highlighting the importance of interactive games and immersive learning experiences in the modern classroom. The study's findings may serve as a valuable resource for educators seeking to integrate game-based learning approaches into their language instruction methods and optimize student engagement and motivation.

2.2 LITERATURE REVIEW

Game-based learning has gained significant attention as an effective pedagogical approach in higher education. The concept is rooted in the idea that students can acquire knowledge and skills more efficiently when engaged in game-like activities that promote active learning (Karagiorgas & Niemann, 2017). This approach has shown promise in various domains, including language acquisition, particularly among adult learners.

2.2.1 Game-Based Learning in Higher Education

According to Jääskä et al. (2022), research has highlighted that game-based learning enhances student engagement, motivation, and active participation. Games provide an immersive and interactive environment that encourages students to explore, experiment, and problem-solve, leading to deeper learning outcomes. Various studies have demonstrated the positive impact of game-based learning on student achievement, knowledge retention, and critical thinking skills across different disciplines (Karakoç et al., 2022).

2.2.2 Language Learning and Game-Based Instruction

Language acquisition, particularly among adult learners, presents unique challenges. Akintolu and Letseka (2021) claimed that traditional lecture-based instruction may not fully engage adult learners or provide them with opportunities for practical application and contextual understanding. Game-based instruction offers an alternative approach that combines entertainment and education to promote language learning (Bygstad et al., 2022). Games provide a supportive and enjoyable environment where learners can practice language skills, apply grammar rules, and develop communicative competence (Yaccob & Yunus, 2019). Several studies have reported positive outcomes of game-based language learning, including improved language proficiency, increased motivation, and enhanced learner autonomy (Breien & Wasson, 2021; Bygstad et al., 2022; Yaccob & Yunus, 2019).

2.2.3 Learning Prepositions and Game-Based Instruction

Learning prepositions is a crucial grammatical aspect of language acquisition, as it contributes to accurate and meaningful communication. Kilimci (2017) stated that traditional approaches to teaching prepositions often involve rote memorization and decontextualized exercises, which may limit students' understanding and retention. Game-based instruction offers a more engaging and interactive way to learn prepositions by providing learners with authentic contexts and opportunities for meaningful practice (Lin et al., 2018). Studies have indicated that game-based instruction can enhance prepositional learning outcomes by promoting active engagement, reinforcing comprehension, and facilitating the application of prepositions in context (Kılıçkaya, 2022; Bahari, 2023).

2.2.4 Student Engagement and Motivation in Game-Based Learning

Engagement and motivation play key roles in effective learning experiences. Game-based learning has the potential to foster higher levels of engagement and motivation due to its interactive and enjoyable nature (Jääskä et al., 2022). Bygstad et al. (2022) stated that games provide immediate feedback, challenge-based tasks, and a sense of autonomy, which can stimulate intrinsic motivation and promote a positive learning experience.

Research has shown that game-based instruction enhances student engagement, persistence, and positive attitudes toward learning, leading to improved academic performance (Adipat et al., 2021). According to Anastasiadis et al. (2018), game-based learning offers an engaging and immersive environment that fosters active participation, motivation, and deep understanding of prepositions. The reviewed studies emphasize the potential benefits of game-based instruction for prepositional learning among adult students.

The current study is filling a research gap in Maltese, which lacks studies demonstrating the importance of game-based learning in Maltese classes, particularly for adult learners. The findings of this study support the idea that interactive games, such as *The Mdina Castle's Treasure Quest*, can improve language learning outcomes and foster positive attitudes toward learning in higher education settings.

2.3 METHODOLOGY

2.3.1 Research Design

This study employed a quasi-experimental design (Thomas, 2022) to investigate the impact of playing *The Mdina Castle's Treasure Quest* on prepositional learning in adult students. A control group design was employed, with the game-based instruction group serving as the experimental group and the traditional lecture-based instruction group as the control group. The study compared the prepositional learning outcomes between the two groups.

2.3.2 Participants

The participants in this study consisted of 63 adult students from diverse international backgrounds. A purposive sampling method (Saunders et al., 2012) was employed to select 32 participants for the game-based instruction group and 31 participants for the traditional lecture-based instruction from the same teacher. These students were all adults aged between 18 and 65 years and enrolled in an A2-level Maltese language course focused on learning Maltese prepositions.

2.3.3 Procedure

The study was conducted over a predetermined three-month period, during which both the game-based instruction and control groups received instruction on Maltese prepositions. The game-based instruction group played *The Mdina Castle's Treasure Quest*, while the control group received instruction through traditional lecture-based methods. The game-based instruction group participated in a series of game sessions where they navigated through the virtual castle, solved puzzles, and overcame challenges to find the hidden medieval gold coin treasure. Throughout the game, prepositions were integrated into the gameplay, requiring students to apply their understanding of prepositions in context. The control group, on the other hand, attended traditional lecture-based classes, where they received instruction on prepositions through lectures, class discussions, and exercises.

2.3.4 Data Collection

Data collection in this study involved multiple measures to assess the impact of game-based instruction on prepositional learning outcomes and students' perceptions and attitudes. Both the game-based instruction and control groups completed the post-instruction assessments, consisting of tasks and questions that assessed their understanding and application of prepositions.

After the two-month instruction period, the game-based instruction group participated in a focus group to collect qualitative data on their perceptions, attitudes, and levels of engagement regarding learning prepositions through game-based instruction.

2.3.5 Data Analysis

The data collected from the post-instruction assessments were analyzed using appropriate statistical methods (Prabhaker et al., 2019) to compare the learning outcomes between the game-based instruction and control groups and to determine the impact of game-based instruction on prepositional learning.

The qualitative data obtained from the focus group interviews were transcribed and analyzed using thematic analysis. Themes and patterns related to students' perceptions, attitudes, and levels of engagement toward game-based instruction were identified and interpreted.

The methodology employed in this study allowed for a rigorous investigation of the impact of game-based instruction on prepositional learning outcomes and students' perceptions and attitudes. The quasi-experimental design with a control group provided a basis for comparing the effectiveness of game-based instruction to traditional lecture-based instruction. By employing purposive sampling, the study ensured the inclusion of adult students from diverse international backgrounds. The combination of quantitative and qualitative data collection methods facilitated a comprehensive analysis of the research questions.

2.4 RESULTS ANALYSIS AND DISCUSSION

2.4.1 Experimental Setup

The group of students was specifically chosen to ensure diversity by recruiting 63 Maltese adult learners from several countries and age ranges. The 63 students were then divided into two subgroups: a control group and an experimental group (see Appendix 2.1). Care was taken to ensure that each group's demographics were as balanced as possible to avoid bias. Table 2.1 provides information on the average age and male-to-female ratios.

Both groups were taught a lesson on Maltese prepositions, with the control group taught using the traditional lecture method and the experimental group taught using a game-based learning platform. The same teacher was in charge of both groups.

TABLE 2.1
The Average Age Was Similar for Both Groups, and the Male-to-Female Ratios Were Nearly Identical

Variable	Control	Experiment	Overall
Average age	27.16	52.38	39.97
Males to females	18–13	19–13	37–26
Gender ratio	72%	68%	70%

Note: However, there was a slight overall bias toward males.

The day after the lesson, students were tested using a multiple-choice questionnaire. The scores ranged from 0 to 100, but there were only 20 questions, resulting in scores ending in 0 or 5 (see Appendix 2.1). If the pass mark for the preposition test is considered to be 50 out of 100, four participants in the experimental group did not pass. They scored as follows: 20 (1 participant), 40 (1 participant), and 45 (2 participants). In contrast, there were six participants in the control group who scored less than 50. Their scores were: 25 (1 participant), 30 (1 participant), 35 (3 participants), and 40 (1 participant). Furthermore, four participants from the experimental group achieved a full mark of 100, while only one participant from the control group scored 100 (see Appendix 2.1).

2.4.2 Data Analysis

The study aimed to investigate the effectiveness of game-based learning compared to traditional methods in teaching Maltese prepositions to adult learners. The independent variable in this study was the instructional method, consisting of two groups: the control group taught using traditional lecture methods, and the experimental group taught using a game-based learning platform. The students' test scores were the dependent variable. To ensure the independence of the groups and avoid knowledge transfer between them, the groups were kept separate. The researcher paid attention to maintaining diversity among the participants by including individuals from different countries and age ranges.

The data analysis began by exporting the test score data from Excel as a CSV file and importing it into RStudio. The test scores for the control and experimental groups were stored in separate data structures named *controlScores* and *experimentScores*, respectively.

```
controlScores <- read.table("D:/StatData/controlScores.csv",
quote="\"", comment.char="")
experimentScores <- read.table("D:/StatData/experimentScores.
csv", quote="\"", comment.char="")
t.test(controlScoresVector,experimentScoresVector)
```

The data were then converted to vector format:

```
controlScoresVector <- controlScores$V1
experimentScoresVector <- experimentScores$V1
```

To compare the means of the two groups, a Welch two-sample t-test was conducted. The t-test is appropriate for comparing the means of two independent groups when the dependent variable is continuous, as it was in this case (Hayes, 2021). The t-test yielded the following results:

```
Welch Two Sample t-test
Data: controlScoresVector and experimentScoresVector
t = -2.7519, df = 60.913, p-value = 0.007794
alternative hypothesis: true difference in means is not equal
to 0
95 percent confidence interval: -23.758843 -3.761319
sample estimates: mean of x mean of y 59.67742 73.43750
```

2.4.3 Interpretation of Quantitative Results

Comparing the average scores, the experimental group had a higher average test score.

The mean test score for the control group was 59.67742, while the mean score for the experimental group was 73.4375. The higher scores obtained by the experimental group suggest that the game-based learning method was more effective than the traditional lecture method.

The negative t-value (−2.7519) indicates that the experimental group's mean score was significantly higher than the control group's mean score. The p-value of 0.007794 indicates that there is a 0.7794% chance of obtaining these results due to random chance alone, suggesting a statistically significant difference between the groups. The standard deviation (SD) was calculated to measure the spread of scores within each group. The control group had an SD of 19.14573, while the experimental group had an SD of 20.53469.

To visualize the differences between the groups, three figures are provided. Figure 2.1 is a box plot generated in Excel, showing the distribution of scores for the control and experimental groups. The box plot indicates that the experimental group had a higher average score compared to the control group.

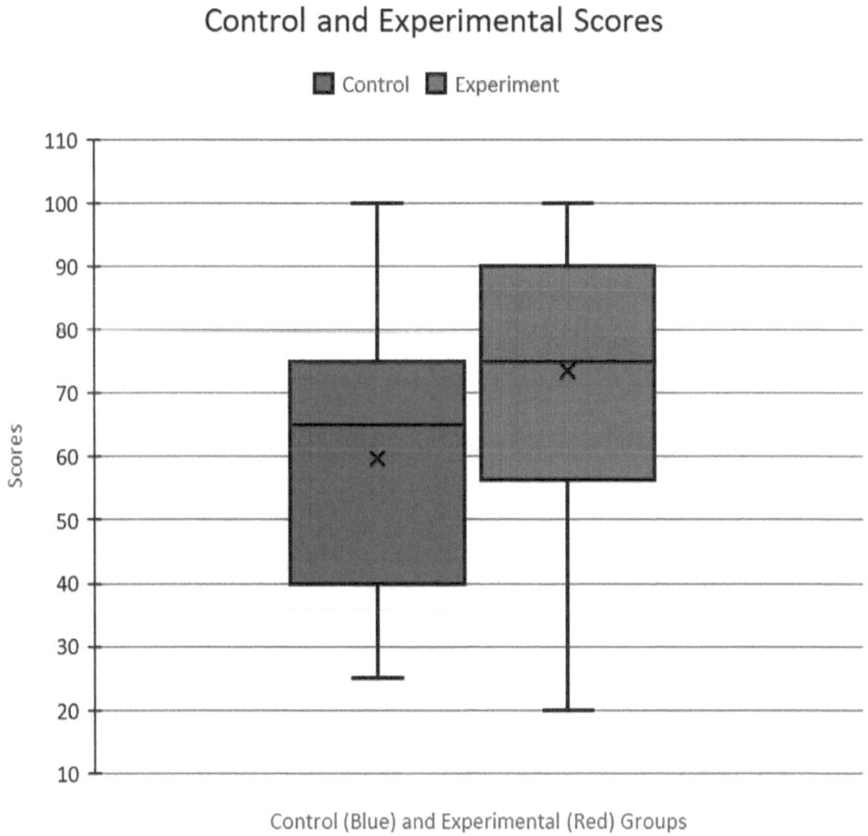

FIGURE 2.1 Box plot comparing scores of the control and experimental groups, showing a higher average score in the game-based method group.

Figures 2.2 and 2.3 illustrate histograms generated in RStudio, showcasing the frequency distribution of scores within the control and experimental groups. These histograms provide additional evidence supporting the observation that the experimental group exhibited a higher concentration of scores toward the upper end, whereas the control group's scores were more clustered in the high 60s range. There is a very good chance (p-value of 0.007794) that adult students who learned Maltese

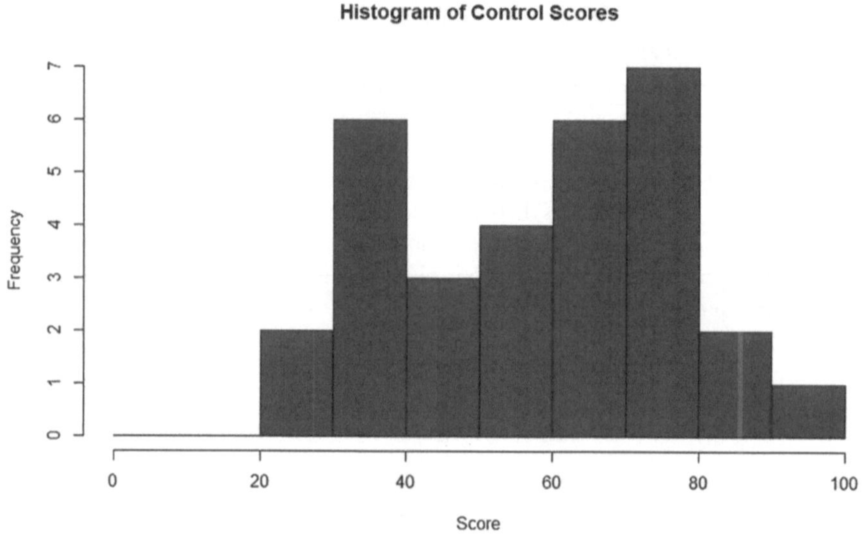

FIGURE 2.2 The frequency distribution of scores in the control group.

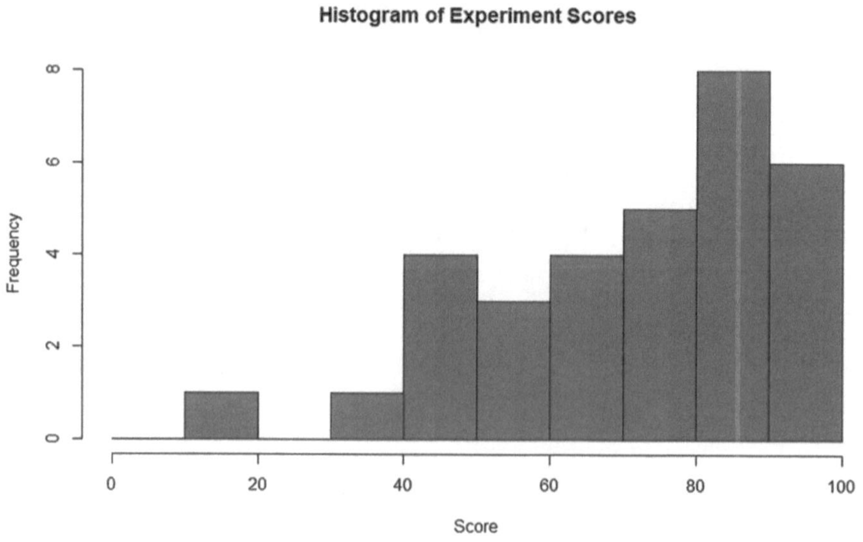

FIGURE 2.3 The frequency distribution of scores in the experimental group.

prepositions through game-based learning did much better on tests than those who learned them through the traditional lecture method. The results suggest a 13% improvement in scores for the experimental group, with an average increase of 13 points out of 100.

The experimental group's higher score than the control group could have been attributed to the intervention of playing the game. *The Mdina Castle's Treasure Quest* game likely provided some form of cognitive stimulation, skill development, or knowledge enhancement that positively impacted the test performance of the experimental group. Therefore, game-based learning can lead to improved outcomes when teaching Maltese prepositions to adult students from diverse backgrounds.

2.4.4 The Qualitative Analysis

The responses from the experimental group's participants were analyzed in a qualitative way to find out how game-based learning affects the learning of Maltese prepositions in adults. The participants' feedback aligns with the aims and results of the study, emphasizing the effectiveness and benefits of the game-based learning approach. Several key themes emerged from the participants' focus group comments, highlighting the positive experiences and outcomes associated with the interactive game *The Mdina Castle's Treasure Quest.*

All 32 participants, including Participant A described the game-based learning approach as "incredibly engaging" and transformed prepositional learning into a "fun challenge" rather than a tedious task. This aligns with the study's aim to investigate the impact of game-based learning on higher education, emphasizing the effectiveness of an engaging learning environment.

Participant B expressed dissatisfaction with traditional lectures, describing them as "monotonous" and lacking practical application. In contrast, the game-based method helped this participant "grasp the concepts much better." This comment reinforces the study's findings, where the game-based learning platform outperformed traditional instruction in terms of prepositional learning outcomes.

> The traditional lectures were quite monotonous. It felt like I was being bombarded with information without really understanding how to apply it. The game-based method helped me grasp the concepts much better.
>
> *Participant B*

Participant C initially expressed skepticism about using games for learning the Maltese prepositions but was pleasantly surprised by the interactive nature of the games. The participant highlighted the ease of remembering and applying the grammatical rules when the Maltese prepositions are combined into articles through the game-based approach. Participant C's comment reinforces the notion that game-based learning provides a unique and effective method for language acquisition.

Participant D emphasized the value of practicing Maltese prepositions in different contexts through the game-based learning platform. By experiencing how prepositions are used in real-life situations, the participant found them more meaningful and easier to remember. This aligns with the study's aim of providing immersive learning experiences and supports the effectiveness of game-based instruction in language acquisition.

Participant E appreciated the collaborative and competitive aspects of the game-based learning, which fostered a "friendly and motivating" environment. His comment highlighted the social dynamics and motivational factors associated with game-based learning, reinforcing the positive attitudes expressed by the participants.

Participant F emphasized the "active participation" and increased engagement experienced with the game-based approach, in contrast to the "limited opportunities" for active involvement in traditional lectures. Participant F's comment aligns with the study's aim of creating immersive learning experiences and highlights the increased motivation and enjoyment associated with game-based instruction.

Participant G reflected on their struggle with Maltese prepositions before engaging with game-based learning. The participant found that the game-based approach "broke down complex concepts" into manageable tasks, allowing for a gradual understanding of Maltese prepositions and improved learning outcomes. Their comment supports the study's findings concerning the effectiveness of game-based instruction in addressing challenging linguistic concepts.

Participant H acknowledged the additional benefits of game-based learning, such as the development of "problem-solving skills" alongside the understanding of prepositions. This claim reinforces the study's emphasis on the immersive and multi-dimensional learning experiences provided by game-based instruction.

Participant I highlighted the "immediate feedback" incorporated into the game-based learning approach. The participant valued the "explanations and corrections" provided within the game, which facilitated learning from errors. This feedback mechanism contributes to the effectiveness of game-based instruction, as observed in the study's results.

Lastly, participant J emphasized the "increased motivation" to study prepositions "outside of class" resulting from the game-based learning method. The participant voluntarily practiced and explored the topic further, indicating a higher level of engagement and self-directed learning. This aligns with the study's focus on the motivational aspects of game-based instruction.

In general, the qualitative analysis of the participants' quotations supported the study's aims and quantitative results, providing evidence for the potential benefits of game-based learning in Maltese language acquisition. The participants' positive experiences, such as increased engagement, motivation, and improved understanding of prepositions, highlight the advantages of using interactive games like *The Mdina Castle's Treasure Quest* in higher education settings. The findings emphasize the importance of immersive learning experiences and offer insights into the effectiveness of game-based language learning.

2.5 LIMITATIONS OF THE STUDY

While this study provided valuable insights into the impact of game-based learning on learning Maltese prepositions in adult learners, it is important to acknowledge its limitations. Firstly, the study was conducted with a relatively small sample size, which may limit the generalizability of the findings to a larger population. Additionally, the study focused specifically on the use of one interactive game, *The Mdina Castle's Treasure Quest*, for learning the Maltese prepositions. The results may vary when different game-based learning platforms or participants are considered.

Furthermore, the study primarily relied on self-reported measures of engagement, motivation, and attitudes toward learning prepositions. While these subjective assessments provided valuable insights, they are susceptible to individual biases and may not fully capture the objective impact of game-based learning on the acquisition of Maltese prepositions. Future studies could incorporate more objective measures, such as performance assessments and cognitive tests, to complement the self-reported data.

Moreover, the study duration was relatively short, with participants engaging in the game-based learning platform for a limited period of two months. The long-term effects and sustainability of the game-based approach were not examined. It would be beneficial to investigate the retention of knowledge and skills over an extended period to assess the long-term effectiveness of game-based learning in language education.

2.6 RECOMMENDATIONS FOR FURTHER STUDIES AND IMPLEMENTATIONS

Based on the findings of this study, several recommendations can be made for further research and potential implementations. It would be valuable to conduct similar studies with larger and more diverse samples to enhance the generalizability of the findings. Including participants from different educational backgrounds, age groups, and cultural contexts would provide a more comprehensive understanding of the effectiveness of game-based learning in various settings.

Additionally, future studies could explore the impact of different game-based learning platforms and instructional designs on language learning outcomes. Comparing the effectiveness of various game mechanics, levels of interactivity, and feedback mechanisms would help identify the key elements that contribute to successful game-based Maltese language instruction.

Furthermore, investigating the long-term effects of game-based learning on Maltese language acquisition is essential. Examining the retention of knowledge, the transferability of skills to real-world contexts, and the sustainability of motivation and engagement over time would provide valuable insights into the durability and practicality of game-based language learning approaches.

In terms of implementation, educational institutions and language instructors could consider integrating game-based learning methodologies into their curricula. Providing opportunities for students to engage in interactive and immersive learning experiences can enhance their motivation, engagement, and language acquisition outcomes. Incorporating game-based elements, such as simulations, puzzles, and challenges, can make Maltese language learning more enjoyable and effective.

2.7 CONCLUSION

This study examined the impact of game-based learning on the acquisition of Maltese prepositions by international adult students in higher education. The findings demonstrate that using the interactive game *The Mdina Castle's Treasure Quest* significantly improved prepositional learning outcomes compared to traditional lecture-based instruction. Participants who engaged with the game reported higher engagement, motivation, and positive attitudes toward learning Maltese prepositions. Game-based learning encourages teamwork, competition, and active engagement, which improves comprehension and memory of linguistic topics. The study suggests that game-based learning can be applied beyond language education, benefiting other areas of education and training. The interactive and immersive nature of games can enhance learning experiences and foster critical thinking, problem-solving, and collaboration skills. As a result, this study gives us useful information about how well game-based learning works for teaching Maltese prepositions to adults. It shows how useful it is for language learning and urges more research and improvement of game-based learning methods.

REFERENCES

Adipat, S., Laksana, K., Busayanon, K., Asawasowan, A., & Adipat, B. (2021). Engaging students in the learning process with game-based learning: The fundamental concepts. *International Journal of Technology in Education* (*IJTE*), *4*(3), 542–552. https://doi.org/10.46328/ijte.169

Akintolu M., & Letseka M. (2021). The andragogical value of content knowledge method: The case of an adult education programme in Kwa-Zulu Natal Province of South Africa. *Heliyon, 7*(9), e07929. https://doi.org/10.1016/j.heliyon.2021.e07929

Anastasiadis, T., Lampropoulos, G., & Siakas, K. (2018). Digital game-based learning and serious games in education. *International Journal of Advances in Scientific Research and Engineering* (*IJASRE*), *4*(12), 139–144. https://doi.org/10.31695/IJASRE.2018.33016

Bahari, A. (2023). CANDA: Computer-assisted nonlinear dynamic approach for the L2 teaching in blended and distance learning. *Interactive Learning Environments, 31*(2), 752–778. https://doi.org/10.1080/10494820.2020.1805774

Breien, F.S., & B. Wasson. (2021). Narrative categorization in digital game-based learning: Engagement, motivation & learning. *British Journal of Educational Technology, 52*(1), 91–111. https://doi.org/10.1111/bjet.13004

Bygstad, B., Øvrelid, E., Ludvigsen, S., & Dæhlen, M. (2022). From dual digitalization to digital learning space: Exploring the digital transformation of higher education. *Computers & Education, 182*, 104463. https://doi.org/10.1016/j.compedu.2022.104463

Hayes, A. (2021). T-test: What it is with multiple formulas and when to use them. *Investopedia*. https://www.investopedia.com/terms/t/t-test.asp

Ismaizam, N. M., Rahman, S. F. A., Ahmad, S. N. S. M., Nazri, N. I. I. M., Idris, N. A. A., Ali, N. A., Rafi, N. F. B. M., Mohamad, S. N. A., Rahim, A. A. A., Rashid, K. K. A., & Aldaba, A. M. A. (2022). An integration of game-based learning in a classroom: An overview (2016–2021). *International Journal of Academic Research in Progressive Education and Development, 11*(1), 1207–1221. https://doi.org/10.6007/IJARPED/v11-i1/12347

Jääskä, E., Lehtinen, J., Kujala, J., & Kauppila, O. (2022). Game-based learning and students' motivation in project management education. *Project Leadership and Society, 3*, 100055. https://doi.org/10.1016/j.plas.2022.100055

Karagiorgas, D. N., & Niemann, S. (2017). Gamification and game-based learning. *Journal of Educational Technology Systems, 45*(4), 499–519. https://doi.org/10.1177/0047239516665105

Karakoç, B., Eryılmaz, K., & Turan Özpolat, E. (2022). The effect of game-based learning on student achievement: A meta-analysis study. *Technology, Knowledge and Learning, 27*(1), 207–222. https://doi.org/10.1007/s10758-020-09471-5

Kılıçkaya, F. (2022) Pre-service language teachers' online written corrective feedback preferences and timing of feedback in computer-supported L2 grammar instruction. *Computer Assisted Language Learning, 35*(1–2), 62–87. https://doi.org/10.1080/09588221.2019.1668811

Kilimci, A. (2017). Integrating cognitive linguistics insights into data-driven learning: Teaching vertical prepositions. *Journal of Language and Linguistic Studies, 13*(2), 681–719.

Lin, C.-J., Hwang, G.-J., Fu, Q.-K., & Chen, J.-F. (2018). A flipped contextual game-based learning approach to enhancing EFL students' English business writing performance and reflective behaviors. *Journal of Educational Technology & Society, 21*(3), 117–131. https://www.jstor.org/stable/26458512

Prabhaker, M., Mani, P.C., Uttam, S., Amit, K., & Mayilvaganan, S. (2019). Selection of appropriate statistical methods for data analysis. *Annals of Cardiac Anaesthesia, 22*(3), 297–301. https://doi.org/10.4103/aca.ACA_248_18

Saunders, M., Lewis, P., & Thornhill, A. (2012). *Research Methods for Business Students.* Harlow: Pearson Education Limited.

Thomas, L. (2022). *Quasi-Experimental Design | Definition, Types & Examples.* Scribbr. https://www.scribbr.com/methodology/quasi-experimental-design/

Yaccob, N.S., & Yunus, M.M. (2019). Language games in teaching and learning English grammar: A literature review. *Arab World English Journal, 10*(1), 209–217. https://doi.org/10.24093/awej/vol10no1.18

APPENDIX 2.1: PARTICIPANT'S DATA

Age	Sex	Country of Origin	Group	Occupation	Score (0–100)
20	F	Albania	Control	Receptionist	75
23	M	Argentina	Control	Salesman	70
18	M	Brazil	Control	Hairdresser	50
33	M	Brazil	Control	Teacher	100
18	M	Brazil	Control	Writer	80
26	F	China	Control	Masseuse	70
24	M	China	Control	Nail technician	70
34	F	Colombia	Control	Maid	25
19	M	Egypt	Control	Builder	75
37	M	Egypt	Control	Carpenter	75
36	F	Hungary	Control	Waitress	55
36	M	India	Control	Carer	55
32	F	India	Control	Nurse	75
28	M	Israel	Control	Online gaming manager	55

(Continued)

(*Continued*)

Age	Sex	Country of Origin	Group	Occupation	Score (0–100)
35	F	Japan	Control	Restaurant chef	35
21	F	Japan	Control	Digital marketing specialist	50
18	F	Kenya	Control	Nun	65
18	M	Lebanon	Control	Builder	85
21	M	Lebanon	Control	Sprayer	65
26	M	Libya	Control	Hotel manager	30
36	M	Morocco	Control	Arabic teacher	70
31	F	Pakistan	Control	Nurse	40
32	F	Pakistan	Control	Nurse	75
35	F	Philippines	Control	Carer	35
27	F	Russia	Control	Sales representative	55
29	F	Russia	Control	Event organizer	35
19	M	Somalia	Control	Builder	85
23	F	Somalia	Control	Builder	45
35	M	Syria	Control	Builder	40
24	F	Venezuela	Control	Maid	35
28	M	Venezuela	Control	Marketing coordinator	75
56	F	Albania	Experiment	Beautician	55
58	M	Albania	Experiment	Customer service representative	75
38	M	Argentina	Experiment	Restaurant chef	65
42	F	Colombia	Experiment	Maid	95
61	M	Colombia	Experiment	Pensioner	20
39	M	Egypt	Experiment	Wall painter	100
51	F	Egypt	Experiment	Builder	75
60	F	Egypt	Experiment	Pensioner	100
58	M	Greece	Experiment	Electrician	85
48	M	Greece	Experiment	Plumber	45
48	M	Greece	Experiment	Salesman	40
62	M	Hungary	Experiment	Pensioner/gardener	60
43	F	Hungary	Experiment	Beautician	85
49	F	India	Experiment	Carer	50
53	F	India	Experiment	Nurse	75
51	M	Israel	Experiment	Software developer	45
44	M	Japan	Experiment	Programmer	75
59	M	Japan	Experiment	Financial analyst	70
61	M	Kenya	Experiment	Pensioner	100
60	M	Lebanon	Experiment	Pensioner	65
65	M	Lebanon	Experiment	Pensioner	90
46	M	Libya	Experiment	Builder	65

(*Continued*)

(*Continued*)

Age	Sex	Country of Origin	Group	Occupation	Score (0–100)
50	F	Libya	Experiment	Plumber	90
60	M	Morocco	Experiment	Pensioner/carpenter	55
44	M	Morocco	Experiment	Truck driver	85
53	F	Pakistan	Experiment	Nurse	95
51	M	Philippines	Experiment	Carer	80
56	M	Philippines	Experiment	Carer	50
44	M	South Africa	Experiment	Delivery man	90
57	M	Syria	Experiment	Builder	100
65	F	Venezuela	Experiment	Housewife	85
44	F	Venezuela	Experiment	Maid	85

3 Gamification and Digitalization in Higher Education

Enhancing Learning and Bridging the Gap in Malaysia via Lean Six Sigma Training

Hasnida Ab-Samat, Nur Amalina Muhammad, Anis Masturina Azhan, and Muhamad Nazrin Muhd Zukri

3.1 INTRODUCTION

Lean Six Sigma has long been established and implemented in some developed countries with the capabilities of developing a decent and complete engineer with lots of skills being applied. One distinctive feature of Lean Six Sigma is systemized knowledge dissemination and institutionalization through a special infrastructure of people within the organization who undergo training and certification known as the belt system. The belt system employs color coding, starting with yellow belts and followed by green and black belts. In the yellow belt, the practitioner learns and gets acquainted with Plan-Do-Check-Act (PDCA), a problem-solving methodology commonly associated with continuous improvement in Lean manufacturing. Ramu (2016) states that the introduction of training can include entry-level employees looking to improve their knowledge or management executives who need an overview of Six Sigma and DMAIC to offer support as sponsors or champions. By taking part in the Lean Yellow Belt (Lean YB) training, the graduates are believed to be familiar with some of the tools and methods used in this problem-solving process.

According to Laureani et al. (2012), implementing Six Sigma motivates employees in an organization. Once implemented, employee engagement becomes a critical factor within the organization, increasing satisfaction levels even further. In this case, graduates that have undergone their Lean YB training during their studies will be given primary exposure to how the training will benefit them in securing a job. Moreover, Svensson et.al. (2015) outlined that the Lean Six Sigma Yellow Belt training was highly interactive, with many exercises and a simulation demonstrating how waste and variation occur in a process and, more importantly, how to eliminate waste

 DOI: 10.1201/9781003400691-3

and variation using Lean and Six Sigma tools. The findings also explored and evaluated the critical success factors such as strategic and visionary leadership, developing organizational readiness, project selection and prioritization, and effective communication at all levels, vertically and horizontally, which is used to overcome the challenges of implementing Lean Six Sigma.

According to the findings of the study, Lean Six Sigma Yellow Belt training has a role in industry and higher education. Alhosani and Tariq (2020) achieved success on the reduction of training preparation time for employees using Lean Six Sigma methodology as they discovered that Six Sigma principles could be used to improve the training process by involving staff at all organizational levels, achieving top management support, creating a culture of continuous learning, and regularly evaluating employee skills to identify gaps and improve resourcing capabilities. Relating this case to graduates with Lean YB training will distinguish themselves to be an important asset to the hiring company as they can apply their knowledge and skills gained from the training without any other training needed to be given.

3.1.1 Objective

1. To assess the impact of digitalization and gamification on student engagement and learning outcomes in higher education settings, with a particular focus on Malaysia.
2. To explore the effectiveness of Lean Six Sigma training as a pedagogical framework for integrating digitalization and gamification into the curriculum, and to identify best practices for enhancing the development of critical skills and competencies required for success in the contemporary workforce.

3.1.2 Research Questions

- Research Question 1: How does the incorporation of digitalization and gamification in higher education settings in Malaysia influence student engagement, motivation, and learning outcomes?
- Research Question 2: What are the most effective strategies for implementing Lean Six Sigma training as a pedagogical framework for integrating digitalization and gamification in the curriculum to enhance the development of essential skills and competencies for employability in the workforce?

3.2 LEAN YB TRAINING

3.2.1 Lean Training

According to Laureani and Antony (2011) since the 1980s, when the terms were first codified and defined, Lean and Six Sigma have taken different paths: Lean was first applied in the Ford plants of Michigan in 1913 and refined in Japan (within the Toyota Production System), while Six Sigma was born in the United States (within the Motorola Research Center). Lean Six Sigma established hierarchies and career paths for quality implementation roles using martial arts. As the success of Six Sigma grew, new terms such as Yellow Belt, Green Belt, and Black Belt were popularized to assist employers in training their employees.

3.2.2 Lean YB Waves

In 2013, Bose Corporation launched its first Malaysian manufacturing plant in Penang, Malaysia, showcasing a commitment to creating local talent. Partnering with Universiti Sains Malaysia (USM), the training program "Lean YB" was introduced for undergraduates aiming to equip students with practical Lean knowledge and enhance their marketability. The Memorandum of Agreement signed in September 2013 solidified this educational alliance. By fostering collaborations, Bose Corporation contributes to manufacturing growth and empowers students with valuable skills for competent future employability. Given the increasing demand for Lean knowledge from industries, USM included Lean training in the curriculum through an elective course.

The path of Lean YB training involves a series of dedicated and comprehensive phases. Due to resource and material constraints, it may not be possible to concurrently provide Lean training to a significant number of individuals. Hence, a selection process comprising an entry test is employed to find 24 students participating in the Lean YB training. This selection process ensures the optimal utilization of resources, including instructors, materials, and equipment, and effective delivery during the training. This course has been in operation for a decade since 2013, with 10 batches of Lean YB students being categorized and referred to as waves.

3.2.3 Lean YB Training – Learning Structure

The students undergo 10 weeks of training, learning the principles, methodologies, and tools of Lean methods. The teaching plan involves theoretical and practical approaches through mixed learning (conventional, gamification, and digitalization). Lean YB training incorporates face-to-face classes to promote practical knowledge acquisition. This approach enables live interactions between students and instructors, supported by printed materials like textbooks, handouts, and flashcards for reference and deeper understanding. The conventional approach is highly effective in this training, offering direct access to instructors for quick access and guidance, making it a preferred method in higher education. Topics covered in Lean YB training include Lean Enterprise Overview, Aligning the Organization, Fundamentals of Problem Solving, Basic Statistics, and Lean Management Systems.

The Lean Enterprise topic encompasses introducing the Lean YB training program, emphasizing hands-on class learning to provide participants with a firsthand experience and make a lasting impression (*Building with LEGO*). Through practical exercises, the training hones problem-solving skills, teamwork, and the application of basic Lean tools. In Aligning the Organization topic, the class delves into cultivating a Lean culture and fostering continuous organizational improvement. Employee engagement and change management are integral to the curriculum, ensuring a well-rounded learning experience. Furthermore, an introduction to core business processes and entertaining minigames that promote active involvement and retention of course material is provided. Then, the students are introduced to the PDCA overview, providing them with a structured framework for continuous improvement. The participants engage in a physical simulation game (*Simulation-Based Learning*) to reinforce their understanding and practical application of PDCA. Students will learn some statistical tools like statistical process control, Pareto Chart (*Pareto Chart with M&M Candies*), and run charts on basic statistics topics and will be introduced to Lean management principles.

3.2.4 Lean YB Training – Certification

Upon reaching the final phase, participants will achieve the esteemed Yellow Belt certification in Lean YB training, marking a significant milestone in their learning journey. Graduation day is a momentous occasion honoring the dedication and hard work. This recognition boosts morale, motivation, and engagement among the students, further strengthening their commitment. Moreover, the attainment of Yellow Belt certification enhances the participant's employability rate, as it demonstrates their proficiency in process and management improvement as well as basic statistics, which makes them highly sought-after candidates in various industries. The acquired skills in Lean YB training empower them to contribute effectively to their organizations, fueling their professional growth and making them valuable assets in the workforce. This Lean YB training aims to foster engaged participants, encouraging active contribution to continuous improvement endeavors, thus positively impacting their careers and the organizations they serve.

3.3 GAMIFICATION IN LEAN YB TRAINING

Gamification is the application of game design elements, mechanics, and principles in non-game contexts, such as Lean YB training in higher education, to enhance their experiences and drive specific behaviors or desired outcomes. The power of gamification in Lean YB training lies in its ability to transform the learning experience, making it more engaging, motivating, and effective in driving innovation. In Lean YB training, gamification is employed in diverse ways, including *Simulation-Based Learning, Building with LEGO: A Hands-On Approach*, and *Pareto Chart with M&M Candies*. Both physical and digital gamification strategies find application in Lean YB training.

3.3.1 Simulation-Based Training

Simulations are dynamic, interactive models replicating real-world processes, systems, or environments. These simulations mimic actual workflows, production lines, or service processes. In the simulation, students engage actively, making decisions and applying Lean methodologies to enhance efficiency and eliminate waste. One of the significant advantages of simulation-based training is the risk-free nature of the learning experience. Students can explore different improvement strategies without fear of negative consequences for the organization. Mistakes made during the simulation offer valuable learning opportunities, allowing trainees to understand the impact of their decisions and learn from their experiences. Students tackle intricate scenarios, diagnosing root causes and optimizing processes through iterative simulations, refining their problem-solving skills continuously.

In Lean YB training, students engage in a physical simulation involving role-playing as emergency and non-emergency helpline operators (Figure 3.1a and b). They handle scenarios like heart attacks, road accidents, homeless individuals, bank robberies, fires, and more. This unique simulation blends real-life situations with Lean tools, offering hands-on learning experiences that enhance understanding and application of Lean tools such as PDCA, visual management, and waste elimination. The simulation emphasizes collaboration and teamwork, providing students with feedback and reflection using graphs drawn by software SPC XL (e.g., run charts, pie charts, and box plots) for self-assessment and progress evaluation.

3.3.2 Building with LEGO: A Hands-On Approach

Building with LEGO: A Hands-On Approach is an engaging method used in Lean YB training to simulate and optimize production lines. Participants construct a physical representation of the production process using LEGO bricks, mimicking the flow of materials and products (Figure 3.2a and b). They run iterative simulations, applying Lean tools like 5S, visual management, pull systems, Heijunka, and SMED. Throughout each run, participants track various performance metrics to assess profitability and efficiency, including the number of operators, work-in-progress, finished products, and shipments. Analyzing the data allows participants to identify effective Lean tools and areas for improvement, fostering experiential learning and data-driven decision-making. The LEGO simulation offers a visual and collaborative learning experience, enabling participants to effectively grasp Lean principles, optimize processes, and embrace continuous improvement.

3.3.3 Pareto Chart with M&M Candies

Pareto Chart is one of the Lean tools to prioritize issues based on frequency and impact. M&M Candies are used for the Pareto Chart to create a hands-on simulation. The assorted colors of M&M candies can represent various categories of defects or issues in a process. Students collect data by counting the occurrences of each colored candy and sorting them in descending order, representing the frequency of different issues (Figure 3.3a and b). A bar chart visually represents the categories, with the most frequent on the left. A cumulative percentage line indicates where the 80/20 rule applies, highlighting the critical few categories contributing most to the issues. Students are encouraged to create a Pareto Chart using SPC XL software to compare it with their manually drawn charts (Figure 3.3c and d). The students demonstrate proficiency in understanding the methodologies and using modern tools and software.

FIGURE 3.1 (a) Student roles play and paricipations and (b) Simulation-based training.

FIGURE 3.2 (a) Hands on activity and (b) Building with LEGO: a hands-on approach.

FIGURE 3.3 (a) A pack of M&Ms candies provided as data set and (b) Drawing of Pareto Chart based on numbers and colors of M&M Candies

(Continued)

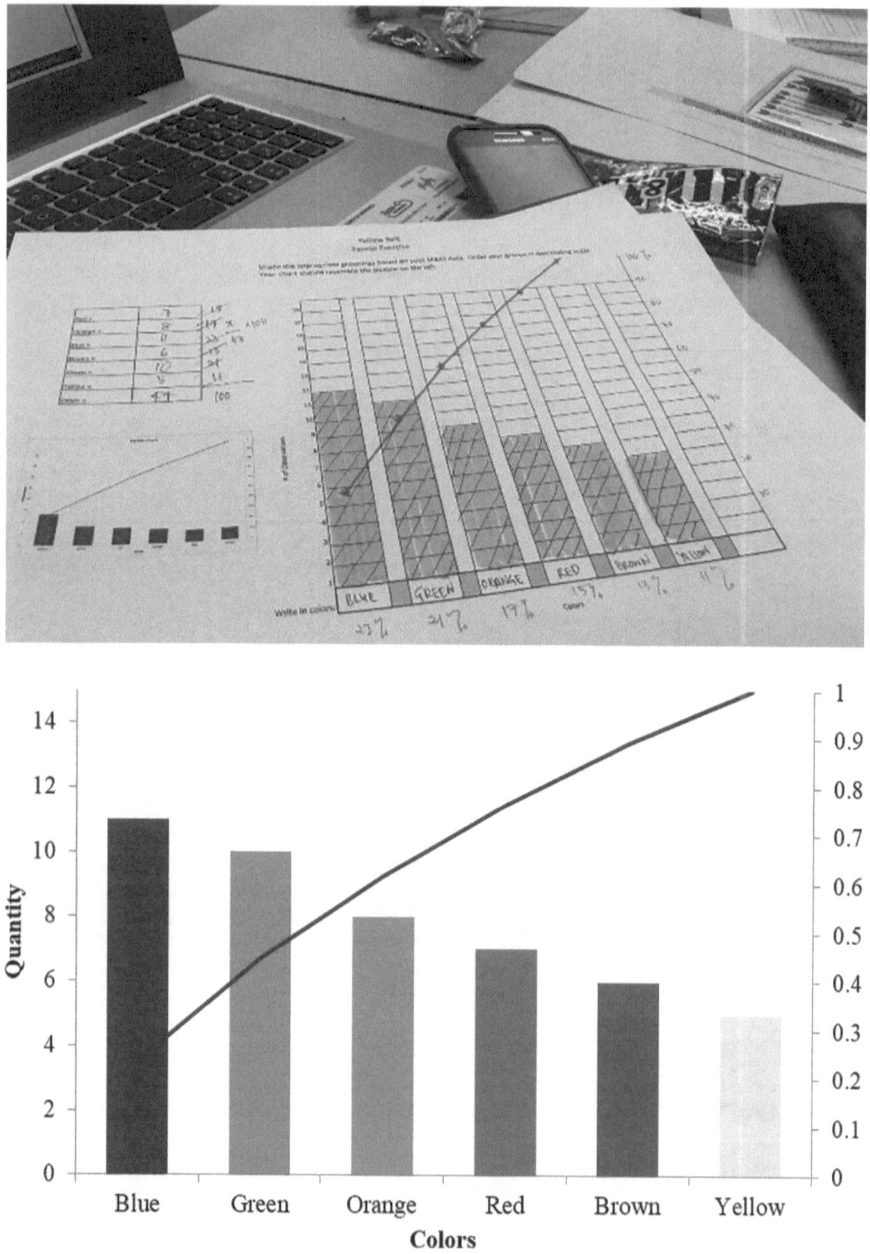

FIGURE 3.3 (*Continued*) (a) A pack of M&Ms candies provided as data set and (b) Drawing of Pareto Chart based on numbers and colors of M&M Candies

3.4 DIGITALIZATION IN LEAN YB TRAINING

3.4.1 Test via Socrative

In the approach toward blended learning, a test was conducted online using the Socrative application. Socrative is an online tool designed to enhance higher education pedagogy through suitable interactive learning experiences. Educators can create real-time quizzes and assessments, gaining immediate insights into students' understanding. The tool offers engaging question formats and activities, promoting active participation and motivation. It is well-suited for remote and hybrid learning environments, providing convenience and flexibility. Through data-driven insights, instructors can make informed decisions to adapt their teaching methods. Socrative supports formative assessment, helping students track their progress and identify areas for improvement. The tool streamlines assessment creation and grading, saving time for personalized instruction. Integration with learning management systems and accessibility options ensures inclusivity and ease of use. Effective integration in Socrative can transform higher education, fostering more interactive and dynamic learning environments.

3.4.2 Ice-Breaking in Padlet

Padlet, an online collaborative tool, proves invaluable as an ice-breaking tool for digitalizing higher education pedagogy. Through virtual bulletin boards, students introduce themselves, fostering a friendly atmosphere and encouraging interaction. Collaborative brainstorming activities promote active participation and student engagement. Icebreakers involving gallery walks and group challenges enhance communication skills and teamwork. Instructors can build connections by sharing personal information.

Furthermore, Padlet facilitates post-activity reflections and feedback, promoting continuous improvement. In the context of flipped classrooms and online discussion forums, Padlet offers a visually appealing alternative for pre-class discussions and multimedia-rich interactions. Ultimately, these ice-breaking techniques create a strong sense of community and enrich the learning experience in the digital higher education landscape.

3.4.3 Class Reflections through Flipgrid

Flipgrid is an online tool that facilitates video reflections, making it ideal for digitalizing higher education pedagogy. Students can record personalized reflections, promoting active participation and critical thinking. The tool empowers introverted students and fosters communication skills. It allows instructors to provide feedback and fosters peer learning and interaction. Integration with learning management systems adds to its versatility. Video reflections can be assessed and compiled into digital portfolios, fostering a learning community. Flipgrid enriches the higher education experience by creating an engaging and inclusive virtual classroom environment. This provides exposure to technology and computer applications in learning. Students have also been encouraged to use social platforms as a tool for learning. Figure 3.4a and b shows the screencaps of the applications used in the Lean YB training for Padlet and Flipgrid.

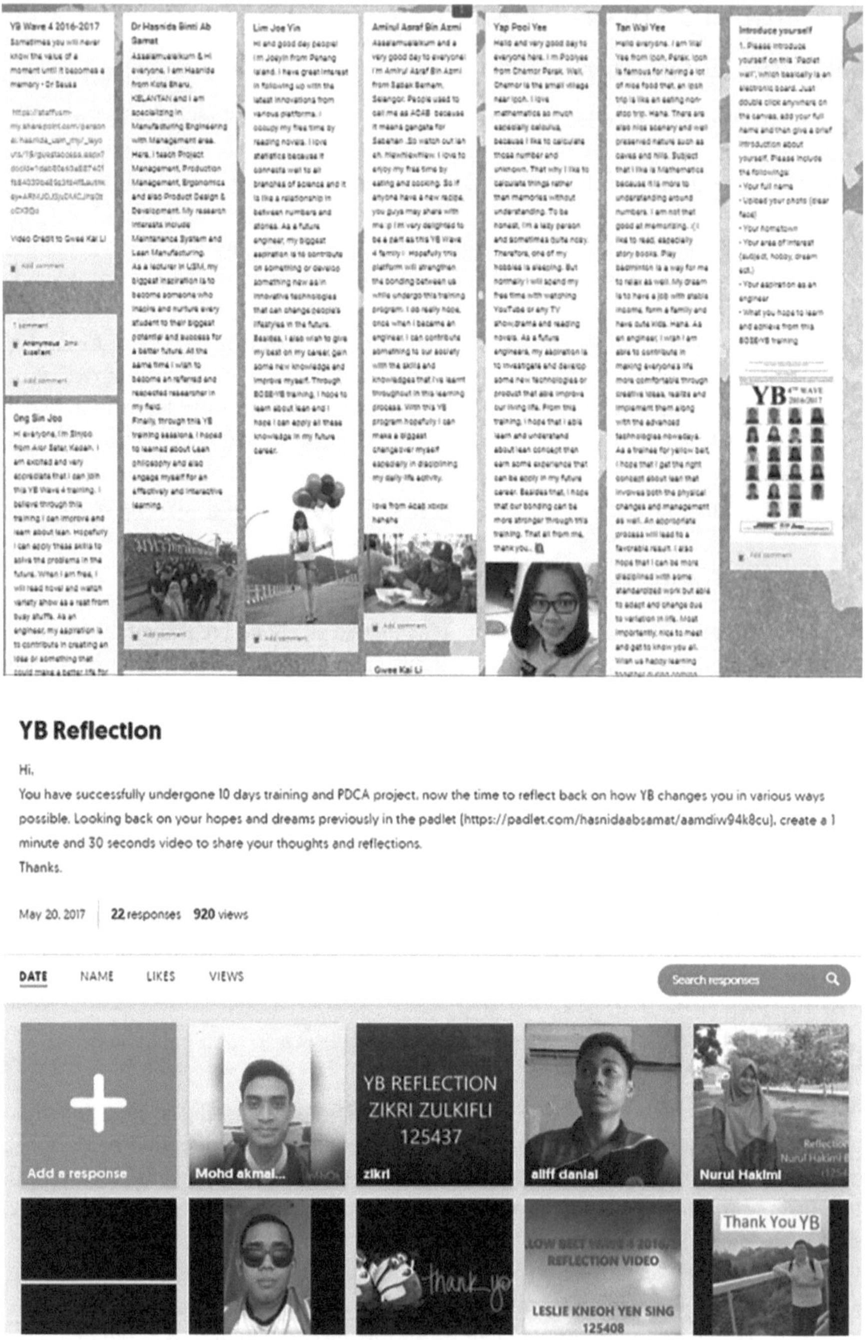

FIGURE 3.4 Screencaps in Padlet and Flipgrid.

3.4.4 Social Engagement in Facebook

In the digitalization of higher education pedagogy, utilizing social media platforms like Facebook can prove beneficial in various aspects. Firstly, Facebook facilitates class discussions and collaboration, creating a virtual learning community where students can actively engage in academic conversations and share their ideas. Secondly, it serves as an efficient learning resource repository, where instructors can organize and share diverse materials, including articles and videos, for easy access by students. Additionally, through private Facebook groups, educators can foster parent and stakeholder engagement, enabling effective communication and updates on class activities and student progress.

Lastly, Facebook offers students a space for reflection and journaling, allowing them to document their thoughts, progress, and learning experiences, thereby promoting a deeper understanding of the subject matter (Figure 3.5). By leveraging these features strategically, educators can enrich the digital learning experience and cultivate a dynamic and collaborative virtual classroom environment. However, it is crucial to consider privacy and accessibility concerns while implementing these practices.

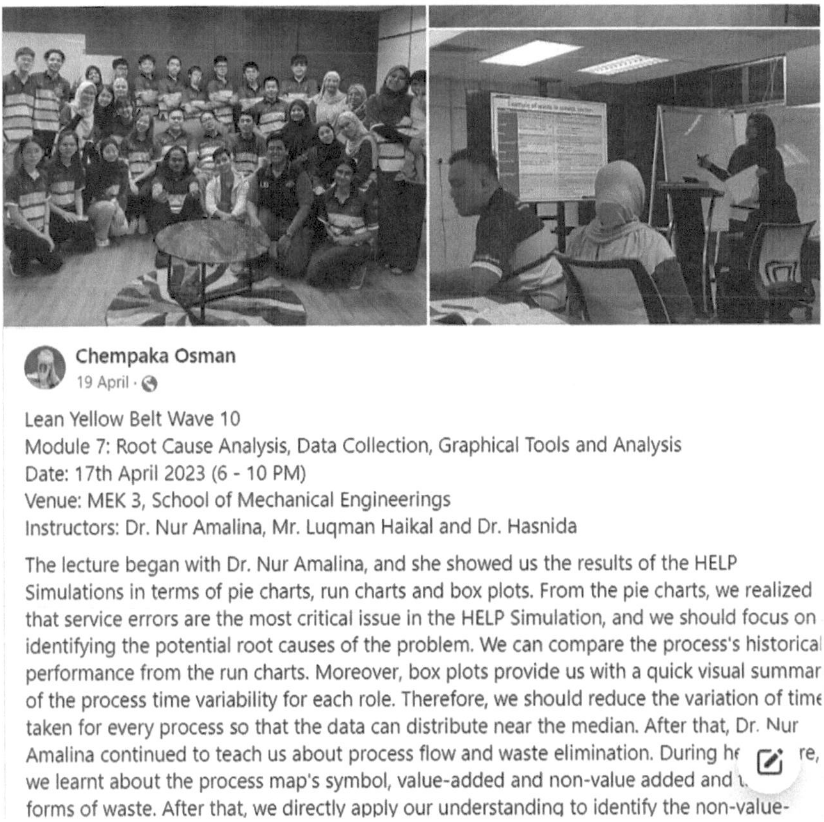

FIGURE 3.5 Screencap of Facebook USM Lean Six Sigma.

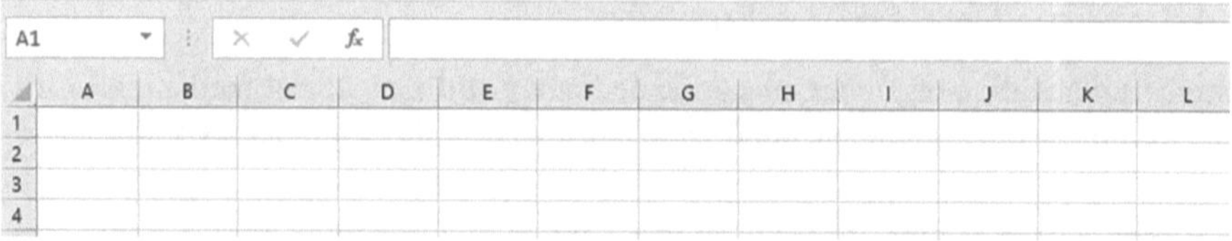

FIGURE 3.6 SPC XL in Microsoft Excel.

3.4.5 Modern Tool Usage

Another example of digitalization in classrooms for higher education is the modern tool and technology usage for learning. During Lean YB training, the classroom utilized SPC XL, an add-in for Microsoft Excel, to provide students with advanced statistical analysis capabilities (Figure 3.6). SPC XL became easily accessible and practical, allowing students to apply statistical concepts in real-world scenarios. SPC XL offers several functions that relate statistical analysis (e.g., correlation matrix, t-test, one-way ANOVA, MSA, etc.), control chart (e.g., XbarR, XbarS, p-chart, c-chart, etc.), process capability analysis (e.g., histogram, Cpk, Pareto Chart, Scatter Plot, etc.), and root cause analysis (e.g., process flow, cause-and-effect analysis, CNX, SOP). Using SPC XL in Lean YB training allows students to create standardization templates for process monitoring and improvement. These templates are structured and consistent formats to analyze data and make informed decisions based on statistical insights.

3.5 RESEARCH METHODOLOGY

To address the first research objective related to the impact of digitalization and gamification on student engagement and learning outcomes, the research methodology involved the design and administration of a survey for data collection. Statistical analysis was conducted using one-way ANOVA and t-tests, facilitated by SigmaZone software, to evaluate the influence of digitalization and gamification strategies on student engagement and learning outcomes in higher education settings in Malaysia. To fulfill the second research objective concerning the effectiveness of Lean Six Sigma training as a pedagogical framework, the research methodology encompassed the development and distribution of a survey to assess the integration of Lean YB training in the curriculum. Statistical analysis, employing one-way ANOVA and t-tests through SigmaZone software, was used to investigate how this training impacts the development of critical skills and competencies relevant to employability and job performance in the contemporary workforce. The target population for this study consisted of graduates who had undergone Lean YB from 2013 to 2022 that includes Wave 1 to Wave 10. This will total up to 218 targeted respondents for this study.

3.6 IMPACT OF LEAN YB TRAINING ON SKILLS

Mamat et al. (2015) mentioned that skills can be categorized into two groups: soft and hard skills. Soft skill includes intrapersonal and interpersonal capabilities linked

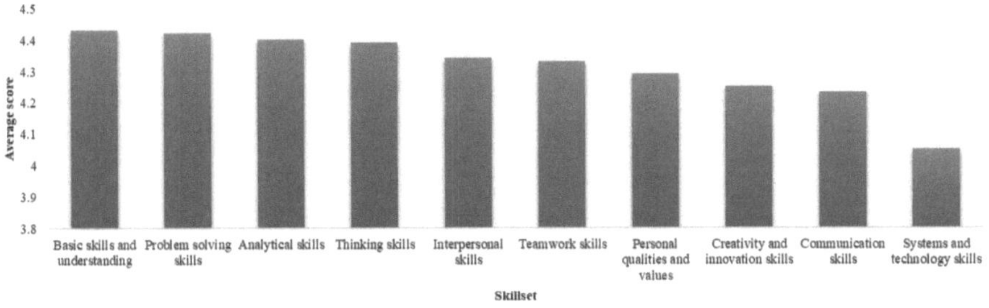

FIGURE 3.7 Skillsets in Lean YB training.

to self-management and communication (Yadav et al., 2021). Meanwhile, hard skill refers to tangible action, including technical knowledge involving equipment, software, data analytics, and other domain expertise such as procedure (Belhadi et al., 2019). Figure 3.7 presents a bar chart illustrating the skillsets imported from Lean YB training. This evaluation is based on the perception and the value added by the graduates after completing Lean YB training. The survey outlines 10 major skillsets, with most respondents expressing strong agreement with the enumerated skillset from the questionnaire. The data reflect the positive impact and effectiveness of the training program in equipping participants with essential skillsets.

Lean YB training gives students ample skillsets, including soft and hard skills, essential for career success. This training provides a conducive environment and effective classroom management as holistic care to ensure a balanced blend of soft and hard skill development. Every classroom layout is flexible and adjustable to accommodate diverse learning activities and teaching approaches. This course uses movable furniture, adjustable seating, and flexible layouts easily reconfigurable to fit different traditional classes, gamification and digital software use, or group discussion. This aligns with the previous study by Lista et al. (2022), where active teaching approaches contribute to soft skills development, regardless of the Lean implementation level. However, perceived learning of hard skills varies with the Lean implementation level.

3.6.1 Lean YB Training Soft Skill

In the context of Lean management, soft skills are behaviors and attitudes gained through experience and practices, reflecting the sociocultural aspects of individuals, teams, or organizations. Referring to Figure 3.7, the feedback data prove that the highest average score for skill is the basic and understanding skillsets. These skills are the most important skill as this skill is the foundation in Lean YB training to pertain to the ability to grasp five principles of Lean management: value, value stream, flow, pull, and perfection. Next, from Figure 3.7, there are another three skills where the average score nearest to the basic skill and understanding skillset are closely related. The skills are problem-solving, analytic skills (technical skills), and thinking skills, respectively. Upon completing Lean YB training, students gain a solid foundation in basic and understanding skills, facilitating their ability to retain the knowledge. This indirectly enables the students to develop and integrate with other skills that link to the foundation, which are problem-solving, analytical, and critical thinking skills, empowering them to face challenges. The survey data reaffirm the importance of these skills, highlighting the Lean YB training effectiveness in cultivating well-rounded Lean Six Sigma practitioners capable of delivering substantial organizational value.

3.6.2 Lean YB Training Technical Skill

Burch and Smith (2019) describe hard skills as an integration of the right technique and relevant technologies to achieve expected results, while Tortorella et al. (2020) defined hard skills as a set of technical expertise that can be easily formalized and shared with other people using resources such as infographics, videos, etc. From Figure 3.7, the highest technical skill is analytical skill. In Lean YB training, analytical skills are critical for the students to effectively analyze data, identify trends, and draw conclusions from information. Some key analytical skills taught in the training module include visualizing problems via process mapping before conducting root cause analysis.

3.6.3 Psychomotor, Cognitive, and Affective

Lean YB training utilizes Psychomotor, Cognitive, and Affective learning, a comprehensive and integrated approach to learning that looks to optimize the educational process and enhance knowledge retention. In Lean YB training, psychomotor learning is observed by the instructor as the student acquires and develops the practical skills necessary to apply methods effectively through hands-on learning and repetition and unlock skill mastery. Cognitive learning in Lean YB training involves gaining knowledge, understanding principles, and developing critical thinking skills related to process improvement and waste reduction elimination while being exposed to lectures, presentations, case studies, and gamification.

This Lean YB training gamification aligned with Deif (2017) idea, where the gamification process in Lean thinking should be coordinated with the learning cycle, often commencing with motivation and progressing to cognitive exercises within a social setting. This supports gamification as an effective mechanism for teaching in higher education. Affective learning plays an important role during hands-on learning sessions, as the students will often engage in group discussions and presentations to address specific problems. This emotional engagement ensures a deeper connection to promoting active participation and knowledge retention.

3.7 IMPACT OF LEAN YB TRAINING ON EMPLOYABILITY RATE

From 2014 to 2023, the impact of Lean YB training on 122 graduates in the School of Mechanical Engineering, USM, yielded noteworthy results: 48% of these graduates secured a job in less than six months, highlighting the immediate employability enhancement offered by the training. While the rest took more than six months to secure a job, the overall outcomes speak volumes about the transformative potential of Lean YB training on the career prospects of the graduates.

Lean YB training equips graduates with skills matching industry requirements, increasing marketability and employability. The training provides a versatile employability skillset (as mentioned in the previous section) applicable across various industry sectors, inducing positive character attitude changes. Industries value graduates with a strong skillset and the ability to learn quickly, key attributes developed through Lean YB training. Thus, graduates who have completed Lean YB training are in demand and have a competitive edge in the job market, even with limited work experience. Their proficiency in Lean methodologies and problem-solving skills sets them apart from other graduates and makes them highly appealing to various industries.

Graduates develop adaptability, taking initiative, accountability, and teamwork skills. They adopt a solution-oriented mindset and commit to continuous improvement through Lean YB training. These attributes make them highly sought-after by employers seeking versatile and proactive professionals. Graduates with Lean YB training possess the qualities necessary to thrive in dynamic work environments, making them stand out in the job market and increasing their chances of securing rewarding career opportunities. Due to their transformed character attitudes, they become valuable assets to organizations, contributing to enhanced productivity and success.

Lean YB training is highly relevant to industries as it equips graduates with valuable Lean methodologies, problem-solving, and continuous improvement skills. This relevancy makes them adept at identifying and resolving industry-specific challenges and inefficiencies, enhancing operational efficiency and productivity. As industries increasingly seek professionals to drive process improvements and streamline operations, graduates with Lean YB training become highly employable. The capacity to provide immediate value and make a positive impact significantly enhances their employability rate, setting them apart in the job market. Industries acknowledge the advantages of hiring trained graduates who can contribute to organizational success, leading to a growing demand for individuals with Lean YB training across diverse sectors.

In Lean YB training, completing a PDCA project in industries is an essential prerequisite for certification. This requirement exposes students to networking opportunities with industry professionals, mentors, and peers, facilitating valuable connections. Engaging with experienced individuals offers insights into real-world challenges and best practices. These connections can lead to job opportunities through referrals or recommendations, providing a competitive edge in the job market. Additionally, building a strong peer network fosters collaboration and knowledge sharing. Enhanced visibility within the industry may lead to access to unadvertised job openings or early career advancements.

3.8 DISCUSSION

Lean Six Sigma is a well-established methodology known for developing skilled engineers and improving processes through continuous improvement. The methodology is disseminated and institutionalized through a belt system, starting with yellow belts and progressing to green and black belts. The training exposes students to problem-solving methodologies like PDCA commonly associated with Lean manufacturing. Implementing Six Sigma in organizations has been found to increase employee motivation and engagement, leading to higher satisfaction levels. Graduates who undergo Lean YB training are better prepared to secure jobs, as they have been exposed to its benefits. The Lean Six Sigma Yellow Belt training is highly interactive, involving exercises and simulations that demonstrate waste and variation in processes and how to eliminate them using Lean and Six Sigma tools. The success factors for implementing Lean Six Sigma include strategic leadership, organizational readiness, project selection, and effective communication at all levels.

3.9 CONCLUSIONS

The study findings indicate that Lean Six Sigma Yellow Belt training has a significant role in both industry and higher education. The application of Six Sigma principles in training preparation for employees has been successful, leading to reduced preparation time and improved resourcing capabilities. Graduates with Lean YB training become valuable assets to hiring companies, as they can apply their knowledge and skills without further training. Therefore, it is important to integrate Lean Six Sigma training in higher education to equip graduates with valuable problem-solving skills, enhancing their employability and contributing to industry success.

REFERENCES

Alhosani, M., S., & Tariq, M.U. (2020). Reduction of Training Preparation Time for Employess using Lean Six Sigma Methodology. *International Journal of Innovation, Creativity and Change,* 14(8), 681–702.

Belhadi, A., Touriki, F. E., & Elfezazi, S. (2019). Evaluation of critical success factors (CSFs) to lean implementation in SMEs using AHP: A case study. International Journal of Lean Six Sigma, *10*(3), 803–829. https://doi.org/10.1108/IJLSS-12-2016-0078

Burch V, R.F., & Smith, B. (2019). Using simulation to teach lean methodologies and the benefits for Millennials. Total Quality Management and Business Excellence, 30(3–4), 320–334. https://doi.org/10.1080/14783363.2017.1303330

Deif, A. (2017). Insights on lean gamification for higher education. International Journal of Lean Six Sigma, *8*(3), 359–376. https://doi.org/10.1108/IJLSS-04-2016-0017

Laureani, A., & Antony, J. (2012). Standards for Lean Six Sigma certification. International Journal of Productivity and Performance Management, *61*(1), 110–120. https://doi.org/10.1108/17410401211188560

Lista, A. P., Tortorella, G. L., Bouzon, M., Thürer, M., & Jurburg, D. (2022). Soft and hard skills development in lean management trainings. International Journal of Lean Six Sigma, 13(5), 1137–1158. https://doi.org/10.1108/IJLSS-06-2021-0116

Mamat, R. C., Deros, M., Nizam, M., Rahman, A., Khalil Omar, M., & Abdullah, S. (2015). Soft lean for successful lean production system implementation in Malaysia Automotive SME Framework. *Jurnal Teknologi, 77*(27), 141–150. www.jurnalteknologi.utm.my

Ramu, G. (2016). *The Certified Six Sigma Yellow Belt Handbook.* Quality Press.

Svensson, C., Antony, J., Ba-Essa, M., Bakhsh, M., & Albliwi, S. (2015). A Lean Six Sigma program in higher education. *International Journal of Quality & Reliability Management, 32*(9), 951–969.

https://doi.org/10.1108/IJQRM-09-2014-0141

Tortorella, G. L., Fettermann, D., Cauchick Miguel, P. A., & Sawhney, R. (2020). Learning organisation and lean production: Empirical research on their relationship. *International Journal of Production Research, 58*(12), 3650–3666. https://doi.org/10.1080/00207543.2019.1633028

Yadav, N., Shankar, R., & Singh, S. P. (2021). Critical success factors for lean six sigma in quality 4.0. International Journal of Quality and Service Sciences, *13*(1), 123–156. https://doi.org/10.1108/IJQSS-06-2020-0099

4 Enhancement of Polya's Problem-Solving Skills by Gamifying Mathematics Lessons

An Experimental Study

Nadira Dayo, Muhammad Mujataba Asad, Fareed Hussain Mangi, Noor-ul-Ain Bhellar, and Unaeza Alvi

4.1 INTRODUCTION

Problem-solving is identified as one of the most required skills to survive in the 21st century (Akinmola, 2014; Barell, 2010; Çetin et al., 2023; Rahman, 2019; Szabo et al., 2020), and it is a prerequisite in the workplace (Limeri et al., 2020). However, it is not possible for teachers to teach solutions to all problems; therefore, instead of teaching solutions, the focus needs to be shifted to teaching problem-solving skills (Eppe et al., 2022; C. Sun et al., 2022). The research on problem-solving and other cognitive skills suggests embedding the teaching of problem-solving from an early age (Council, 2012). For several compelling reasons, it is of utmost importance to teach problem-solving skills from an early age, mainly, in the pre-university period as it is a crucial developmental period in which cognitive capacities such as comprehension, interpretation, abstract reasoning, critical thinking, and problem-solving skills develop significantly (Erol & Çırak, 2022). Strong problem-solving skills at this level help in a successful transition to higher education, serve as a foundation to learn advanced mathematics problem-solving, and enable students to navigate challenging coursework with solid analytical and strategic mindset (Conley, 2007; Dias & Amaral, 2014). Early development of problem-solving skills prepares students to flourish in higher education and thrive in a variety of professions that require complex cognitive skills (Carneiro et al., 2007).

Typically, systematic problem-solving is involved in Science, Technology, Engineering, and Mathematics (STEM)-related disciplines and these fields cannot exist without problem-solving (Priemer et al., 2020). However, most students avoid solving mathematics problems (Khasawneh et al., 2021) in general, and this problem can be reduced by introducing innovative teaching methods and resources.

 DOI: 10.1201/9781003400691-4

Problem-solving is the center of mathematics education (Baumanns & Rott, 2019). In addition, effective mathematics problem-solving involves a willingness to solve, perseverance to keep going, awareness about problem-solving strategies, valuing solutions, understanding multiple ways to solve a problem, making and implementing plans to solve problems, and verifying the solutions (Aydoğdu & Ayaz, 2008; Gurat, 2018; Sinaga et al., 2023). Learners need to see the application of mathematics in real life; this way, accuracy and proficiency in mathematics problem-solving develop (García et al., 2019; Sorby et al., 2022).

To make mathematics problem-solving easier, studies have suggested the following steps such as understanding the problem, choosing a strategy, putting the plan to use, and assessing the answer (Ibrokhimovich, 2022; Polya, 1957; Simamora et al., 2017). Problem-solving skills as suggested by George Polya are regarded as appropriate as they are divided into four stages, helping learners develop a logical and organized framework to direct their thinking and decision-making (Okafor, 2019). Polya's framework is suitable for teaching mathematics problem-solving as it recognizes the non-linearity and complexity built into the problem-solving process (Ortiz, 2016). Polya's problem-solving steps encourage the acquisition of a richer understanding of problem-solving notions, creative thinking, and adaptability by pushing students to seek numerous possibilities, reflect on their thinking, and engage in recursive problem-solving (Amit & Neria, 2008; Yapatang & Polyiem, 2022).

More study has been done in recent years on how computer games improve math learning. People who play computer games are often involved in using their brains to do tasks and solve problems (Garris et al., 2002; Hewett et al., 2020; Reynaldo et al., 2021). There is a close linkage between Polya's problem-solving steps and digital game-based mathematics learning, for example, Setyaningrum et al. (2018) used Polya's problem-solving procedure to teach problem-solving through digital game-based learning (DGBL). Similarly, Chadli et al. (2019) conducted a study to learn about the process of problem-solving in serious games by using Polya's suggested interrelated skills. Research also suggests certain game characteristics (e.g., challenges, badges, points, rewards, surprising events, feedback, and competition) support learning and mathematics problem-solving (Green & Bavelier, 2007; Prensky, 2006; Wouters et al., 2017).

Though a considerable number of studies have analyzed the effects of gamified lessons on learning (Chen et al., 2015; Chen & Ren, 2013; Demirel & Karakus Yilmaz, 2019; Lee et al., 2004; Li & Tsai, 2013; Setyaningrum et al., 2018; Spires et al., 2011; Yeh & Ting, 2023), very few have adopted the theoretical framework to study the effects of gamified lessons on mathematical problem-solving skills (Dai et al., 2023; Guan et al., 2022; Li & Tsai, 2013). Given this dearth of credible empirical information on the practical use of a learning framework to teach mathematics problem-solving through lessons, the current study intends to examine the impacts of gamified mathematics lessons on pre-university students with a major focus on Polya's problem-solving skills. To confirm the validity of the conclusions, the reliability of the results, and a thorough knowledge of the influence of gamified lessons on mathematics problem-solving, a comparison with non-gamified mathematics was also made.

4.1.1 Research Questions

Q1: What is the effect of gamified and non-gamified mathematics lessons on Polya's problem-solving skills of pre-university students?

Q2: Is there a statistically significant difference in Polya's problem-solving skills of pre-university students who are taught by using gamified and non-gamified mathematics lessons?

4.1.2 Hypotheses

H_0: Gamified mathematics lessons do not have a significant effect on Polya's problem-solving skills of pre-university students.

H_1: Gamified mathematics lessons have a significant effect on Polya's problem-solving skills of pre-university students.

4.2 LITERATURE REVIEW

With its ability to uniquely adapt to students' interests, digital games provide a new perspective on the culture of learning (Whitton, 2009). DGBL is one of the practical and engaging methods for higher education (Chu & Chang, 2014; Prensky, 2003). In the realm of mathematics problem-solving, gamified mathematics lessons have several benefits for students' learning. The review of the literature reveals the following benefits of gamified mathematics lessons:

- Engagement (L. Sun et al., 2022)
- Motivation and persistence (Dayo et al., 2020; Zapata-Cáceres & Martín-Barroso, 2021)
- Active learning (Ting et al., 2019)
- Individualized and personalized learning (Augustin et al., 2010; Bang et al., 2023)
- Social interaction and communication (Al-Washmi et al., 2013; Sjöberg & Brooks, 2022)
- Real-world application (Abarca, 2023; Nilimaa, 2023)
- Assessment and feedback (Bellotti et al., 2013; Garris et al., 2002; Shute & Ke, 2012)
- Learning retention (Divjak & Tomić, 2011; Sarsa, 2013; Ting et al., 2019)
- Reduce mathematics problem-solving anxiety (Bouzid et al., 2021; Dondio et al., 2023; Vanbecelaere et al., 2020)
- Transferrable skills (Gee, 2003; Van Eck, 2015)
- Multimodal learning (Ke & M. Clark, 2020; Whittington, 2010)
- Progress tracking (Dichev et al., 2018; Shute & Ventura, 2013)
- Growth mindset (Araya et al., 2019)
- Ubiquitous learning (Burke, 2016; Haleem et al., 2022; Hou & Li, 2014)

All of the above-mentioned skills improve problem-solving skills as well. According to studies, students need to learn particular problem-solving techniques to master problem-solving skills (Albay, 2019). George Polya, in 1945, created a problem-solving model. Because the problem-solving process is broken down into different stages and emphasizes carefully the skill requirements at each level, the problem-solving abilities proposed by Polya are emphasized. Students may tackle hard jobs meticulously and effectively using this disciplined approach (Chadli et al., 2019; Kaur, 2019). They may develop a deeper knowledge and mastery of problem-solving by improving their abilities at each level, which will lead to improved results and iterative learning. Additionally, this approach stimulates teamwork, is adaptable to many different fields, and helps people skillfully handle plenty of challenges. The model includes four skills for solving the problem effectively:

Understanding the problem: Students should comprehend the meaning of any given problem, recognize the known and the unknown, and the association between them, and they must know the concepts that may be used to solve problems.

Devising a plan: Students must explicate the relations between different parts of the problem, use prior knowledge to construct and develop solutions to any problem, and develop a plan accordingly.

Carrying out the plan: Once the students have plans, they must be able to carry out the plans. At this step, students do the computations that were previously listed in the plan.

Verifying the solution: This step involves verifying and examining the solution to see if the solution makes sense. In an effort to see whether there are any further approaches to problem resolution, students carefully review the answer and the plan they used.

Gee (2003) describes that the learning scenarios are available in games through simulations, which expands the creative mind of students and gives them a chance to think, comprehend, get ready, and execute activities; these highlights of games are in line with the four skills of problem-solving recommended by Polya (1945). Moreover, Baek (2013, pp. 388–389) makes clear that the game's rules and structure are essentially the same as Polya's four-step approach to problem-solving: (a) comprehension of the problem, (b) creation of a plan, (c) execution of the plan, and (d) assessment of the solution. When it comes to games, Polya's method of problem-solving entails first comprehending the issue or the state of the game, then coming up with a plan, and finally making well-informed decisions. This approach creates a structure that runs parallel to the mechanics of the game and complements problem-solving abilities.

In Table 4.1 an illustration of interlinkage between Polya's four problem-solving steps and games can be seen.

Soancatl et al. (2010) proposed a method for solving mathematical problems based on Polya's four-phase method, through question-led learning. Intending to incorporate it into a game-based learning program, they created a database to hold information about problems and learner data. Furthermore, Karatas and Baki (2013) used Polya's method of problem-solving to promote math achievement and problem-solving skills among students. The experimental group significantly improved in solving problems, while the control group did not demonstrate any substantial

TABLE 4.1
Interlinkage between Polya's Four Problem-Solving Steps and Games

Polya Steps	Understanding of the Problem	Understanding of the Game
Understanding the problem	Identification of problem data	Comprehension of game requirements
Devising a plan	Comparison with similar problems	Comparison of the game to similar ones and making strategies
Carrying out a plan	Execution of the plan	Assessment of devised strategy
Verifying the solution	Generalization of the strategy to different problems	Generalization of strategy for different game conditions

Source: Adapted from Piñero Charlo (2020).

improvement. Lester et al. (2014) developed a game-based learning environment using Polya's problem-solving model to enhance the problem-solving effectiveness and efficiency of students, and the findings revealed significant progress from pre-test to post-test, indicating improved problem-solving skills. In another study by Yang et al. (2016) it was investigated how peer tutoring and tablet PCs may help students improve their math communication skills. They applied Polya's method of problem-solving, which led to more accurate mathematical representations and explanations. Previous research has shown that game-based learning that is in line with Polya's approach to problem-solving has a good impact on students' problem-solving skills. Likewise, Ortiz Rojas et al. (2017) investigated how gamification with badges affected student learning outcomes in a computer programming course. It was predicted that the gamified group would exhibit greater intrinsic drive, self-efficacy, and engagement versus the control group. To address the programming environment, the researchers adapted Polya's four problem-solving techniques. Results showed that gamified conditions significantly increased student engagement. Similarly, Chadli et al. (2019) presented a practical educational game based on Polya's problem-solving model that aims to improve second-graders' problem-solving skills by breaking the process down into steps and helping them comprehend key elements and solution links. Recently, Chacón-Castro et al. (2021) assessed the effectiveness of using escape rooms with 38 first-grade high school students taking mathematics to improve group collaboration, problem-solving, and logical thinking. Bruner's theory served as the foundation for the strategy, which employed George Polya's problem-solving techniques for mathematical challenges in a virtual classroom. By encouraging entertaining and hands-on learning in the classroom, the gamification method enhanced student engagement and enthusiasm throughout mathematics lessons. The importance of Polya's approach comes from the fact that few studies on game-based learning have used theoretical underpinnings (Li & Tsai, 2013; Wu et al., 2012). For instance, Wu et al. (2012) found that adding learning theories into game design resulted in favorable effects in a study of 567 published studies, but unexpectedly, the majority of studies did not take learning theories into account. However, in a different study, Qian and Clark (2016) discovered that constructivism was the most popular learning theory, along with several other learning theories, and that 76% of 29 articles directly included recognized learning theories in their research or game design. Young et al. (2012) claim that for game-based learning to be successful, modern learning theories must be carefully considered since it is not enough to just provide students with games and hope for improved motivation and information acquisition. According to past studies, game-based learning that exists using Polya's problem-solving technique improves students' problem-solving skills. As a result, it's essential to include learning theories in the development of serious games that are appropriate for improving students' problem-solving skills.

4.3 RESEARCH METHOD

4.3.1 Research Design

The current study employed an experimental research design involving two groups: an experimental group and a controlled group. Specifically, true experimental research design was used to examine the effects of gamified and non-gamified mathematics lessons on Polya's problem-solving skills of pre-university students. True experimental research design is also called randomized experimental research since participants are randomly assigned to controlled and treatment groups. Stone-Romero (2002) explains the following four characteristics of true experimental research design: (a) manipulation of one or more independent variables, (b) random assignment of participants to ensure internal validity of the research, (c) measurement of the effects of manipulated condition on dependent variables, and (d) controlling the confound variables. All these four characteristics were considered and met while conducting this study.

4.3.2 Participants

Sixty-four pre-university (grade eight) students of a private school were recruited as the sample of the current study by using a random sampling method. The sample was randomly divided into two experimental and control groups with 32 students in each group. The private school was chosen as it had a well-established technological infrastructure, and the use of technology was welcomed in all subjects, including mathematics, unlike other schools which restricted the use of computer labs to teach computer literacy only. Regarding sampling, it was random as there were four sections of eighth grade in the selected school, the two sections were high performing while the other two were average; therefore, a computer-generated random list of students was chosen to select 64 sample of students out of 143 students in total.

4.3.3 Treatment

To prove the research hypothesis, an experiment was conducted in which pre-post Polya's Mathematics Problem-solving Skills Survey was taken. The treatment was developed by consulting with the subject teacher of mathematics; he proposed teaching two topics, e.g., operations on sets and solving quadratic equations.

Following this, digital mathematics games were searched, which align with the curricular objectives of the school's mathematics syllabus. An alignment of digital mathematics games, mathematics topics, and Polya's problem-solving skills was made; an example can be seen in Table 4.2.

TABLE 4.2
Alignment of Type of Digital Game, Mathematics Topics, and Polya's Problem-Solving Skills

Topics	Type of Digital Mathematics Game	Polya's Problem-Solving Skills
	Operations on sets	
Union, intersection, and compliment on three sets	Paint and puzzle	1. Understanding the problem: Players identify the concepts of intersections, set unions, and Venn diagrams to solve the problem. To effectively depict the connections between sets, they decide which areas to paint and in which colors. 2. Devising a plan: Players come up with a plan for assessing the sets and their intersections, choosing which areas to paint first and in what sequence. They think carefully about the logical connections between sets. 3. Carrying out a plan: Using the selected strategy, players will paint specific areas of the Venn diagrams by using an understanding of set theory. 4. Verifying the solution: While working on the diagrams, players evaluate their progress and accuracy at any time by using the "Check" button in the game. If they make a mistake, they can evaluate their decisions and correct them by using the ndo" button.
	Solving quadratic equations	
Factorizing quadratic equation	Math balloon quadratic equation game (quiz style arcade game)	1. Understanding the problem: Players understand the quadratic equation and the task/goal given in the game. 2. Devising a plan: Players will make a strategy to achieve the game objective (solve a quadratic equation) 3. Carrying out a plan: To solve, players will execute the plan, calculate, and manipulate the quadratic equations. 4. Verifying the solution: After finishing a level or challenge, players may be prompted to review their strategy, admit any errors they made, and reflect on different approaches.

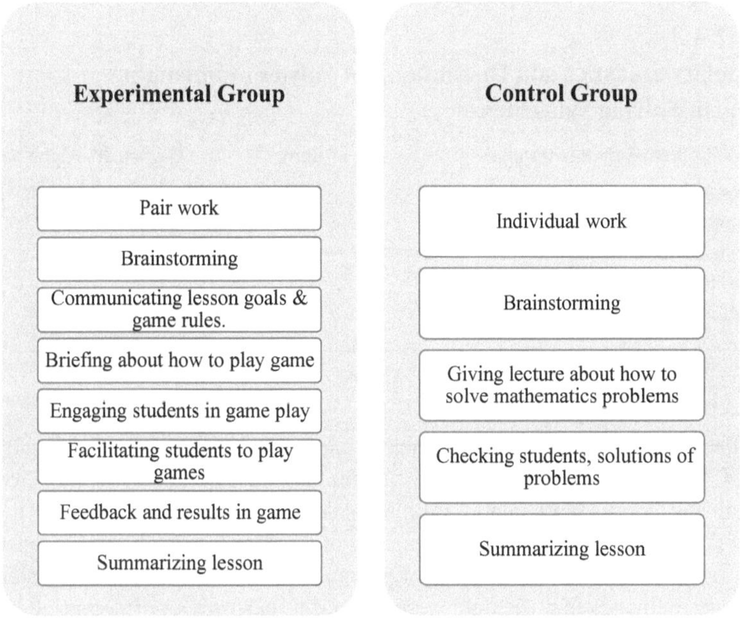

FIGURE 4.1 Instructional strategy of experimental and control groups.

The experimental group learned both topics in a computer lab where they were involved in playing digital mathematics games, whereas the control group was taught the same content without gamifying the mathematics lessons. Each group had the same learning objectives, problem-solving techniques, and instructional hours for learning. The experiment was conducted for a month. Figure 4.1 illustrates the procedures of both the experimental and control groups.

4.3.4 Data Collection Tool

To analyze the difference in Polya's mathematics problem-solving skills of students after gamifying mathematics lessons, a survey based on Polya's problem-solving skills was administered among students. The survey consisted of 15 items, and they all were positive because the sample was pre-university students; so it was anticipated that they might not understand or inappropriately fill the reverse questions. The number of items of each sub-category of Polya's problem-solving skills is shown in Table 4.3.

The responses to the survey were recorded on 5-point Likert scales, i.e., 1 = never, 2 = rarely, 3 = sometimes, 4 = often, and 5 = always. The validity of the survey was ensured by getting it reviewed by three subject experts, whereas the reliability of the survey was calculated by using Statistical Package for the Social Sciences (SPSS) 24; the overall Cronbach's alpha value of the survey is 0.847, which means the survey was reliable, whereas the Cronbach's alpha value of each of the five sub-categories of Polya's problem-solving skills is shown in Table 4.3.

TABLE 4.3
Reliability Statistics and Distribution of Polya's Mathematics Problem-Solving Skills Survey

Polya's Problem-Solving Strategies	No. of Items	Cronbach's Alpha Value
Understanding the problem	1,2,3	0.825
Devising a plan	4,5,6,7,8,9	0.808
Carrying out the plan	10,11,12,13	0.827
Evaluating the solution	14,15	0.868
Overall survey items	15	0.847

4.3.5 Data Analysis

The collected data of both the experimental and control groups were analyzed using SPSS 24. Several types of data analysis techniques were used. Firstly, the Cronbach alpha value of Polya's mathematics problem-solving skills was calculated for the reliability of the survey. Secondly, the normality of the data was analyzed by generating a histogram and Shapiro-Wilk's test of normality; the value of p was greater than 0.05, which indicates that the data were normal to perform any inferential analysis techniques. Hence, along with descriptive statistics, an independent samples t-test was calculated to determine the significance of the difference between the control and treatment groups. Finally, Cohen's d was calculated as a measure of effect size by using its formula.

4.4 RESULTS AND DISCUSSION

Tokac et al. (2019) discuss that minimal rigorous studies have been conducted to explore the effects of digital games on mathematics skills development. Chadli et al. (2019) used Polya's problem-solving model to determine whether students gain from game-based learning. A similar objective has been fulfilled in this study.

Table 4.4 shows the mean, standard deviation, and minimum and maximum scores of both groups in four skills (to understand the problem, to devise a plan, to execute the plan, and to verify the solution) included in Polya's problem-solving skills survey. The survey results depict that the experimental group's mean score in all four Polya's problem-solving skills was more than that of the control group. In the first Polya's suggested problem-solving skill regarding problem understanding, the mean score in the pre-survey of the experimental group was 3.32, and in the post-survey, it drastically raised to 4.11. Similarly, in the pre-survey, the mean score of the control group was 3.12, and in the post-survey, the score increased to 3.30, but the increase in mean score from pre-survey to post-survey was less than in the experimental group.

In terms of plan devising, the maximum score difference has been observed; in the pre-survey, the mean score is 2.78, whereas in the post-survey, it is 3.80, which is an increase of more than one point. The control group's mean score increased from pre-survey ($M=2.64$) to post-survey ($M=3.03$), but again, the increase in average

TABLE 4.4
Descriptive Statistics of Pre-Post Polya's Mathematics Problem-Solving Survey

Polya's Problem-Solving Skills	Time	Group	Mean	SD
Problem understanding	Pre-survey	Experimental	3.32	0.80
		Control	3.12	0.85
	Post-survey	Experimental	4.11	0.60
		Control	3.30	0.82
Plan devising	Pre-survey	Experimental	2.78	0.50
		Control	2.64	0.59
	Post-survey	Experimental	3.80	0.69
		Control	3.03	0.54
Plan executing	Pre-survey	Experimental	3.02	0.89
		Control	2.74	0.71
	Post-survey	Experimental	3.61	0.71
		Control	3.28	0.73
Solution verifying	Pre-survey	Experimental	2.83	0.72
		Control	3.04	0.62
	Post-survey	Experimental	3.26	0.88
		Control	3.09	0.80

Source: N = 64 (32 experimental and 32 control).

mean scores from pre-survey to post-survey of the control group is less than the experimental group. Next, in the other two problem-solving skills regarding plan execution and solution verifying, a significant difference in the mean scores from pre-survey to post-survey in the experimental and control group is observed, and that is also slightly better in the experimental group than that of the control group's score.

The mean score in plan executing skill in the pre-survey of the experimental group was 3.02, and in the post-survey, it increased to 3.61, while the pre-survey score of the control group was 2.74 and post-survey it was 3.28. Finally, in the last Polya's problem-solving skill, solution verifying, the mean score of the experimental group in pre-survey was 2.83, and in post-survey, it raised to 3.26, and the pre-survey score of the control group was 3.04, and in post-survey, it increased very slightly to 3.09. All these data reveal that Polya's suggested problem-solving skills of students who were taught through digital mathematics games improved more than the students who did not learn through digital mathematics games. These results are aligned with the study of Cankoy and Darbaz (2010), who found that games influence the problem-solving skills of students. Since all scales of Polya's mathematics problem-solving survey are designed to measure the fundamental construct of mathematics problem-solving, a change in any scale score can be interpreted as a change in mathematics problem-solving skills.

The results of the independent samples t-test are reported in Table 4.5 to determine whether the difference in scores between the experimental and control groups is significant.

TABLE 4.5
Results of Independent-Sample t-Test of Polya's Mathematics Problem-Solving Skills Survey

Polya's Problem-Solving Skills Survey	Group	Mean	SD	t	Sig (2-Tailed)	Cohen's d
Pre-survey	Experimental	2.98	0.57	0.76	0.44	0.58
	Control	2.88	0.44			
Post-survey	Experimental	3.70	0.48	4.25	0.00	4.62
	Control	3.17	0.49			

Source: N=64 (32 experimental and 32 control).

Table 4.5 reports the statistical results of the independent samples t-test; it can be seen in the table that the mean scores of experimental and control groups in the pre-survey are approximately the same with no statistically significant difference. The post-survey results after one month of the gamified mathematics intervention showed considerable improvements in problem-solving skills, the mean scores in the post-survey of the experimental group (M=3.70, SD=0.48) were statistically significant than those of the control group (M=3.17, SD=0.49), t=4.25, p=0.00. Although the "Sig (2-tailed)" score for the t-test is 0.00 (showing a statistically significant difference between the groups), the large Cohen's d value demonstrates that the difference between the means is both statistically and practically significant. For the pre-survey, the computed Cohen's d value was around 0.5817, indicating a modest to medium effect size. The post-survey Cohen's d value was around 4.62, which denotes a large effect size. This indicates that, following the intervention, there was a significant difference in the mean problem-solving scores between the experimental and control groups. The gamified mathematics lessons significantly improved students' problem-solving skills.

The study's overall conclusions show how effective the digital mathematics intervention was in boosting students' problem-solving skills concerning operations with sets and quadratic equations. The post-survey's substantial effect size highlights the intervention's impact importance in practical terms. The findings back up the use of online games and tools for mathematics instruction as a successful pedagogical strategy for enhancing students' mathematical problem-solving skills. A similar conclusion was reached by Chang et al. (2012), who found higher problem-solving abilities in the experimental group after embedding Polya's problem-solving model in digital games. Likewise, Kenna (2016) in a qualitative case study research concluded that digital games have a positive effect on students' problem-solving skills. These results indicate that the problem-solving skills of students who learn through digital mathematics games are enhanced more than those who learn without their use. Based on the findings of the studies the null hypothesis is rejected and the alternate hypothesis is accepted.

4.5 CONCLUSION

The importance of problem-solving skills cannot be overlooked. Gamified mathematics lessons offer an immersive experience where players interact with problems as if they are physically there in the game environment, focusing on various components of problem-solving such as critical thinking and decision-making. Through difficult challenges, iterative learning, instant feedback, cooperative play, realistic simulations, time-sensitive tasks, resilience, and creativity, gamified mathematics lessons provide an ideal platform for developing problem-solving skills. Gamified lessons are a valuable approach for improving problem-solving skills because of their entertaining character, which keeps learners motivated for lifelong learning. Furthermore, Polya's problem-solving skills encourage learners to utilize a structured and systematic approach to solving mathematics problems. Given this, the primary purpose of the current study was to examine the effects of gamified mathematics lessons on Polya's problem-solving skills. To achieve this purpose an experiment of one month was implemented in which operations on sets and solving quadratic equations were taught by using gamified mathematics lessons to an experimental group and without gamified mathematics lessons to the control group. The findings were similar to the study of Chadli et al. (2019) who also found the positive effect of adopting Polya's problem-solving skills on word problem-solving procedure. Likewise, the results of the study are also aligned with the studies of Okafor (2019) and Yapatang and Polyiem (2022) who found that gamified lessons using Polya's problem-solving method had positive effects on the academic achievement of senior students and mathematical problem-solving ability, respectively. By breaking down the problem-solving process into steps, Polya's problem-solving model focuses on helping students comprehend the semantics and context of mathematical concepts. Additionally, the stages at which mistakes happen when a student confronts challenges may be a useful aid for student support. Based on the findings of the study the alternate hypothesis that gamified mathematics lessons have a significant effect on problem-solving skills of pre-university students is accepted and null is rejected.

The significance of the study lies in its ability to provide empirical evidence of how gamified mathematics intervention improves problem-solving skills before they enter higher education. The large effect size in Polya's mathematical problem-solving skills survey indicates that gamified mathematics lessons have the potential to be an innovative method of teaching mathematics. For educators and institutions looking for research-based educational techniques to foster problem-solving skills, the findings are extremely insightful. The beneficial findings of the study may also be used to design curriculum and higher education policies, enhancing students' readiness for the complexity of the modern workforce and society. The study also advances the use of digital resources for 21st-century teaching and learning, adding to the body of knowledge on evidence-based strategies to enhance student outcomes.

However, it is crucial to note the limitations that can compromise the validity and generalizability of the results. First of all, because it was conducted on pre-university students, it is unclear what would be the results for students in other grades. The fact that the experiment was only conducted for one month added another restriction. Additionally, the experiment was carried out using already existing games. The study was only completed once; a longitudinal study to see if gamified mathematics lessons promote problem-solving skills would have been more advantageous. Therefore, future research should take these things into account.

REFERENCES

Abarca, A. D. (2023). *A Link Between Worlds: The Special Educator's Guide to Teaching Algebra 1 Through Games and Play.* Saint Mary's College of California.

Akinmola, E. (2014). Developing mathematical problem solving ability: A panacea for a sustainable development in the 21st century. *International Journal of Education and Research, 2*(2), 1–8.

Al-Washmi, R., Hopkins, G., & Blanchfield, P. (2013). Investigating collaborative games to teach mathematics based problem solving in the classroom. In *7th European Conference on Games Based Learning (Proc. ECGBL 2013),* Porto, Portugal.

Albay, E. M. (2019). Analyzing the effects of the problem solving approach to the performance and attitude of first year university students. *Social Sciences & Humanities Open, 1*(1), 100006.

Amit, M., & Neria, D. (2008). "Rising to the challenge": Using generalization in pattern problems to unearth the algebraic skills of talented pre-algebra students. *Zdm, 40,* 111–129.

Araya, R., Arias Ortiz, E., Bottan, N. L., & Cristia, J. (2019). Does Gamification in Education Work? Experimental Evidence from Chile. Inter-American Development Bank (IDB), Washington, DC.

Augustin, T., Hockemeyer, C., Kickmeier-Rust, M. D., & Albert, D. (2010). Individualized skill assessment in digital learning games: Basic definitions and mathematical formalism. *IEEE Transactions on Learning Technologies, 4*(2), 138–148.

Aydoğdu, M., & Ayaz, M. F. (2008). The importance of problem solving in mathematics curriculum. *Physical Sciences, 3*(4), 538–545.

Baek, Y. (2013). *Cases on Digital Game-Based Learning: Methods, Models, and Strategies: Methods, Models, and Strategies.* IGI Global.

Bang, H. J., Li, L., & Flynn, K. (2023). Efficacy of an adaptive game-based math learning app to support personalized learning and improve early elementary school students' learning. *Early Childhood Education Journal, 51*(4), 717–732.

Barell, J. (2010). Problem-based learning: The foundation for 21st century skills. *21st Century Skills: Rethinking How Students Learn, 8,* 175–199.

Baumanns, L., & Rott, B. (2019). Is problem posing about posing "problems"? A terminological framework for researching problem posing and problem solving. In *2018 Joint Conference of ProMath and the GDM Working Group on Problem Solving.* Germany

Bellotti, F., Kapralos, B., Lee, K., Moreno-Ger, P., & Berta, R. (2013). Assessment in and of serious games: An overview. *Advances in Human-Computer Interaction*, 2013, 1–1.

Bouzid, T., Kaddari, F., Darhmaoui, H., & Bouzid, E. G. (2021). Enhancing Math-class experience throughout digital Game-based learning, the case of Moroccan Elementary Public Schools. *International Journal of Modern Education & Computer Science, 13*(5), 1–13.

Burke, B. (2016). *Gamify: How Gamification Motivates People to do Extraordinary Things.* Routledge.

Cankoy, O., & Darbaz, S. (2010). Effect of a problem posing based problem solving instruction on understanding problem. *Hacettepe University Journal of Education, 38,* 11–24.

Carneiro, P., Crawford, C., & Goodman, A. (2007). The impact of early cognitive and non-cognitive skills on later outcomes. Centre for the Economics of Education, London School of Economics, London.

Çetin, I., Erümit, A. K., Nabiyev, V., Karal, H., Temel, K., & Kokoc, M. (2023). The effect of gamified adaptive intelligent tutoring system Artibos on problem-solving skills. *Participatory Educational Research, 10*(1), 344–374.

Chacón-Castro, M., Aimacaña-Espinosa, L., & Jadán-Guerrero, J. (2021). Escape rooms: Mathematical challenges available to educators. In *Perspectives and Trends in Education and Technology: Selected Papers from ICITED 2021* (pp. 195–205), Springer.

Chadli, A., Tranvouez, E., & Bendella, F. (2019). Learning word problem solving process in primary school students: An attempt to combine serious game and Polya's problem solving model. In A. Tlili & M. Chang (Eds.), *Data Analytics Approaches in Educational Games and Gamification Systems*, 139–163.

Chang, K.-E., Wu, L.-J., Weng, S.-E., & Sung, Y.-T. (2012). Embedding game-based problem-solving phase into problem-posing system for mathematics learning. *Computers & Education, 58*(2), 775–786.

Chen, H.-R., Liao, K.-C., & Chang, J.-J. (2015). Design of digital game-based learning system for elementary mathematics problem solving. In *2015 8th International Conference on Ubi-Media Computing (UMEDIA),* Colombo, Sri Lanka.

Chen, M.-P., & Ren, H.-Y. (2013). Designing a RPG game for learning of mathematic concepts. *In 2013 Second IIAI International Conference on Advanced Applied Informatics.*

Chu, H.-C., & Chang, S.-C. (2014). Developing an educational computer game for migratory bird identification based on a two-tier test approach. *Educational Technology Research and Development, 62*, 147–161.

Conley, D. T. (2007). Redefining college readiness. *Educational Policy Improvement Center (NJ1).* Vol (3), Eugene, OR.

Council, N. R. (2012). *Education for Life and Work: Developing Transferable Knowledge and Skills in the 21st Century*, National Academies Press.

Dai, C.-P., Ke, F., Pan, Y., & Liu, Y. (2023). Exploring students' learning support use in digital game-based math learning: A mixed-methods approach using machine learning and multi-cases study. *Computers & Education, 194*, 104698.

Dayo, N. A., Alvi, U., & Asad, M. M. (2020). Mechanics of digital mathematics games for learning of problem-solving: An extensive literature review. In *2020 International Conference on Emerging Trends in Smart Technologies (ICETST),* (pp. 1–6). IEEE.

Demirel, T., & Karakus Yilmaz, T. (2019). The effects of mind games in math and grammar courses on the achievements and perceived problem-solving skills of secondary school students. *British Journal of Educational Technology, 50*(3), 1482–1494.

Dias, D., & Amaral, A. (2014). Assessment of higher education learning outcomes (AHELO): An OECD feasibility study. In *Quality Assurance in Higher Education: Contemporary Debates* (pp. 66–87), Springer.

Dichev, C., Dicheva, D., & Irwin, K. (2018). Gamification driven learning analytics. In *Proceedings of the 13th International Conference on e-Learning* (pp. 70–76).

Divjak, B., & Tomić, D. (2011). The impact of game-based learning on the achievement of learning goals and motivation for learning mathematics-literature review. *Journal of Information and Organizational Sciences, 35*(1), 15–30.

Dondio, P., Gusev, V., & Rocha, M. (2023). Do games reduce maths anxiety? A meta-analysis. *Computers & Education, 194*, 104650.

Eppe, M., Gumbsch, C., Kerzel, M., Nguyen, P. D., Butz, M. V., & Wermter, S. (2022). Intelligent problem-solving as integrated hierarchical reinforcement learning. *Nature Machine Intelligence, 4*(1), 11–20.

Erol, O., & Çırak, N. S. (2022). The effect of a programming tool scratch on the problem-solving skills of middle school students. *Education and Information Technologies, 27*(3), 4065–4086.

García, T., Boom, J., Kroesbergen, E. H., Núñez, J. C., & Rodríguez, C. (2019). Planning, execution, and revision in mathematics problem solving: Does the order of the phases matter? *Studies in Educational Evaluation, 61*, 83–93.

Garris, R., Ahlers, R., & Driskell, J. E. (2002). Games, motivation, and learning: A research and practice model. *Simulation & Gaming, 33*(4), 441–467.

Gee, J. P. (2003). What video games have to teach us about learning and literacy. *Computers in Entertainment (CIE), 1*(1), 20–20.

Green, C. S., & Bavelier, D. (2007). Action-video-game experience alters the spatial resolution of vision. *Psychological Science, 18*(1), 88–94.

Guan, X., Sun, C., Hwang, G.-j., Xue, K., & Wang, Z. (2022). Applying game-based learning in primary education: A systematic review of journal publications from 2010 to 2020. *Interactive Learning Environments*, 32(2), 534–556.

Gurat, M. G. (2018). Mathematical problem-solving strategies among student teachers. *Journal on Efficiency and Responsibility in Education and Science, 11*(3), 53–64.

Haleem, A., Javaid, M., Qadri, M. A., & Suman, R. (2022). Understanding the role of digital technologies in education: A review. *Sustainable Operations and Computers, 3*, 275–285.

Hewett, K. J., Zeng, G., & Pletcher, B. C. (2020). The acquisition of 21st-century skills through video games: Minecraft design process models and their web of class roles. *Simulation & Gaming, 51*(3), 336–364.

Hou, H.-T., & Li, M.-C. (2014). Evaluating multiple aspects of a digital educational problem-solving-based adventure game. *Computers in Human Behavior, 30*, 29–38.

Ibrokhimovich, F. J. (2022). Teaching mathematics in elementary school: Issues and solutions. *Eurasian Journal of Learning and Academic Teaching, 4*, 84–87.

Karatas, I., & Baki, A. (2013). The effect of learning environments based on problem solving on students' achievements of problem solving. *International Electronic Journal of Elementary Education, 5*(3), 249–268.

Kaur, B. (2019). The why, what and how of the 'Model' method: A tool for representing and visualising relationships when solving whole number arithmetic word problems. *Zdm, 51*(1), 151–168.

Ke, F., & Clark, K.M. (2020). Game-based multimodal representations and mathematical problem solving. *International Journal of Science and Mathematics Education, 18*, 103–122.

Kenna, A. L. (2016). Problem solving and interactive educational games: A case study of year 6 children. In G. Adams (Ed.), *Proceedings of the British Society for Research into Learning Mathematics,* 36(1), 42. British Society for Research into Learning Mathematics.

Khasawneh, E., Gosling, C., & Williams, B. (2021). What impact does maths anxiety have on university students? *BMC Psychology*, *9*(1), 1–9.

Lee, J., Luchini, K., Michael, B., Norris, C., & Soloway, E. (2004). More than just fun and games: Assessing the value of educational video games in the classroom. CHI'04 extended abstracts on Human factors in computing systems (pp. 1375–1378).

Lester, J. C., Spires, H. A., Nietfeld, J. L., Minogue, J., Mott, B. W., & Lobene, E. V. (2014). Designing game-based learning environments for elementary science education: A narrative-centered learning perspective. *Information Sciences, 264*, 4–18.

Li, M.-C., & Tsai, C.-C. (2013). Game-based learning in science education: A review of relevant research. *Journal of Science Education and Technology, 22*, 877–898.

Limeri, L. B., Choe, J., Harper, H. G., Martin, H. R., Benton, A., & Dolan, E. L. (2020). Knowledge or abilities? How undergraduates define intelligence. *CBE-Life Sciences Education, 19*(1), ar5.

Nilimaa, J. (2023). New examination approach for real-world creativity and problem-solving skills in mathematics. *Trends in Higher Education, 2*(3), 477–495.

Okafor, T. U. (2019). Effect of Polya's problem solving technique on the academic achievement of senior secondary school student in physics. *European Journal of Physics Education, 10*(1), 38–48.

Ortiz, E. (2016). The problem-solving process in a mathematics classroom. *Transformations, 1*(1), 4–13.

Ortiz Rojas, M. E., Chiluiza, K., & Valcke, M. (2017). Gamification in computer programming: Effects on learning, engagement, self-efficacy and intrinsic motivation. In *11th European Conference on Game-Based Learning (ECGBL)* (pp. 507–514). Acad Conferences LTD.

Piñero Charlo, J. C. (2020). Educational escape rooms as a tool for horizontal mathematization: Learning process evidence. *Education Sciences, 10*(9), 213. https://www.mdpi.com/2227-7102/10/9/213

Polya, G. (1957). *How to Solve It*. 2nd ed. Princeton, NJ: University Press.

Prensky, M. (2003). Digital game-based learning. *Computers in Entertainment (CIE), 1*(1), 21–21.

Prensky, M. (2006). *Don't Bother Me, Mom, I'm Learning!: How Computer and Video Games are Preparing Your Kids for 21st Century Success and How You can Help!* Paragon House St. Paul, MN.

Priemer, B., Eilerts, K., Filler, A., Pinkwart, N., Rösken-Winter, B., Tiemann, R., & Zu Belzen, A. U. (2020). A framework to foster problem-solving in STEM and computing education. *Research in Science & Technological Education, 38*(1), 105–130.

Qian, M., & Clark, K. R. (2016). Game-based learning and 21st century skills: A review of recent research. *Computers in Human Behavior, 63*, 50–58.

Rahman, M. M. (2019). *21st century skill "problem solving": Defining the concept.* Asian *Journal of Interdisciplinary Research, 2*(1), 64–74.

Reynaldo, C., Christian, R., Hosea, H., & Gunawan, A. A. (2021). Using video games to improve capabilities in decision making and cognitive skill: A literature review. *Procedia Computer Science, 179*, 211–221.

Sarsa, H. (2014). *Does Gamification Work? A Literature Review of Empirical Studies on Gamification.* In 2014 47th Hawaii International Conference on System Sciences (pp.3025–3034). IEEE. Waikoloa, USA.

Setyaningrum, W., Pratama, L. D., & Ali, M. B. (2018). Game-based learning in problem solving method: The effects on students' achievement. *International Journal on Emerging Mathematics Education, 2*(2), 157.

Shute, V., & Ventura, M. (2013). *Stealth Assessment: Measuring and Supporting Learning in Video Games.* The MIT Press.

Shute, V. J., & Ke, F. (2012). Games, learning, and assessment. In *Assessment in Game-based Learning: Foundations, Innovations, and Perspectives* (pp. 43–58). Springer.

Simamora, R., Sidabutar, D., & Surya, E. (2017). Improving Learning Activity and Students' Mawaddah. *Erlinawaty Simanjuntak Perbedaan Kemampuan Komunikasi Matematis Siswa Melalui Pendekatan Pembelajaran Open Ended Dengan Pendekatan Pembelajaran Differentiated.*

Sinaga, B., Sitorus, J., & Situmeang, T. (2023). The influence of students' problem-solving understanding and results of students' mathematics learning. *Frontiers in Education* (Vol. 8, p. 1088556). Frontiers Media, SA.

Sjöberg, J., & Brooks, E. (2022). Collaborative interactions in problem-solving activities: School children's orientations while developing digital game designs using smart mobile technology. *International Journal of Child-Computer Interaction, 33*, 100456.

Soancatl, V., León, A., Martínez, C., & Torres, L. (2010). Leading students to solve math problems using question-led learning. In *Proceedings of the 4th European Conference on Games Based Learning* (pp. 368–374). Copenhagen, Denmark.

Sorby, S. A., Duffy, G., & Yoon, S. Y. (2022). Math instrument development for examining the relationship between spatial and mathematical problem-solving skills. *Education Sciences, 12*(11), 828.

Spires, H. A., Rowe, J. P., Mott, B. W., & Lester, J. C. (2011). Problem solving and game-based learning: Effects of middle grade students' hypothesis testing strategies on learning outcomes. *Journal of Educational Computing Research, 44*(4), 453–472.

Stone-Romero, E. F. (2002). The relative validity and usefulness of various empirical research designs. *Handbook of Research Methods in Industrial and Organizational Psychology*, 77–98.

Sun, C., Shute, V. J., Stewart, A. E., Beck-White, Q., Reinhardt, C. R., Zhou, G., Duran, N., & D'Mello, S. K. (2022). The relationship between collaborative problem solving behaviors and solution outcomes in a game-based learning environment. *Computers in Human Behavior, 128*, 107120.

Sun, L., Ruokamo, H., Kangas, M., & Siklander, P. (2022). Effects of collaborative digital gameplay on students' three dimensions of engagement in mathematics. *International Journal of Game-Based Learning (IJGBL), 12*(1), 1–16.

Szabo, Z. K., Körtesi, P., Guncaga, J., Szabo, D., & Neag, R. (2020). Examples of problem-solving strategies in mathematics education supporting the sustainability of 21st-century skills. *Sustainability, 12*(23), 10113.

Ting, F. S. T., Lam, W. H., & Shroff, R. H. (2019). Active learning via problem-based collaborative games in a large mathematics university course in Hong Kong. *Education Sciences, 9*(3), 172.

Tokac, U., Novak, E., & Thompson, C. G. (2019). Effects of game-based learning on students' mathematics achievement: A meta-analysis. *Journal of Computer Assisted Learning, 35*(3), 407–420.

Van Eck, R. N. (2015). SAPS and digital games: Improving mathematics transfer and attitudes in schools. *Digital Games and Mathematics Learning: Potential, Promises and Pitfalls*, 141–173.

Vanbecelaere, S., Van den Berghe, K., Cornillie, F., Sasanguie, D., Reynvoet, B., & Depaepe, F. (2020). The effects of two digital educational games on cognitive and non-cognitive math and reading outcomes. *Computers & Education, 143*, 103680.

Whittington, J. L. (2010). *Serious Games: How Instructional Design and Game Experts Design Multimodal Learning Environments,* Capella University.

Whitton, N. (2009). *Learning with Digital Games: A Practical Guide to Engaging Students in Higher Education*, Routledge.

Wouters, P., van Oostendorp, H., ter Vrugte, J., Vandercruysse, S., de Jong, T., & Elen, J. (2017). The effect of surprising events in a serious game on learning mathematics. *British Journal of Educational Technology, 48*(3), 860–877.

Wu, W.-H., Chiou, W.-B., Kao, H.-Y., Hu, C.-H. A., & Huang, S.-H. (2012). Re-exploring game-assisted learning research: The perspective of learning theoretical bases. *Computers & Education, 59*(4), 1153–1161.

Yang, E. F., Chang, B., Cheng, H. N., & Chan, T.-W. (2016). Improving pupils' mathematical communication abilities through computer-supported reciprocal peer tutoring. Journal of Educational Technology & Society, *19*(3), 157–169.

Yapatang, L., & Polyiem, T. (2022). Development of the mathematical problem-solving ability using applied cooperative learning and Polya's problem-solving process for grade 9 students. *Journal of Education and Learning, 11*(3), 40–46.

Yeh, Y. c., & Ting, Y. S. (2023). Comparisons of creativity performance and learning effects through digital game-based creativity learning between elementary school children in rural and urban areas. *British Journal of Educational Psychology, 93*(3), 790–805.

Young, M. F., Slota, S., Cutter, A. B., Jalette, G., Mullin, G., Lai, B., Simeoni, Z., Tran, M., & Yukhymenko, M. (2012). Our princess is in another castle: A review of trends in serious gaming for education. *Review of Educational Research, 82*(1), 61–89.

Zapata-Cáceres, M., & Martín-Barroso, E. (2021). Applying game learning analytics to a voluntary video game: Intrinsic motivation, persistence, and rewards in learning to program at an early age. *IEEE Access, 9*, 123588–123602.

5 Applying Gamification as an Innovative Teaching Model with Quizalize for EFL Learners in Higher Education

Muthmainnah Muthmainnah, Ahmad Al Yakin, and Muhammad Mujtaba Asad

5.1 INTRODUCTION

The 21st century and the continuous influence of ongoing technological advances on people's lives mean that it is undeniable that today's society is undergoing a process of digitization and modernization (Marieiev et al., 2023). This condition is illustrated by the projections it makes and by the main role of information and communication technology (ICT) or IoT in various activities ranging from social activities to education and citizen employment.

One of the most prominent ICT applications is in the field of education. This educational transformation is supported by the evolution of the training process that benefits from continuous innovation generated across the educational spectrum, from the shift in the role of the main agents (teachers and students) to the emergence of new methodologies packed with digital pedagogy, resources, IoT-based tools, and locations to develop teaching and learning processes (Davronovich and Mansurjonovich, 2023; Okoye et al., 2023).

This factor is the driving force behind the educational transformation that takes place in the classroom, the development of educational technology, and innovative techniques to be strong and effective (Serdyukov, 2017). As a result of all the accessible digital methods, resources, and tools to support the formative process, the incorporation of these new ways of learning has produced various attitude-type academic indicators, such as motivation, attitude, engagement, learning autonomy, and student perceptions (Ferrer et al., 2020; Tang et al., 2023).

Along with various modernizations of learning models, of course, this supports better academic opportunities because of changes in conventional teaching and learning methods and learning time. Their motivation and attitude toward learning (Tang et al., 2023), engagement (Bringula and Atienza, 2023), self-efficacy (Aguilera, 2020), interaction socialization (Abascal and Balcells, 2023), autonomy

DOI: 10.1201/9781003400691-5

(Barbosa and Ferreira 2023), and even control of learning at the personal level have all been improved as a result of this methodology.

Along with the emergence of the millennial generation, whose lives are heavily influenced by digital technology and who enjoy learning in a fun, dynamic, comfortable, and interesting environment, gamification in education has become one of the unavoidable trends in modern learning models. Researchers and stakeholders in education are starting to pay attention to gamified learning media and technologies such as Quizalize (España, 2023), which incorporate specific game components including leaderboards, memes, and online quiz reports. All of these things work together to make learning more fun, active, and stress-free for undergraduate students in higher education. Educators have a responsibility to prepare the next generation for the Fifth Industrial Revolution by encouraging the development of essential skills such as digital fluency, problem-solving skills, creative and critical thinking, social interaction, learning autonomy, and commitment to their own learning.

Modern, applicable, and efficient game-based learning uses interactive games to provide the fundamental knowledge needed to maintain an efficient modern way of life, which is heavily influenced by ICT interventions or the metaverse, i.e., interactions in cyberspace. As a result, Indonesia is updating its education system to equip its youth with the essential skills they need to thrive in today's fast-paced digital era through 21st-century learning, which emphasizes the integration of IoT and self-directed learning.

The foundation of the Merdeka curriculum is to free instructors to carry out learning innovations, design creative digital humanist classes, one of which is game-based, as a result of the contemporary climate where educators are "forced" to adapt their pedagogical approaches to the increasing influence of technology through mobile interventions, be flexible, and be open to new experiences where online games are more popular among young adults today.

Quizizz, Kahoot!, Quizlet, and Plickers are just a few of the popular gamified learning platforms that have emerged from recent academic research. Students can learn at their own pace and acquire the digital literacy required by today's employers using interactive online games as a classroom tool that can help enhance digital literacy and provide a competitive learning experience. Zhang and Hasim (2023) and Tawafak, et al. (2023) argue that by placing students at the center of the learning process, a game-based approach can make education easier, more interesting, and more productive. This is in line with the goals of 21st-century education, which prioritizes the individual needs of students and does not prohibit or discourage the use of creative and interesting learning methods.

A number of studies have shown that gamifying the learning process can improve learning outcomes as well as students' interest and motivation for the course. However, local research on the use of the Quizalize game application is to improve micro (vocabulary and grammar) and macro-English learning (listening, reading, speaking, and writing) and how Quizalize affects foreign language acquisition (EFL) in Indonesia. In fact, learning English opens up a whole new world of information for learners to explore. The fact is that English is one of the subjects that does not interest students in college. To overcome this worry, the game-based learning model through Quizalize online interactive games in EFL basic English classes offers useful new results in which learning English becomes more interesting and increases their self-confidence and motivation, especially in terms of educating generations in the digital era.

The main purpose of this study is to explore how the innovation of the English learning model using the Quizalize application affects the students' English skills. Research using quantitative studies aims to describe the effectiveness of quiz-based English learning. We recognize that language acquisition has given rise to several theories and has been a topic of concern from time to time, with age, IQ, aptitude, attitude, motivation, personality, and context highlighted as determinants of language learning (Zainal, 2023). A teacher must be adept at creating engaging course materials as well as implementing a variety of instructional strategies that cater to different types of learning and create a collaborative, fun, student-centered, and interactive learning environment. This study is anticipated to provide answers to the problem of learning English. Considering that there are still many teachers who have not created interesting and up-to-date game media, especially in the digital era, this research can provide solutions to the problems of learning English in universities.

In the light of the importance of learning English encouraging the acquisition of English among EFL students in undergraduate students, the following research questions are discussed in this study:

1. Does using Quizalize's game-based teaching model help undergraduate students better improve their English learning outcomes?
2. Does using innovative game-based teaching models make them interested in learning English and actively involved?

5.2 THE IMPORTANCE OF INNOVATIVE TEACHING MODEL IN EDUCATION

Learning English must be continuously updated with teaching methodologies that follow trends by increasing teaching resources, educational facilities, and infrastructure, including learning media, in line with advances in information technology (Rudolph et al., 2023). Learning resources are very important, especially when learning a foreign language. Learning media and technology can be used for undergraduate students. Several studies have also shown the advantages and potential of using media and technology to learn English (Cheng, et al., 2023). However, there are still many local teachers who do not use educational media as a tool that can help students learn English because using media and technology or innovative learning models takes a lot of time, requires money and effort. Because lecturers don't want to take many of these risks, extensive language learning makes students bored and not interested in learning English.

We recognize that computer- or smartphone-assisted English learning has several features that encourage student immediacy, full engagement on their own devices, and the potential for a more collaborative and fun learning environment. The advantage of using media in the learning process for students is that the teaching system will attract more students' attention, even though a teaching material has a clearer meaning so that learning is easy to understand, saves time, and maximum feedback responses can be obtained (Kacetl and Klímová, 2019; Luo and Watts, 2022).

In addition, the learning model will also be more varied and creative so that students can do more learning activities because they not only listen to the teacher's explanation but also carry out activities that stimulate their psychomotor as well as other activities such as observing, doing, and asking questions. Learning is an educational activity that results from teaching or imparting knowledge, and teaching is a process that involves a person having information, knowledge, or skills ready and willing to pay attention or attend classes and some other exercises that may be required. According to Plucker et al. (2023) and Tu and Akhter (2023), creative and innovative teachers often use technology and design learning models that aim to enhance subject representation, adaptation, presentation, and interaction for the 21st-century learning environment.

Learning model innovations can be carried out by integrating the media and technology used to design teaching materials that can be visualized efficiently. The use of these media in formal settings where students work alone, or the teacher interacts with other groups of students makes them essential for effective teaching. Lecturers use game quiz apps to determine how well their undergraduate students have mastered content, and teachers can use them to determine how well students have learned it. In other words, you can evaluate your mastery of the material by taking quizzes. We recognize that quizzes are used by teachers as a reflection tool to improve their teaching or as a guide for choosing the best approach. Innovative language game-based learning models are offered so that learning becomes fun and interesting and reflects a student-centered approach to learning. The game-based learning model is carried out at the end or after the learning process. The purpose of the review is to measure how well students learn and engage in interactions with the latest technology, enabling a direct combination between teaching and evaluation that is more enjoyable and enabling formative assessment not only through conventional methods but also through technology (Dilekçi & Karatay, 2023). We recognize that feedback, especially when it comes to learning, should always be a two-way conversation. If learning is constant, a feedback culture increases managerial effectiveness, especially with regard to learning (Piatnychuk et al., 2022).

The Quizalize application used in this study offers the advantage of helping participants learn in a meaningful and memorable way by collaborating and participating in a competitive, fun, and interesting learning environment. This allows the language teacher community to quickly gain a clear understanding of technological approaches and their outcomes (Abdul, 2021). In this regard, Quizalize is a program that is part of the 4.0 learning revolution, also known as artificial intelligence. There are similarities between Quizalize and the currently popular Quizalize app for assessing students, particularly in learning games during online classes (Johansson, 2021). In the field of learning revolution 4.0, which is the same as artificial intelligence, Quizalize is one of them, even though it has switched to face-to-face teaching. According to Troussa et al. (2020), the media functions as a messenger (educational module) from a teacher to his students in order to achieve educational goals and as a tool for assessing educational outcomes. Quizalize is one of the e-learning-based educational tools used in hybrid classrooms, comparable to the Quizizz and Kahoot programs.

5.3 TEACHING MODEL MIXED QUIZALIZE

Since there are so many technologies available today, it is important to examine and determine which tools can deliver the desired results with the help of gamification. Gamification involves the use of game-learning mechanics and design principles. The game assigns experience levels to scores, and shared leaderboards include contest setups for novice players. To stimulate learners, encourage engagement, and incentivize them to develop the knowledge acquisition and skill development behaviors that designers need to engage in, this notion has been used in contexts that are not just for games (Mohd et al., 2023). However, gamification is not intended to replace traditional teaching methods in modern learning models with video games that combine instruction with aspects of game play. To engage students in learning and teaching based on the problems and priorities mentioned above, the teaching and learning process can involve developing game-like experiences. To help undergraduate students improve their English skills by imitating a play environment, the researchers were interested in creating a teaching and learning model that used Quizalize as their starting point in relation to the idea of gamification. When students complete a given mission, they will be awarded points or prizes.

To keep students' attention in online classes, teachers must conduct research and learn how to use new tools (Jurado et al., 2023). The teacher must examine the success of learning and relate subjects to internet sources. A teacher must be aware of the effects of visual, aural, and tactile awareness and how most students will interpret any technological instrument used for classroom instruction taking into account the student's learning style, and this game can embrace all those learning styles just like verbal learning styles (using speech and writing), physical learning styles (doing practice), auditory learning styles (listening to sounds, music, and rhythms), logical learning (relying on reasoning to get the big picture), group learning (collaborative work in teams), and independent learning (learning alone).

Apart from imparting knowledge, educational games can help players develop critical qualities including problem-solving, teamwork, and communication (Almulla and Al-Rahmi, 2023). Gamification is a method of incorporating game components such as points, leaderboards, and badges into situations unrelated to the game. The use of games in class increases the enthusiasm and involvement of students, develops their visual abilities, and increases their motivation to interact and work together with peers (Caiza, 2022).

Game-based teaching models should ideally promote fun, engagement, and competition, as well as opportunities for social interaction and self-expression in a relaxed or formal setting. It is important to investigate the possibilities of gamified educational software to simplify teaching and improve learning outcomes. Yamani (2021) notes that "Gamification aims to increase participation and motivate users through the use of game elements such as points, leaderboards, and direct feedback, among others". Quiz is a fun multiplayer game platform or application-based class activity for interactive quiz games, according to Costello (2020) and Szeto et al. (2021) "where students become active controllers of the pace of their game class activities". Leaderboards, time limits, points, and social connections are just a few aspects of the gamification that Quizalize has implemented to increase user engagement, motivation, and a sense of empowerment in completing the task at hand.

The use of Quizalize in education has been the subject of much research, and the findings are encouraging. To incorporate fun learning, Oulaich (2020) looks at using Quizalize in English classes. The study's authors claim that students love Quizalize, and the addition of fun and interactive features helps maintain their interest and motivation in learning. The leader board, which motivates them to compete and aim for better performance, is called the students' favorite Quizalize feature. Research by Capinding (2022) looks at Quizalize as a game-based assessment learning tool in English classes. The research findings indicate that using Quizalize in English classes has a high probability of being beneficial, as it has been shown to increase students' attention span and level of engagement in understanding the subject matter. The research findings by Bratel et al. (2021) and Çekiç and Bakla (2021) revealed that students consider Quizalize the right choice for online quiz web applications. Although Quizalize has been shown to increase interest and motivation, there has not been sufficient research to determine whether it is an effective and appropriate tool for language teaching and education for micro-skills and macro-skills. The purpose of this research is to investigate the impact of Quizalize to support the claim that Quizalize is an effective tool for improving language teaching, provides comprehensive data to all stakeholders in the field of education, and deserves to be considered as an innovative teaching model suitable for today's 21st century.

5.4 METHOD

The method used in this study was pre-experimental with a single-group pre-test and post-test design. Before and after treatment, pre- and post-tests were carried out. Assessments are given to produce results, which will be used to compare student performance between genders and evaluate student achievement both before and after the intervention. The population of this study was 350 undergraduate students who were selected purposefully. A total of 34 undergraduate students in the Pancasila and citizenship education study program became the research sample, with 13 elementary school students, 19 years; 14 people aged 18 years; 4 people aged 22 years; 2 people aged 21 years; and 1 person aged 20 years. As for gender, 13 males and 21 females were selected as samples.

Purposive sampling was used to select students who had difficulty learning English, were not motivated to learn, felt insecure, and lacked vocabulary. The aim is to determine how the teaching model with the Quizalize application affects the interest and proficiency of the respondents toward the acquisition of the targeted language. The University of Al Asyariah Mandar, West Sulawesi, was used as a place for this research. In addition, the survey was conducted during the first period of the first semester, when respondents were formally taught micro- and macro-English. The main experiment of this study was started after the pre-test by eight meetings with the following stages:

1. Inform undergraduate students to prepare smartphones and download the Quizalize application, then lecturer explains the purpose and function of the application to the students.

2. The lecturer gives learning instructions. The teaching materials in this study are introducing yourself and friends, providing key vocabulary on how to introduce yourself and friends, providing sample texts, providing introductory dialogues, playing self-introduction and other videos, completing paragraphs, and compiling your own role-play scripts practiced in each unit. During the first 15 min, the undergraduate students worked in groups on a predetermined topic, using their smartphones to map the vocabulary and definitions of new vocabulary words they encountered in the English conversational dialogue. They are permitted to use Google Translate and screen translation using their smartphone at any time during the session if needed.
3. The lecturer then spends around 40 min teaching and explaining the dialogue to students using Canva, covering vocabulary, grammatical structures, phrases, and expressions. The lecturer asks students to form groups of three and then to listen to a video introducing themselves and others for about 20 min, asking them to imitate and recite vocabulary to the fullest with their group mates.
4. Students then compose their own dialogue script and practice the conversation skills they have learned in class for an additional 20 min. They then made presentations and conversed with students in the groups they had previously arranged.
5. The lecturer goes around the class helping students in need while students practice the dialogues given. EFL students often participate in role-playing exercises to practice what they have learned in class. However, students must visualize the situation they are in when they participate in the classroom role-play exercise. This challenges students to experience real feelings in real situations and understand the usefulness of lessons in class.
6. The students also spent 20 min in the environment recording their presentations using smartphones, correcting each other's conversation skills that they had learned in class, and uploading them to their respective YouTube accounts.
7. The lecturer then shares the Quizalize link via the WhatsApp application and asks them to work on the quiz according to the specified time. Quizalize's integrated learning instruction seeks to achieve two main goals: (a) to increase the learner's capacity to retain and use new expressions and vocabulary; and (b) to deepen the relevance and application of these expressions in the classroom.
8. The lecturer closes the lesson.

5.5 COLLECTING DATA

Results from the survey and test scores were both aggregated for the current investigation. To investigate the variation in mean scores between pre- and post-stages, paired sample t-tests were run. The questionnaire has 10 questions about students' views on using Quizalize and 10 questions about their experience of learning languages using this teaching model. The survey was conducted using a 5-point Likert scale

(1 = strongly disagree, 2 = disagree, 3 = slightly disagree, 4 = agree, and 5 = strongly agree). Given the low average scores for each problem, the survey analysis concentrated on percentage comparisons rather than interpreting student answers in more detail. Two open-ended questions discussing the advantages and disadvantages of using Metaverse are also included. Replies from students were examined, classified into important ideas, and their frequency and percentages calculated.

By recording some data that occurred during lectures, the data collection approach also includes direct observation with students and observing lecture conditions specifically mentioned in Quizalize practice for one semester. It is well recognized that pre-test and post-test learning engagement, direct observation, and conducting surveys are methods that quantitative research relies on to gather information.

5.6 DATA ANALYSIS

By comparing each respondent's pre-test and post-test scores, the first research question was resolved. The results are then converted into percentages for both exams. The information gathered to answer the first study question is presented using inferential statistics. Using the SPSS Statistics Version 26 application, the average sample score and response to the game-based teaching model were determined. Before discussing undergraduate students' responses to their learning experience with Quizalize, a normality test was performed to help choose the best test based on the data collected. Analysis of the data in this study presents the findings of the normality test, homogeneity test, and partial *t*-test on both scores.

5.7 RESULTS AND DISCUSSION

The purpose of this study is to determine how the innovative English learning model using the Quizalize application affects students' English skills. In this section, quantitative data will be disclosed. Below, we will provide details of the learning outcomes of undergraduate students and the results of responses regarding the Quizalize application applied while learning English as a component of the relationship between this Quizalize application and learning outcomes. You can see the respective values from the test and questionnaire results in Table 5.1.

Based on the initial test data for English skills, a pre-test was carried out, and the average test results for the initial test data for English skills with a pre-test score of 70.32 were in the sufficient category. After the results of the pre-test are

TABLE 5.1
Statistical Description of Pre-test and Post-test

One-Sample Statistics

	N	Mean	Std. Deviation	Std. Error Mean
Pre-test	34	70.3235	2.62529	.45023
Post-test	34	82.9118	7.15782	1.22756

known, a trial is conducted on the use of learning media and technology, namely the Quizalize application with the stages shown in Figures 5.1–5.7. The final post-test is given after learning quizzes using SPSS version 26 and the average post-test score is 82.9. The data in Table 5.1 show a significant difference in learning outcomes after being given treatment using Quizalize. The results of pre-test and post-test data can be explained through Table 5.2.

Based on the data in Table 5.2, it can be seen that as many as 34 undergraduate students have an average value of 70.32, where the minimum value for the pre-test is 66.00 and the maximum value is 74.00. Then the increase in post-test scores is indicated by the lowest score of 70.00 and the highest score of 90.00.

Based on the output of the "test statistics" (Table 5.3), it is known that Asymp. Sig. (2-tailed) is worth 0.000. Because the value of 0.000 is less than <0.05, it can be concluded that there is a difference between the results of learning English using the Quizalize application, so it can be concluded that there is an effect of using Quizalize media and technology on English micro- and macro-skills.

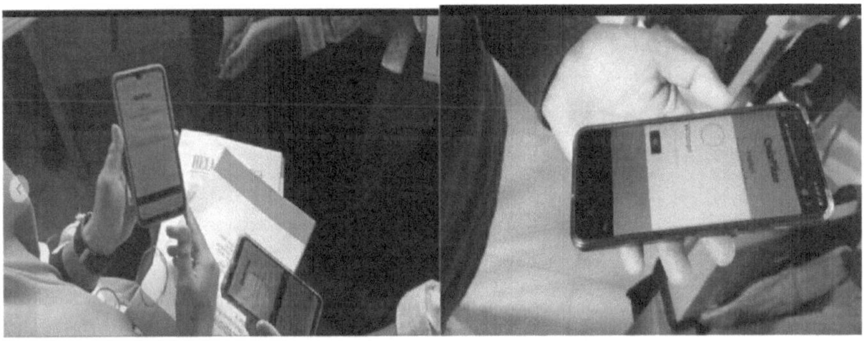

FIGURE 5.1 Download Quizalize.

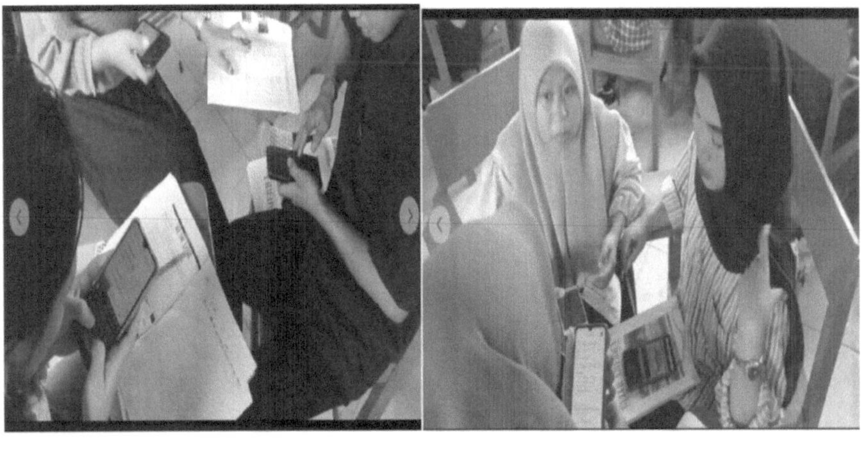

FIGURE 5.2 Group discussion through Google Translate.

FIGURE 5.3 Listening from YouTube.

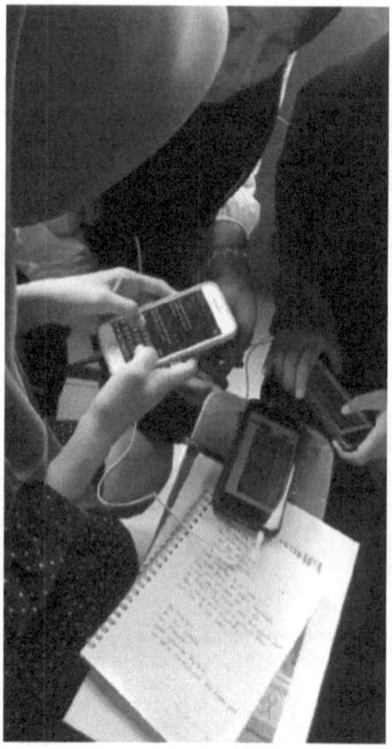

FIGURE 5.4 Construct sentences and dialogue.

From Table 5.4, the results of the post-test scores are in the range of 80–100, which is included in the very good category. The presentation of undergraduate student response data regarding their learning experience using the Quizalize application can be seen in Table 5.5.

Table 5.5 shows the statistical results of obtaining undergraduate student responses through the use of Quizalize, which achieved an average score of 109.52. In this case, using an interval scale with a length of 30 and a maximum value of 155, the use of the Quizalize application gets a positive response with the following very good categories.

FIGURE 5.5 Practice with group team.

FIGURE 5.6 Students project via YouTube.

Based on the survey results in Table 5.6 the Quizalize application received a positive response. The responses of undergraduate students stated that this application was easy to operate, they enjoyed learning English, and the presence of this technology trained micro- and macro-skills well. They can correct errors in vocabulary, grammar, reading (completing vocabulary), writing, speaking, and listening.

FIGURE 5.7 Quizalize in action.

TABLE 5.2
Statistical Description of the Students' Score

	N	Minimum	Maximum	Mean	Std. Deviation
Pre-test	34	66.00	74.00	70.3235	2.62529
Post-test	34	70.00	90.00	82.9118	7.15782
Valid *N* (listwise)	34				

TABLE 5.3
Statistic Paired Sample *t*-Test

One-Sample Test						
	Test Value = 0					
	t	df	Sig. (2-tailed)	Mean Difference	95% Confidence Interval of the Difference	
					Lower	Upper
Pre-test	156.193	33	.000	70.32353	69.4075	71.2395
Post-test	67.542	33	.000	82.91176	80.4143	85.4092

Expressions of words, sentences, and dialogues are part of learning activities with the application of this technology. After obtaining data from learning responses with Quizalize and learning outcomes, the two data are analyzed into product moment correlation calculations. Correlation calculations are shown in Table 5.7.

TABLE 5.4
Frequency Distribution of Student Learning Outcomes

No.	Interval	Category
1.	80–100	Very good
2.	70–79	Good
3.	60–69	Fairly
4.	50–59	Low

TABLE 5.5
Statistical Data on Responses Using Quizalize

Descriptive Statistics of Quizalize Questionnaire

	N	Minimum	Maximum	Mean	Std. Deviation
Quizalize	34	44.00	150.00	109.5294	23.86482
Valid *N* (listwise)	34				

TABLE 5.6
Students Interest toward Quizalize in Learning

No.	Interval	Category
1.	121–150	Very good
2.	91–120	Good
3.	61–90	Fairly
4.	31–60	Low
5.	0–30	Very low

The most used correlation index is the Pearson correlation coefficient (r). The range of r values is between –1 and +1, where a positive value indicates a positive relationship, a negative value indicates a negative relationship, and a value of 0 indicates there is no linear relationship between the two variables. To interpret the strength of the correlation, the correlation level is generally used as follows:

Table 5.8 presents the calculated rxy value (correlation between Quizalize use and final English test score) of 0.59, indicating a positive relationship between the two variables. If these findings are interpreted using the r table for correlation based on Pearson's guidelines, a value of 0.59 indicates that there is a correlation between X and Y, and this correlation is included in the moderate to strong group. So, it can be concluded that the teaching model with Quizalize has a significant relationship to improving learning outcomes in a hybrid environment.

The histogram image after applying Quizalize application shows the range of values obtained by students from 60 to 100, where the histogram bar shows the highest

TABLE 5.7
Descriptive Statistics of Student's Responses to Quizalize

Descriptive Statistics					
	N	Minimum	Maximum	Mean	Std. Deviation
1. Accessing Quizalize's application environment was easy.	34	1.00	5.00	3.6471	1.29994
2. Accessing the Quizalize application using a mobile device or smartphone was easy.	34	1.00	5.00	3.5000	1.05169
3. The Quizalize application was easy to use.	34	1.00	5.00	3.8824	1.17460
4. I did not need assistance to use the Quizalize application.	34	1.00	5.00	3.6471	1.12499
5. I found the various functions in the Quizalize application well integrated.	34	1.00	5.00	3.6471	1.22802
6. I did not need assistance before I could get going with the Quizalize application.	34	2.00	5.00	3.7353	1.10943
7. I felt very confident using the Quizalize application in English	34	1.00	5.00	3.7059	1.19416
8. I feel comfortable learning English using the Quizalize online application.	34	1.00	5.00	3.4706	1.13445
9. The time display of the activities in the Quizalize application facilitates time management.	34	1.00	5.00	3.7059	1.08793
10. I had fun learning English while using the Quizalize application.	34	1.00	5.00	3.5588	1.13328
11. I had an increase in learning while using the Quizalize application.	34	1.00	5.00	3.6176	1.23128
12. Timely questions in quizzes and application activities increase students' interest.	34	1.00	5.00	3.7647	1.07475
13. The Quizalize application helps students correct misinformation and understand the mistake.	34	1.00	5.00	3.7647	0.98654
14. The Quizalize application helps students correct answer in realtime.	34	1.00	5.00	3.3235	1.24853
15. The scoring system of the Quizalize application increases students' ambition.	34	1.00	5.00	3.5882	1.20900
16. The Quizalize application improves the rapid-thinking abilities of students.	34	1.00	5.00	3.6176	1.25565

(Continued)

TABLE 5.7 (*Continued*)
Descriptive Statistics of Student's Responses to Quizalize

Descriptive Statistics

	N	Minimum	Maximum	Mean	Std. Deviation
17. The Quizalize application provides an active and relaxing environment by helping students improve their performance.	34	1.00	5.00	3.5588	1.18555
18. The Quizalize application helps me answer questions better than in the classroom.	33	1.00	5.00	3.4545	1.32502
19. The Quizalize application is educational and informative for English.	33	1.00	5.00	3.6364	0.96236
20. Quizalize's application scores answers instantly.	34	2.00	5.00	3.6471	0.91725
21. The Quizalize application displays my progress in English.	34	1.00	5.00	3.6765	1.06517
22. I have been able to self-evaluate my learning process using the Quizalize application.	34	2.00	5.00	3.5882	1.01854
23. The Quizalize application gives me the opportunity to see the questions. I got it wrong with the correct answers.	34	1.00	5.00	3.4706	1.28477
24. I can invite a friend to join English by using the Quizalize application platform.	34	2.00	5.00	3.2647	0.93124
25. The Quizalize application enables me to access assessments anywhere.	34	1.00	5.00	3.5000	1.10782
26. The Quizalize application enables error-free scoring.	34	1.00	5.00	3.6176	1.07350
27. The Quizalize application was difficult to use.	34	1.00	5.00	3.1471	1.13170
28. Internet connection problems affected my access to the Quizalize application environment.	34	1.00	5.00	3.4118	1.28199
29. There was difficulty selecting the options in the Quizalize application.	34	1.00	5.00	3.0882	1.33411
30. It was difficult to read from the dashboard of the Quizalize application.	34	1.00	5.00	3.1765	1.33645
31. The Quizalize application is user friendly for English.	34	1.00	5.00	3.3235	1.29616
Valid *N* (listwise)	32				

TABLE 5.8
The Correlation Quizalize and Learning Outcome

Correlations

		Quizalize	English Score
Quizalize	Pearson correlation	1	–0.110
	Sig. (2-tailed)		0.535
	N	34	34
English score	Pearson correlation	–0.110	1
	Sig. (2-tailed)	0.535	
	N	34	34

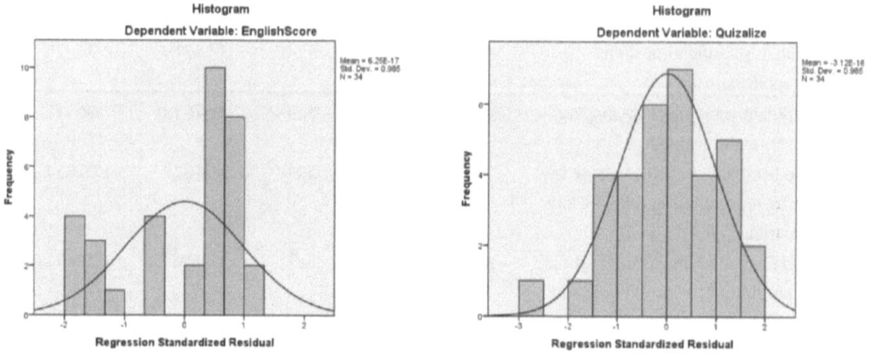

FIGURE 5.8 Histogram of the data.

value is 90, so it is assumed that the mean value of students' English skills is 90 (Figure 5.8).

5.8 DISCUSSION

Based on the analysis of the pre-test and post-test data, it is known that there are significant differences in learning outcomes before and after implementing the teaching model using the Quizalize application, which is strengthened by statistical data from the survey results of undergraduate students in Table 5.6. The findings of this study modify the theory regarding the importance of media integration and technology as an innovation teaching model with a learning syntax that involves undergraduate students learning more actively with Quizalize, which is described in Figures 5.1–5.6.

Designing a teaching model in accordance with the right instructions so that the benefits of this technology are right on target, namely micro- and macro English skills (Table 5.2) at universities, is not just limited to game use but also correlates to learning outcomes (Table 5.7) according to the reference table of Pearson (Table 5.9). Broadly speaking, this study states that there is an effect of the teaching model using Quizizz learning media and technology on English learning outcomes, as seen from the response shown by the respondents through the data from

TABLE 5.9
Pearson Category

No	Interval	Category
1	0.00–019	Very weak correlation or no relationship
2	0.20–0.39	Weak to moderate correlation
3	0.40–0.59	Moderate to strong correlation
4	0.60–0.79	Strong to very strong correlation
5	0.0–1.00	Very strong correlation

the questionnaire results in Table 5.6, which is strengthened by the support of several relevant research results such as Zhussupova and Zhumabek (2021).

The use of Quizalize media can provide benefits for undergraduate students and lecturers in tertiary institutions. Some of the benefits for students that the researcher can summarize are that there is an increase in learning concentration, an increase in learning outcomes, an increase in motivation, an increase in understanding, and a fostering of an attitude of thinking, critical thinking, creating a learning environment that is fun, interesting, and comfortable, and increasing learning autonomy by using this application not only in the classroom but also as an exercise in honing English skills outside the classroom (Figure 5.7).

In addition, the benefits obtained by lecturers in tertiary institutions are that the learning process becomes more effective, fair assessments, explanations, and work instructions can be easily conveyed, which assists in conveying learning material, and it is easy to evaluate learning outcomes anywhere and anytime, either directly or online. In addition to the various benefits that undergraduate students and lecturers get, there are also various conveniences brought about by using Quizalize media to train undergraduate students' computational and digital literacy skills. Several other conveniences are easy installation on smartphones (Table 5.6 and Figure 5.1) and ease of managing applications. In addition, using media and technology, Quizalize can create a comfortable learning atmosphere. It is more active and fun, fosters interaction, learning becomes effective and efficient, cannot be limited by space and time, and learning materials can be presented more attractively by using videos, music, games, pictures, and animations.

Quizalize is one example of an online learning platform that should be adopted in universities to keep up with 21st-century teaching and learning standards. English skills are enhanced, supporting learning autonomy, by using the online evaluation tool Quizalize, described as a fun multiplayer activity by Kuzmina (2018), which allows all students to practice together using their computers, smartphones, or iPads. It is a great way for students to measure their understanding and growth, and university teachers can use it to assess students' language skills and knowledge across disciplines through learning activities in a hybrid mode environment.

This research looked at how a Quizalize intervention that incorporated gaming could affect student engagement and performance. An interactive quiz with points and team standings served as the gamification intervention. As was hypothesized, gamification's indirect impact on students' acquisition of practical skills during a

Quizziz classroom intervention was favorable. Students' successes in the learning process moderate this impact. In addition, gamification boosts intrinsic motivation and satisfies the demand for social relatedness but has no discernible impact on satisfying the desire for competence. We adopted a true experimental design, which sets us apart from most of the prior research in the subject. Our findings lend credence to the concept of gamified education put out by Beltran (2018).

The results of this study indicate that undergraduate students' active learning during the learning process with the teaching model innovation of Quizalize is known to receive quick feedback at the task level thanks to gamified interventions. Krystalli and Arvanitis (2018) and Qoura (2021) both found that using the Quizalize model as a learning medium and technology improved student performance during the intervention. Although we found an indirect effect, the impact of gamification on skill-based knowledge in this study was very good, even though the micro-grammar skills and macro-listening skills were still of minimum effect. The findings from this study corroborate the findings of Abdul (2021) that gamification may have a direct impact on the learning process itself. A teaching model with gamification improves practical skills significantly by considering the design of learning activities and statistics on how students learn. To some extent, the impact of a teaching model with Quizalize in hybrid classes on students' overall educational rankings can explain the direct effect of gamification on application-oriented knowledge in the experimental group that was exposed to innovative and new teaching methods that seek to increase student engagement during class time.

Time constraints meant rather brief interventions and brief instruments with few items were required because of the incorporation of studies into the curriculum; however, this approach has the benefit of high ecological validity. Our research was carried out in an environment that is representative of a regular college classroom in hybrid mode. This research can answer questions about the long-term consequences of the intervention, both emotionally and practically. In addition, the level of preparation significantly affects the effectiveness of the learning process and the acquisition of practical knowledge presented in this study, namely learning syntax with students centered on learning elements to maximize Quizalize. We found that student readiness, the teaching model, and lecturer readiness are important elements for the success of Quizalize. Teachers in higher education need to ensure their students are ready for a hybrid classroom environment by providing them with high-quality learning assignments and individualized audiovisual materials (Nguyen, 2022). It has been suggested that gamification can be used as an intervention to help students prepare for exams and other assignments outside the classroom.

Table 5.6 and Figures 5.1–5.6 show the survey results. During the learning process, we observed that the teaching model with Quizalize quizzes with points and team leaderboards significantly increased their confidence in learning English when they initially felt anxious and unmotivated. This finding is consistent with previous studies on in-class quiz gamification (Kuzmina, 2019) and this study, which demonstrated the beneficial effects of gamification in addition to learning motivation as findings (Chuquín et al., 2020). This study found widely that the impact of the teaching model with Quizalize in Table 5.6 is comportable, confident, fun, increasing students' interest, ambition, self-correction, rapid thinking, relaxing atmosphere, digital literacy, and user-friendly in higher education. The result consistently meets psychological needs in this case. Consistent with previous research (Capinding, 2022), we found that the teaching model with gamification significantly improves English performance skills (Figure 5.6), which is presented on social media on YouTube. Team leaderboards, which encourage healthy competition among members, can satisfy participants' desire for self-correction so that presentation scripts can be maximized.

Our findings show that the teaching model innovation has a visible impact on students' satisfaction with their competency demands and does not rule out the possibility that this teaching model can be adopted with other learning media and technologies (besides Quizalize). The digital era and technology adoption require optimal challenges for lecturers so that the teaching model is not monotonous through the design of various teaching material components, learning instructions, game interventions, and gamified quizzes according to their previous knowledge and skills. This is very important for satisfying the desire for competency, and we present considerable challenges in our research. Based on these findings, it is important to think about how the gamified teaching model's syntax interacts with the feedback provided to lecturers, students, and technology. Our research is centered on the idea of a teaching model with quiz-gamification functions as a mediator, implying a moderate effect on instructional learning.

REFERENCES

Abascal, J., & Civit Balcells, A. (2023, March). Mobile communication for older people: new opportunities for autonomous life. In *Workshop on Universal Accessibility of Ubiquitous Computing: Providing for the Elderly (2000)*. Spain.

Abdul Samea Qoura, A. (2021). Guidelines for EFL teacher education and development in the Humachine Era. *Sohag University International Journal of Educational Research*, *3*(3), 1–19.

Abdul Samea Qoura, A. (2021). Guidelines for EFL teacher education and development in the Humachine Era. *Sohag University International Journal of Educational Research*, *3*(3), 1–19.

Aguilera-Hermida, A. P. (2020). College students' use and acceptance of emergency online learning due to COVID-19. *International Journal of Educational Research Open*, *1*, 100011.

Almulla, M. A., & Al-Rahmi, W. M. (2023). Integrated social cognitive theory with learning input factors: the effects of problem-solving skills and critical thinking skills on learning performance sustainability. *Sustainability, 15*(5), 3978.

Barbosa, M. W., & Ferreira-Lopes, L. (2023). Emerging trends in telecollaboration and virtual exchange: A bibliometric study. *Educational Review, 75*(3), 558–586.

Beltran, J. C. (2018). Examining the Effects of the Application of a Student Response System in Teaching Media & Information Literacy to Senior High School Students: A Micro-Research Proposal. *Online Submission*.

Bratel, O., Kostiuk, M., Bratel, S., & Okhrimenko, I. (2021). Student-centered online assessment in foreign language classes. *Linguistics and Culture Review, 5*(S3), 926–941.

Bringula, R. P., & Atienza, F. A. L. (2023). Mobile computer-supported collaborative learning for mathematics: A scoping review. *Education and Information Technologies*, *28*(5), 4893–4918.

Caiza Imbaquingo, M. N. (2022). *Gamificación en el desarrollo de la expresión oral del idioma extranjero* (Bachelor's thesis, Quito: UCE).

Capinding, A. T. (2022). Utilization of 'Quizizz'a game-based assessment: an instructional strategy in secondary education science 10. *European Journal of Educational Research,* 11(4), 1959–1967.

Capinding, A. T. (2022). Utilization of 'Quizizz'a game-based assessment: an instructional strategy in secondary education science 10. *European Journal of Educational Research, 11*(4), 1959–1967.

Çekiç, A., & Bakla, A. (2021). A review of digital formative assessment tools: features and future directions. *International Online Journal of Education and Teaching, 8*(3), 1459–1485.

Cheng, Y. P., Lai, C. F., Chen, Y. T., Wang, W. S., Huang, Y. M., & Wu, T. T. (2023). Enhancing student's computational thinking skills with student-generated questions strategy in a game-based learning platform. *Computers & Education, 200*, 104794.

Chuquín, M. Á. A., Paguay, J. A. N., & Mora, S. X. O. (2020). Impact of the digital didactic material on students' reading and grammar acquisition. *UTE Teaching & Technology* (*Universitas Tarraconensis*), vol. 1, pp. 22–36.

Costello, R. (Ed.). (2020). Gamification Strategies for Retention, Motivation, and Engagement in Higher Education: Emerging Research and Opportunities: Emerging Research and Opportunities. IGI global.

Davronovich, A. D., & Mansurjonovich, J. M. (2023). Important advantages of organizing the educational process in a digital technology environment. *Galaxy International Interdisciplinary Research Journal, 11*(2), 149–154.

Dilekçi, A., & Karatay, H. (2023). The effects of the 21st century skills curriculum on the development of students' creative thinking skills. *Thinking Skills and Creativity, 47*, 101229.

España-Delgado, J. A. (2023). Kahoot, Quizizz, and Quizalize in the English class and their impact on motivation. *HOW Journal, 30*(1), 65–84.

Ferrer, J., Ringer, A., Saville, K., A Parris, M., & Kashi, K. (2020). Students' motivation and engagement in higher education: The importance of attitude to online learning. *Higher Education, 83*, 317–338.

Johansson, C. (2021). El rol de la instrucción ante el uso de herramientas digitales en el aula de ELE: Un estudio sobre el efecto de la instrucción antes del uso de la herramienta digital Quizalize.

Troussas, C., Krouska, A., Alepis, E., & Virvou, M. (2020). Intelligent and adaptive tutoring through a social network for higher education. *New Review of Hypermedia and Multimedia, 26*(3–4), 138–167.

Jurado-Castro, J. M., Vargas-Molina, S., Gómez-Urquiza, J. L., & Benítez-Porres, J. (2023). Effectiveness of real-time classroom interactive competition on academic performance: a systematic review and meta-analysis. *PeerJ Computer Science, 9*, e1310.

Kacetl, J., & Klímová, B. (2019). Use of smartphone applications in English language learning – A challenge for foreign language education. *Education Sciences, 9*(3), 179.

Krystalli, P., & Arvanitis, P. (2018). Self-assessment and immediate feedback in language learning. In *ICERI2018 Proceedings* (pp. 2066–2072). IATED.

Kuzmina, J. (2018). Use of mobile applications teaching English grammar at tertiary level. In *ICERI2018 Proceedings* (pp. 4212–4218). IATED.

Kuzmina, J. (2019). Selecting mobile and web-based applications to teach English grammar at tertiary level. In *INTED2019 Proceedings* (pp. 7473–7480). IATED.

Luo, Y., & Watts, M. (2022). Exploration of university students' lived experiences of using smartphones for English language learning. *Computer Assisted Language Learning*, volume 1 —(26).

Marieiev, D., Marieieva, T., Yaroshevska, L., Kaminskyy, V., & Viesova, O. (2023). Modern teacher education in Ukraine and EU countries: transformation, vectors of development. *Journal of Higher Education Theory & Practice*, volume *23*(5), 5–15.

Mohd, C. K., Nuraini, C. K., Mohamad, S. N. M., Sulaiman, H., Shahbodin, F., & Rahim, N. (2023). A review of gamification tools to boost students' motivation and engagement. *Journal of Theoretical and Applied Information Technology*, volume *101*(7), 2771–2782.

Nguyen, T. H. (2022). The effects of using online applications to teach vocabulary to English learners of HUFI in Ho Chi Minh city. *International Journal of TESOL & Education*, *2*(3), 32–42.

Okoye, K., Hussein, H., Arrona-Palacios, A., Quintero, H. N., Ortega, L. O. P., Sanchez, A. L., ... & Hosseini, S. (2023). Impact of digital technologies upon teaching and learning in higher education in Latin America: an outlook on the reach, barriers, and bottlenecks. *Education and Information Technologies*, *28*(2), 2291–2360.

Oulaich, S. (2020). Pedagogy in the digital age: Making learning effective and engaging for students. In *Innovations in Smart Cities Applications Edition 3*: *The Proceedings of the 4th International Conference on Smart City Applications 4* (pp. 168–182). Springer International Publishing.

Piatnychuk, I., Boryshkevych, I., Tomashevska, A., Hryhoruk, I., & Sala, D. (2022). Online tools in providing feedback in management. Journal of Vasyl Stefanyk Precarpathian National University, 9(3), 6–17.

Plucker, J. A., Meyer, M. S., Karami, S., & Ghahremani, M. (2023). Room to run: using technology to move creativity into the classroom. In *Creative Provocations*: *Speculations on the Future of Creativity, Technology & Learning* (pp. 65–80). Cham: Springer International Publishing.

Qoura, D. A. A. S. (2021). The teaching and learning of English language in the digital transformation era. *The International Journal of E-Learning Technology1*(3), 11–27.

Rudolph, J., Tan, S., & Tan, S. (2023). War of the chatbots: Bard, Bing Chat, ChatGPT, Ernie and beyond. The new AI gold rush and its impact on higher education. *Journal of Applied Learning and Teaching*, *6*(1) page 365-389.

Serdyukov, P. (2017). Innovation in education: what works, what doesn't, and what to do about it?. *Journal of Research in Innovative Teaching & Learning*, *10*(1), 4–33.

Szeto, M. D., Strock, D., Anderson, J., Sivesind, T. E., Vorwald, V. M., Rietcheck, H. R., ... & Dellavalle, R. P. (2021). Gamification and game-based strategies for dermatology education: Narrative review. *JMIR Dermatology*, *4*(2), e30325.

Tang, T., Abuhmaid, A. M., Olaimat, M., Oudat, D. M., Aldhaeebi, M., & Bamanger, E. (2023). Efficiency of flipped classroom with online-based teaching under COVID-19. *Interactive Learning Environments*, *31*(2), 1077–1088.

Tang, T., Abuhmaid, A. M., Olaimat, M., Oudat, D. M., Aldhaeebi, M., & Bamanger, E. (2023). Efficiency of flipped classroom with online-based teaching under COVID-19. *Interactive Learning Environments*, *31*(2), 1077–1088.

Tawafak, R. M., Al-Obaydi, L. H., Klimova, B., & Pikhart, M. (2023). Technology integration of using digital gameplay for enhancing EFL college students' behavior intention. *Contemporary Educational Technology*, *15*(4), ep452.

Tu, J. J., & Akhter, S. (2023). Exploring the role of entrepreneurial education, technology and teachers' creativity in excelling sustainable business competencies. *Economic Research-Ekonomska Istraživanja*, *36*(1), 2119429.

Yamani, H. A. (2021). A conceptual framework for integrating gamification in elearning systems based on instructional design model. *International Journal of Emerging Technologies in Learning* (*Online*), *16*(4), 14.

Zainal, N. (2023). University students' perspectives on gamification in vocabulary learning: a literature review. *Malaysian Journal of Social Sciences and Humanities* (*MJSSH*), *8*(5), e002314.

Zhang, S., & Hasim, Z. (2023). Gamification in EFL/ESL instruction: A systematic review of empirical research. *Frontiers in Psychology*, *13*, 1030790.

Zhussupova, R., & Zhumabek, G. (2021). Benefits of using online assessment applications in foreign language learning. Proceedings of the International Scientific and Practical Conference. Volume Part 1. Ekibastuz, 2021Publisher: Branch of the Federal State Budgetary Educational Institution of Higher Professional Education "Kuzbass State Technical University in Prokopyevsk (pp. 426–431).

6 Developing 3D Games for Foreign Language Education

PV-Hunt (Phrasal Verbs Hunt)

D. Kalay and K. Erarslan

6.1 INTRODUCTION

Digital technologies have always played a crucial role in people's lives across various fields, including engineering, medicine, education, and architecture. They are easily accessible to people worldwide. In the context of education, the sudden outbreak of the COVID-19 pandemic has further underscored the significance of digital technologies, such as computers, in supporting teaching and learning, particularly with the transition to online and remote education.

Computers are employed in various ways in classrooms. However, in many schools, the prevailing approach is characterized by "*making the computer teach the child*" (Papert, 2020), which does not fully harness the potential and benefits of computer skills in the educational process. Consequently, there exists a lack of understanding among people regarding the broader advantages that computer skills can bring to education, which may also stem from the fact that, while individuals interact with computers from a very early age, these interactions often remain limited to gaming purposes (Bucksch et al., 2016), failing to realize their full potential (Toker & Baturay, 2016).

Furthermore, financial constraints may limit the availability and use of computers in classrooms, which, in turn, restricts the integration of computer skills into the curriculum. In other words, having access to computers, updated software, and hardware is crucial for the successful adoption and integration of technology into classrooms (Buabeng-Andoh, 2012). To address these challenges, it is essential for both students and teachers to gain experience in effectively using computers for educational purposes. By incorporating computer skills into the curriculum and providing training and support for teachers, schools can empower students to develop essential digital literacy and computer skills (Giannikas, 2022).

Educational games and simulations have become integral components of traditional education, with extensive research exploring their impact on learning outcomes (Chiang et al., 2011). In recent years, the use of digital and web-based games for educational purposes, particularly in the realm of online education, has gained considerable attention from the scientific and academic communities, including

DOI: 10.1201/9781003400691-6

educators, students, and game designers. As technological advancements continue, there is growing interest among educators and policymakers in incorporating innovative tools like video games, virtual worlds, and Massive Multiplayer Online Games (MMPOGs) into the learning process (Buckless et al., 2014; Gómez, 2014).

Computer games have produced a variety of outcomes in several dimensions, including the perspectives of instructors, student performance, engagement, and motivation for learning. Nonetheless, existing research has primarily focused on specific disciplines, leaving a gap in the literature when it comes to a comprehensive framework for online games designed based on augmented reality, to be integrated across academic language teaching programs. Overall, the primary objective of this chapter is to create and introduce a first-person shooter (FPS) online computer game specifically aimed at English language teaching. This game is expected to provide not only language instructors but also educators from diverse disciplines with a deeper understanding of the benefits and limitations of using such online games aligned with their pedagogical objectives.

6.2 LITERATURE REVIEW

6.2.1 Need for Computers in Language Education

In recent years, substantial research efforts have emerged, focusing on various approaches to Computer-Assisted Language Learning (CALL). These approaches encompass teaching language elements and skills in various ways, in addition to creating, delivering, and assessing the target language (Mak, 2011). Specialized associations such as the Computer-Assisted Language Instruction Consortium (CALICO), European Association of Computer-Assisted Language Learning (EUROCALL), and International Association for Language Learning Technology (IALLT) have been established to dedicate their attention to the integration of technology into language education (Chun et al., 2016). Furthermore, numerous educators have recognized the computer as an effective tool for language learning. Employing computers in language education has demonstrated its capacity to not only engage learners but also yield beneficial and motivating outcomes for both teachers and students (Mehring, 2016).

Digital technology serves as an educational tool to support language learners in enhancing their language skills, supplementing traditional teaching methods, and creating an immersive and linguistically diverse learning environment. As contemporary education is increasingly based on e-learning with the integration of modern technological applications, the current focus is on equipping students with continuous learning skills that go beyond the mere transfer of knowledge, enabling them to access information and take on a more active role in the classroom, which transforms them into engaged, motivated, and active learners rather than passive recipients (Berk, 2009).

Active learner participation in the language acquisition process is essential, particularly as a foreign language differs from learners' native language (Tosuncuoğlu, 2019). Active learning, which is based on student-centered education, is promoted by computer-based instructional technologies. These technologies create a learning environment with visually stimulating and authentic course materials, such as news articles, movies, and music, catering to various learning styles. They also facilitate online communication, cultural comparisons in the target language, and opportunities to develop language skills and elements that support second-culture competency (Ghasemi et al., 2011).

From the perspective of instructors, who are in constant need of additional and appropriate teaching resources, digital technologies offer invaluable tools to foster a dynamic learning environment. These technologies assist teachers in providing timely and efficient feedback to their students, monitoring their progress, and identifying areas that need improvement (Hunukumbure et al., 2021). In addition to textbooks, instructors can also benefit from numerous resources such as interactive whiteboards, listening audios/videos, software tools (e.g., word processing and multimedia authoring (Ghasemi et al., 2011)), and online/offline games.

6.2.2 The Concept of Gamification

Recently, scholars, educators, and policymakers have shown increasing interest in exploring gamification in higher education, which encompasses various forms such as educational games (Çankaya & Karamete, 2009), digital game-based learning (DGBL) (Yang et al., 2012), and applied games (van Roessel & van Mastrigt-Ide, 2011). Some scholars may even categorize these games as interactive exercises (Mueller et al., 2003), video games (Biddiss & Irwin, 2010), and next-generation video games (Bausch et al., 2008). In the context of web-based online games, digital platforms implementing game codes include desktops/laptops and consoles (Salen & Zimmerman, 2003), as well as web browsers on mobile phones and tablets (Willoughby, 2008).

However, despite the diversity of game types, there seems to be no clear and shared terminology among researchers and educators, often leading to what is referred to as "terminological ambiguity" (Klabbers, 2009). Nevertheless, it is crucial to establish shared terminology when referring to the various types of computer games within the educational process. While scholars and game developers may use different taxonomies to categorize games, the following seven genres (Gros, 2007) are commonly recognized:

a. **Action Games:** Games that require quick responses and reflexes from the player.
b. **Adventure Games:** Games where players solve problems and navigate through levels in a virtual world.
c. **Fighting Games:** Games involving combat against computer/player-controlled characters.
d. **Role-Playing Games:** Games where players take on the roles of fictional characters and immerse themselves in their roles.
e. **Simulations:** Games that model natural or man-made systems or phenomena, challenging players to achieve specific objectives.
f. **Sports Games:** Games based on various sports, allowing players to participate in athletic activities virtually.
g. **Strategy Games:** Games with historical or fictional scenarios where players frame strategies to achieve their objectives.

The effectiveness of game-based learning in curricular education has been extensively investigated (Hwang & Wu, 2012), with findings showing that educational games offer several benefits for learners (Tsekleves et al., 2014; Huang et al., 2010). First, knowledge and skills acquired and practiced through gaming are more likely to transfer to real-world situations compared to practicing them in isolation. Mastery of these skills is further reinforced through overlearning, leading to automaticity and better retention in memory, allowing learners to focus more consciously on the understanding and application of new information (Gentile & Gentile, 2008). Second, educational gaming aims to promote intrinsic motivation by integrating learning with engaging experiences (Egenfeldt-Nielsen, 2006), which helps raise interest in learning. Such games, in addition to the virtual worlds specifically designed for teaching/learning, highlight the potential of digital technologies to motivate and engage learners beyond mere recreational activities (Anderson et al., 2009). Third, educational games create a more personalized learning experience for learners by providing different levels of difficulty, various types of activities, and diverse content, catering to individual differences and various learning styles (Putz et al., 2020).

Moreover, these games place learners in the position of decision-makers, guiding them through increasingly challenging scenarios and fostering learning through trial-and-error methods (Mikalef et al., 2012). Additionally, games offer the advantage of providing immediate feedback on students' actions and decisions, promoting a sense of exploration and experimentation (Kirriemuir, 2002). Finally, the use of game-based technology for educational purposes provides support and facilitates teaching/learning in addition to assessment/evaluation (Connolly et al., 2012).

The Cognitive Theory of Multimedia Learning (CTML) accounts for the efficacy of educational games in fostering learning (Huang et al., 2010). The fundamental premise is drawn from Paivio's (2013) Dual Coding Theory (DCT) in that a learner's cognitive system (memory) comprises two distinct channels for processing pictorial and verbal information. By presenting instructional materials that cater to both these channels, learners can enhance their understanding by integrating information from both sources (Mayer, 2001).

Although many scholars have investigated the effectiveness of educational games in learning experiences, the scarcity of clear empirical evidence (Farrington, 2011) remains a challenge (Vlachopoulos & Makri, 2017). This is primarily because the field of education is extensive and requires further analysis. Moreover, governments and educational institutions often lack official policy frameworks or endorsed guidelines regarding the integration of games into teaching, as is the case in many American, European, and Australian countries, where the decision to incorporate games into curricular education falls upon individual instructors or institutions (Vlachopoulos & Makri, 2017).

Despite these challenges, games have already been, to some extent, integrated into the teaching process to achieve diverse learning objectives, even in the absence of comprehensive policies (Connolly, 2012). The primary motivation for this chapter appears to be the need for practical step-by-step implementations to guide stakeholders (i.e., researchers, educators, teachers, and students) in formulating such policies.

Overall, this chapter primarily aims to introduce an online computer game with the objective of providing deeper insights into the field of language education.

6.3 A 3D GAME: PV-HUNT

As for the main purpose of this chapter, PV-Hunt (Phrasal Verbs Hunt) is developed as an online computer game to review specific target vocabulary previously taught. In particular, it focuses on phrasal verbs, which are known to be "notoriously challenging" elements, especially for English as a Second Language (ESL)/English as a Foreign Language (EFL) students (Kalay & Keçik, 2023).

The fundamental components of the gamification approach include a scenario created by a field expert, scenes, and graphic materials designed according to this scenario, and program scripts that drive the action. To bring the events to life, PV-Hunt was developed by creating a training scenario, relevant graphic designs, visual materials, and C# code. The game is designed using a versatile game engine called UNITY, which is particularly prominent for crafting mobile games for platforms like iOS and Android. It has been used for popular mobile games such as Pokémon Go and Call of Duty. This game engine is also employed for the creation of both two-dimensional (2D) and three-dimensional (3D) online games, incorporating interactive simulations and immersive experiences (*Unity Real-Time Development Platform | 3D, 2D, VR & AR Engine, n.d.*).

PV-Hunt is designed as an online FPS game using UNITY 3D. In simple terms, students are tasked with navigating a designated path, locating target phrasal verb-related questions, and responding to these questions from a first-person perspective. The game's mechanics will be discussed in the following section.

6.3.1 Scenario

PV-Hunt has been developed as a digital resource that supports and complements the educational materials used in language classrooms by assessing students' knowledge of phrasal verbs. Given that phrasal verb structures are among the most challenging language components, additional materials that employ visualization or memory aids are often necessary (Kalay & Keçik 2023). Leveraging the gamification approach to enhance the learning process, PV-Hunt aims to familiarize students with phrasal verbs and test their knowledge in a dynamic virtual environment.

As per the game scenario, students initiate the PV-Hunt application at the starting point of a specially designed terrain. The user, or game player (i.e., the student), first encounters information boards that provide an overview of the game, control keys, and task instructions. The primary objective of the game is to locate phrasal verb questions at five different stations along the path, answer them correctly, and while progressing along the path, collect coins. Each correct answer earns the player +10 points, while incorrect answers result in a loss of −5 points. Additionally, there are 15 coins scattered along the path, each worth +2 points, providing an additional element of excitement to the game. The game concludes once all the questions are answered and all the coins have been collected.

6.3.2 Design

In this chapter, as mentioned previously, the UNITY 3D Game Engine is utilized to develop PV-Hunt (*Unity Real-Time Development Platform | 3D, 2D, VR & AR Engine, n.d.*). It is one of the most widely used game engines for various purposes, including serious games, gamification, and virtual/augmented reality applications. Scenes comprising the game's elements can be created within UNITY 3D, and designs from software such as AutoCAD, 3DSMax, and Blender can also be imported. Unity allows the development of applications for multiple platforms, including PC-based, web-based, Android/IOS systems, consoles, and VR/AR headsets. In this study, a serious game was developed that can be used for both PC-based and web-based platforms.

Based on the scenario, a terrain with a spacious area for movement has been created, incorporating five question stations related to phrasal verbs, as illustrated in Figure 6.1.

Initially, the student comes across the task descriptions, instructions, and game control explanations, as demonstrated in Figures 6.2–6.4.

When players traverse the wooden bridge (boernie97, 2023) spanning the river on the path, they arrive at the station section. The graphical assets for each station along the path have been created to align with the context of the target phrasal verb question. Additionally, the question boards have been designed and placed adjacent to the stations within the scenes, as depicted in Figures 6.5–6.9.

In addition to each station on the path, all the target PV questions directed to the students are presented in Figure 6.10.

The total score is kept on a leaderboard available in the upper right corner of the screen, as demonstrated in Figure 6.11. The total score is out of 80. After completing the path, the player can end the game by pressing the ESC key on the keyboard.

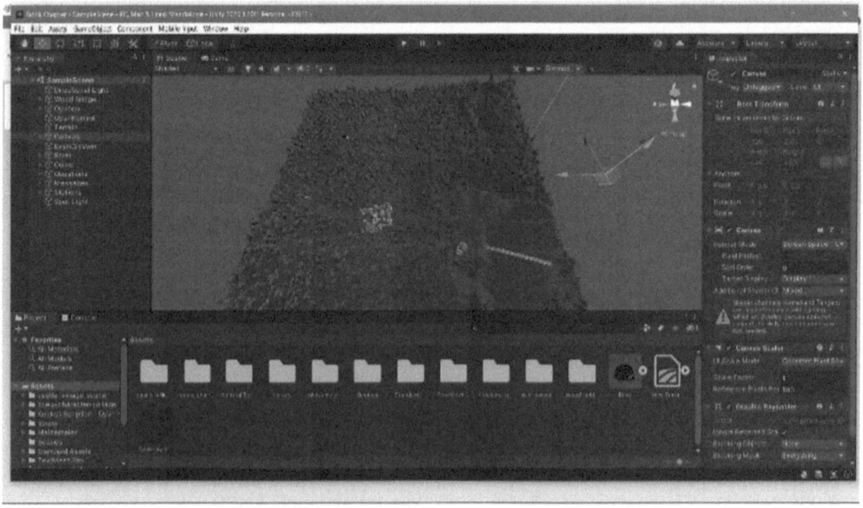

FIGURE 6.1 Game platform.

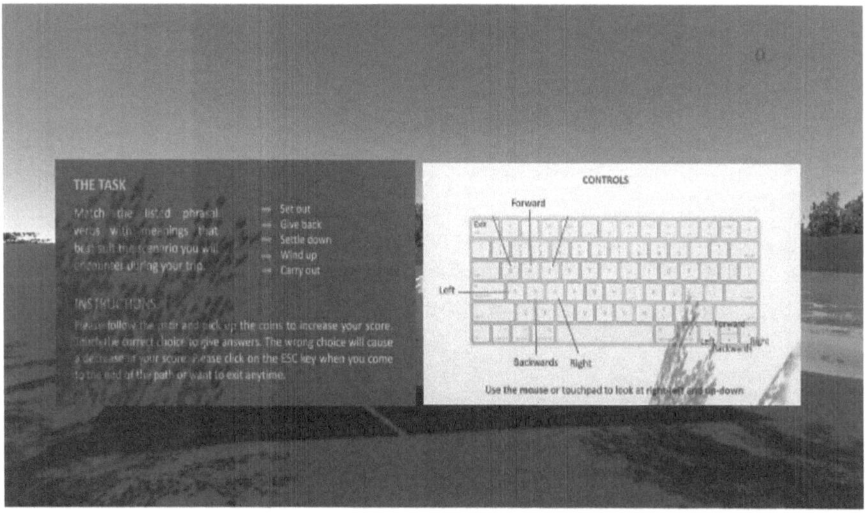

FIGURE 6.2 Start-up point.

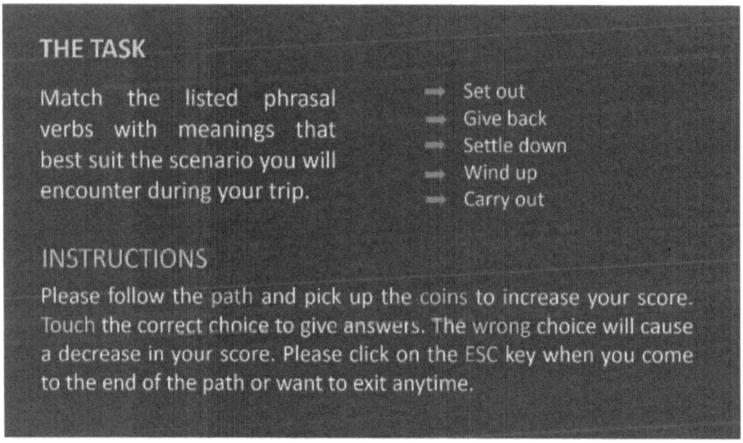

FIGURE 6.3 Task and instructions.

6.3.3 Code Scripting

After designing the stage, in line with the first-person perspective/FPS concept, three code files are created using the C-Sharp (C#) programming language to manage movements within the terrain, answering questions, and scoring. These codes are interconnected components of the overall scenario. They allow for movement in the horizontal directions using designated keys and for viewing in all directions through mouse movements. Points are accumulated by answering questions on the track and collecting coins. The *CameraControl.cs* code file enables a 360-degree view via mouse control. The *PlayerControl.cs* file contains the codes responsible for

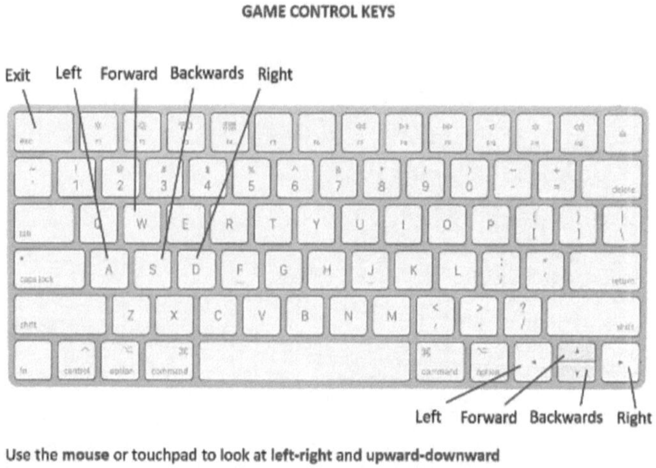

FIGURE 6.4 Game control keys.

FIGURE 6.5 Station 1; PV question 1.

horizontal movement assigned to specific keys. Figure 6.12 demonstrates the presence of codes controlling the score texts in the *GameControl.cs* file.

6.3.4 Implementation

UNITY 3D, known as a cross-platform program, can produce outputs in various formats, such as exe, html, apk etc. To clarify, the final output can be saved in the form of a PC-based (exe) application, web-based (HTML), Android/IOS mobile application (APK), game console application (PS4, PS5, Xbox), and virtual and augmented reality, as illustrated in Figure 6.13.

FIGURE 6.6 Station 2; PV question 2.

FIGURE 6.7 Station 3; PV question 3.

Overall, in the context of the PV-Hunt game, players are expected to run the EXE or HTML application on an assigned computer and complete the target task by locating the relevant questions and providing answers within a limited time, all under the instructor's supervision. The faster players reach the finish scene, the higher their scores will be.

6.4 DISCUSSION

6.4.1 Possible Challenges

As a matter of fact, players may encounter several challenges and obstacles while playing the PV-Hunt game. These potential issues can be outlined as follows:

FIGURE 6.8 Station 4; PV question 4.

FIGURE 6.9 Station 5; PV question 5.

a. **Navigational Challenges:** Players might face difficulties in effectively navigating the terrain, making it harder to locate designated questions or stations.
b. **Time Management:** Balancing the need to answer questions quickly while collecting coins within the allocated time could prove challenging for players.
c. **Distractions:** Environmental distractions or sudden changes in the game environment could divert players' attention, making it more challenging to focus on the questions.
d. **Technical Glitches:** Technical issues stemming from various factors, such as students' computer hardware or software, or internet connectivity problems like lag, freezing, or crashes, could disrupt gameplay and frustrate players.

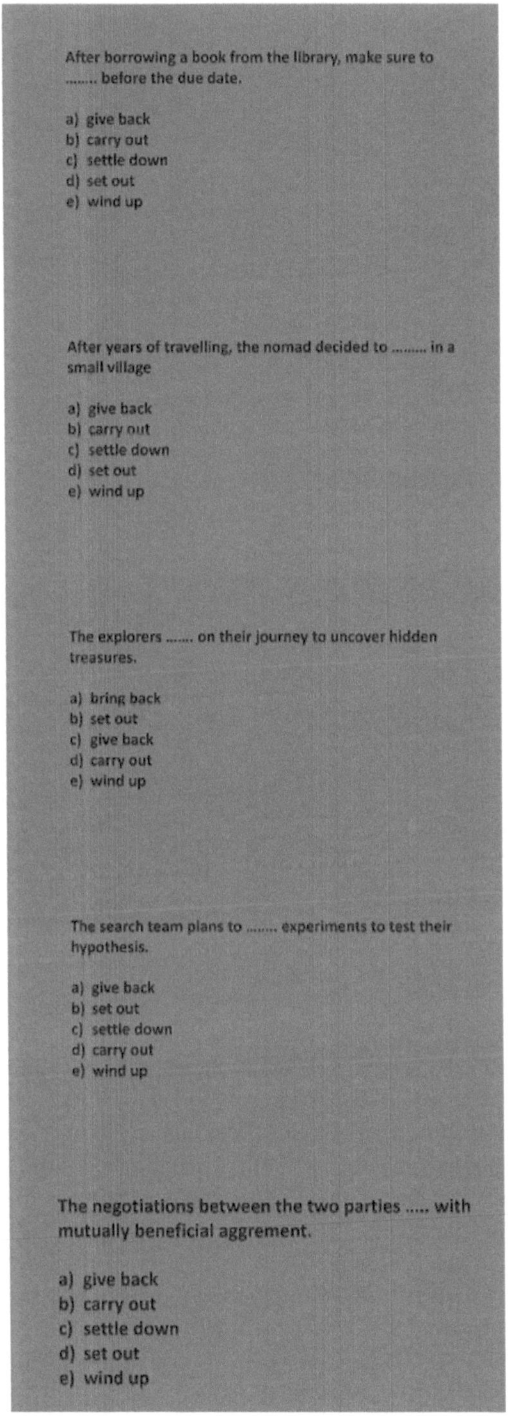

After borrowing a book from the library, make sure to .... before the due date.

a) give back
b) carry out
c) settle down
d) set out
e) wind up

After years of travelling, the nomad decided to .... in a small village

a) give back
b) carry out
c) settle down
d) set out
e) wind up

The explorers .... on their journey to uncover hidden treasures.

a) bring back
b) set out
c) give back
d) carry out
e) wind up

The search team plans to .... experiments to test their hypothesis.

a) give back
b) set out
c) settle down
d) carry out
e) wind up

The negotiations between the two parties ..... with mutually beneficial aggrement.

a) give back
b) carry out
c) settle down
d) set out
e) wind up

FIGURE 6.10 All PV questions.

FIGURE 6.11 Coins and scores.

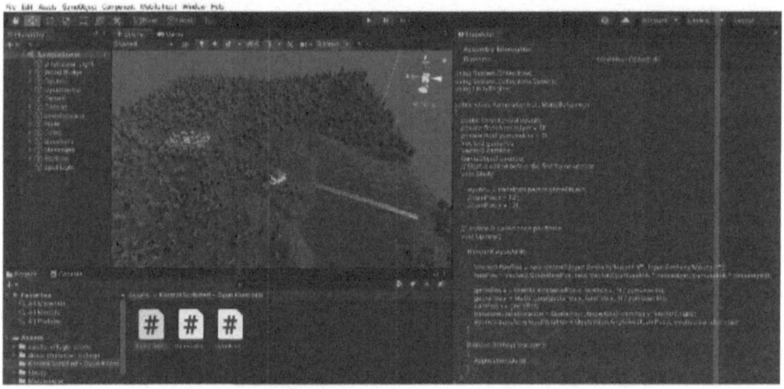

FIGURE 6.12 C# script files.

e. **Lack of Familiarity with Controls:** Players unfamiliar with the game controls or mechanics might struggle to perform actions efficiently.
f. **Misinterpretation of Instructions:** Misunderstanding game instructions or task descriptions could lead to confusion and incorrect actions.
g. **Language Barrier:** Non-native English speakers or players with lower language proficiency might encounter difficulties understanding and responding to PV questions.
h. **Visual Clues Misinterpretation:** Players might misinterpret visual cues or context clues related to PV questions, leading to incorrect answers.
i. **Loss of Direction:** Players could become disoriented in the game world or lose track of their progress along the path. This could be due to a lack of game design experience or because the game's challenges are either too complex or too easy.

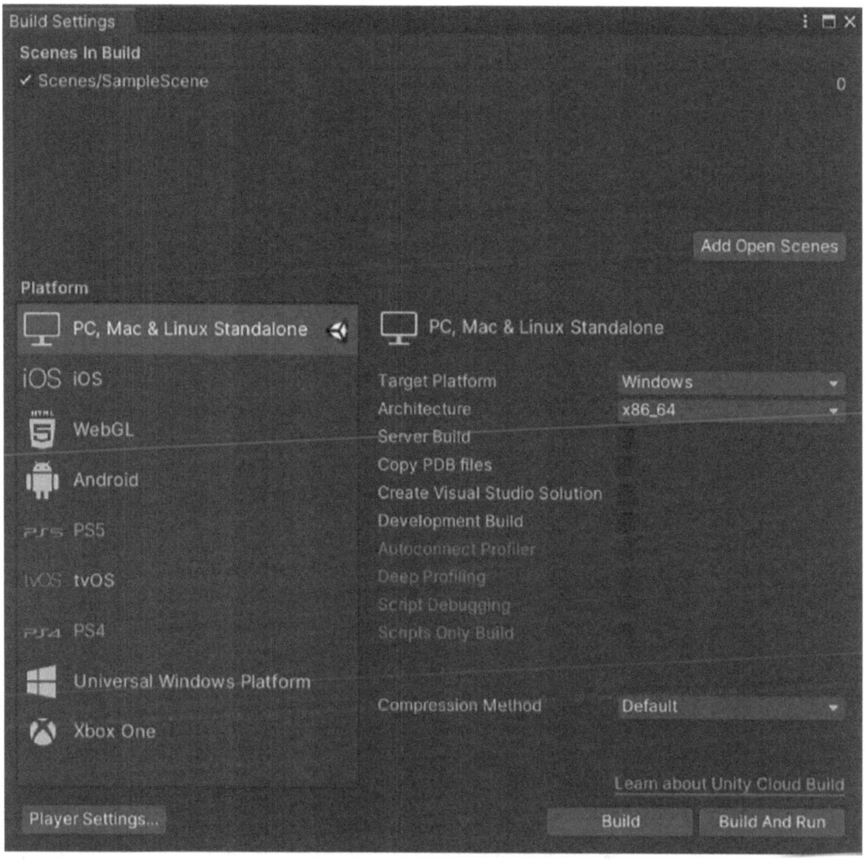

FIGURE 6.13 Cross-platform property of UNITY 3D.

j. **Performance Pressure:** The competitive aspect of the game or the desire to achieve a high score could lead to performance anxiety, impacting decision-making.

All in all, recognizing these potential challenges is crucial, and it's essential to implement strategies to mitigate them. By anticipating and addressing these possible issues, educators and policymakers can fine-tune the PV-Hunt game to create a positive, enriching, balanced, and enjoyable learning experience. This will encourage students to overcome challenges, improve their language skills, and derive meaningful learning outcomes from their interaction with the game.

6.4.2 Implications

In addition to the challenges that may be encountered, the game PV-Hunt holds significant implications for both educators and policymakers in the field of language teaching:

a. PV-Hunt offers a novel approach to language teaching, providing a gamified, engaging, and motivating language learning experience for learners. Instructors can utilize PV-Hunt's interactive and immersive nature to spark learners' interest and maintain their engagement during language classes.
b. PV-Hunt fosters a dynamic and experiential teaching/learning context. Educators can integrate the game as a supplementary tool alongside traditional language teaching methods, enhancing students' learning experiences and deepening their understanding of target language structures.
c. PV-Hunt may guide language instructors in promoting digital literacy skills among students. Furthermore, since technology has become an integral part of education, policymakers can recognize the significance of cultivating digital competence in language learners to prepare them for the digital age.
d. PV-Hunt's adaptable nature allows instructors to tailor the game's difficulty levels to accommodate learners' language proficiency levels, promoting personalized learning where students can progress at their own pace. Policymakers can endorse these technology-enhanced strategies to address diverse learner needs.
e. The tracking and scoring features of PV-Hunt serve as assessment mechanisms that generate valuable data on learner performance and progress. Educators and policymakers can analyze this data to gain insights into students' strengths and areas for improvement, informing instructional decisions and curriculum development.
f. Educators and policymakers can collaborate to integrate PV-Hunt into the language curriculum. Such integration contributes to a comprehensive and modern curriculum that reflects the evolving needs of language learners.
g. PV-Hunt opens avenues for further research on game-based language learning and its effectiveness. Educators and policymakers can collaborate on designing research projects to investigate the efficacy of web-based games like PV-Hunt and similar tools on language proficiency and overall learning outcomes.

In summary, the integration of online games like PV-Hunt into language education as an innovative pedagogical practice encourages educators to explore creative and effective ways to incorporate technology into their teaching methods, enhancing language learning outcomes. Furthermore, policymakers may consider such a web-based game as an example of how educational technology can align with language learning objectives, underscoring the potential of technology to support and achieve broader educational goals.

6.5 CONCLUSION

In conclusion, the creation and exploration of PV-Hunt, an innovative educational game designed for language learning, underscore the transformative capabilities of gamification in contemporary education. This chapter has traversed a multifaceted terrain, delving into the convergence of digital technology, language teaching methods, and gamified learning experiences. In doing so, it has revealed the promising opportunities that PV-Hunt and similar gamified approaches offer to learners, educators, policymakers, and game developers.

In a broader context, PV-Hunt represents a departure from conventional didactic approaches, signaling a shift toward comprehensive, immersive, and student-centered educational experiences. It reinforces the notion that educational games are not just tools but transformative vehicles that bridge the gap between pedagogical theory and practical application. As digital technologies continue to evolve and education adapts to new horizons, PV-Hunt is poised to become a symbol of the possibilities that await those who embrace innovative approaches to teaching and learning.

To summarize, the journey through the development of PV-Hunt has illuminated the potential for gamification to revolutionize language education. By harnessing the power of technology and embracing the principles of active learning, engagement, and motivation, PV-Hunt invites educators and policymakers to embark on a transformative journey toward a future where learning is not merely a pursuit but an immersive, enjoyable, and empowering experience. As the impact of PV-Hunt resonates through classrooms and educational institutions, it serves as a guiding light of innovation, leading the way to a more dynamic and effective educational landscape.

REFERENCES

Anderson, E. F., McLoughlin, L., Liarokapis, F., Peters, C., Petridis, P., & De Freitas, S. (2009). Developing serious games for cultural heritage: A state-of-the-art review. *Virtual Reality, 14*(4), 255–275. https://doi.org/10.1007/s10055-010-0177-3

Bausch, L., Beran, J., Cahanes, S., & Krug, L. (2008). Physiological responses while playing Nintendo Wii sports. *Journal of Undergraduate Kinesiology Research,* 3, 19–25.

Berk, R. A. (2009). Multimedia teaching with video clips: TV, Movies, YouTube, and mtvU in the college classroom. *International Journal of Technology in Teaching & Learning, 5*(1), 1–21.

Biddiss, E., & Irwin, J. D. (2010). Active video games to promote physical activity in children and youth. *Archives of Pediatrics and Adolescent Medicine, 164*(7), 664–672. https://doi.org/10.1001/archpediatrics.2010.104

Buabeng-Andoh, C. (2012). Factors influencing teachers' adoption and integration of information and communication technology into teaching: A review of the literature. *International Journal of Education and Development Using Information and Communication Technology, 8*(1), 136–155. https://files.eric.ed.gov/fulltext/EJ1084227.pdf

Buckless, F. A., Krawczyk, K., & Showalter, D. S. (2014). Using virtual worlds to simulate Real-World audit procedures. *Issues in Accounting Education, 29*(3), 389–417. https://doi.org/10.2308/iace-50785

Bucksch, J., Sigmundová, D., Hamrik, Z., Troped, P. J., Melkevik, O., Ahluwalia, N., Borraccino, A., Tynjälä, J., Kalman, M., & Inchley, J. (2016). International trends in adolescent screen-time behaviors from 2002 to 2010. *Journal of Adolescent Health,* 58(4), 417–425. https://doi.org/10.1016/j.jadohealth.2015.11.014

Chiang, Y. T., Lin, S. S. J., Cheng, C., & Liu, E. Z. F. (2011). Exploring online game players' flow experiences and positive affect. *The Turkish Online Journal of Educational Technology, 10*(1), 106–114.

Chun, D. M., Smith, B. R., & Kern, R. (2016). Technology in language use, language teaching, and language learning. *The Modern Language Journal, 100*(S1), 64–80. https://doi.org/10.1111/modl.12302

Connolly, T., Boyle, E., MacArthur, E., Hainey, T., & Boyle, J. M. (2012). A systematic literature review of empirical evidence on computer games and serious games. *Computers & Education, 59*(2), 661–686. https://doi.org/10.1016/j.compedu.2012.03.004

Çankaya, S., & Karamete, A. (2009). The effects of educational computer games on students' attitudes towards mathematics course and educational computer games. *Procedia – Social and Behavioral Sciences, 1*(1), 145–149. https://doi.org/10.1016/j.sbspro.2009.01.027

Egenfeldt-Nielsen, S. (2006). Overview of research on the educational use of video games. *Digital Kompetanse, 1*(3), 184–214. https://doi.org/10.18261/issn1891-943x-2006-03-03

Farrington, J. (2011). From the research: Myths worth dispelling: Seriously, the game is up. *Performance Improvement Quarterly, 24*(2), 105–110. https://doi.org/10.1002/piq.20114

Gentile, D. A., & Gentile, J. R. (2008). Violent video games as exemplary teachers: A Conceptual analysis. *Journal of Youth and Adolescence, 37*(2), 127–141. https://doi.org/10.1007/s10964-007-9206-2

Ghasemi, B., Hashemi, M. M., & Bardine, S. H. (2011). The capabilities of computers for language learning. *Procedia – Social and Behavioral Sciences, 28*, 58–62. https://doi.org/10.1016/j.sbspro.2011.11.012

Giannikas, C. N. (2022). Developing students' digital literacy skills. *Structural Learning*. https://www.structural-learning.com/post/developing-students-digital-literacy

Gómez, M. (2014). Ciencias Sociales y gamificación, ¿una pareja con futuro? In J. Pagès (Ed.), *Una mirada al pasado y un proyecto de futuro: investigación e innovación en didáctica de las ciencias sociales* (pp. 257–262). Barcelona: Universitat Autónoma de Barcelona; Asociación Universitaria de Profesores de Didáctica de las Ciencias Sociales.

Gros, B. (2007). Digital games in education: The design of games based learning environments. *Journal of Research on Technology in Education, 40*(1), 23–39. https://files.eric.ed.gov/fulltext/EJ826060.pdf

Hwang, G., & Wu, P. (2012). Advancements and trends in digital game-based learning research: A review of publications in selected journals from 2001 to 2010. *British Journal of Educational Technology, 43*(1), E6-E10. https://doi.org/10.1111/j.1467-8535.2011.01242.x

Huang, W. D., Huang, W., & Tschopp, J. (2010). Sustaining iterative game playing processes in DGBL: The relationship between motivational processing and outcome processing. *Computers & Education, 55*(2), 789–797. https://doi.org/10.1016/j.compedu.2010.03.011

Hunukumbure, A. D., Horner, P. J., Fox, J., & Thakerar, V. (2021). An online discussion between students and teachers: A way forward for meaningful teacher feedback? *BMC Medical Education, 21*(1). https://doi.org/10.1186/s12909-021-02730-8

Kalay, D., & Keçik, İ. (2023). Metaphorical insights: Phrasal verb knowledge gains in the light of conceptual metaphor theory. *European Journal of Foreign Language Teaching, 7*(3). https://dx. doi.org/10.46827/ejfl.v7i3.5070

Kirriemuir, J. (2002). The relevance of games and gaming consoles to the higher and further education learning experience. *Jisc Techwatch*, https://doi.org/10.13140/rg.2.2.34765.03044

Klabbers, J. (2009). Terminological ambiguity. *Simulation & Gaming, 40*(4), 446–463. https://doi.org/10.1177/1046878108325500

Mak, S. H. (2011). Tensions between conflicting beliefs of an EFL teacher in teaching practice. *RELC Journal, 42*(1), 53–67. https://doi.org/10.1177/0033688210390266

Mayer, R. E. (2001). *Multimedia Learning*. Cambridge: Cambridge University Press.

Mehring, J. (2016). Present research on the flipped classroom and potential tools for the EFL classroom. *Computers in the Schools, 33*(1), 1–10. https://doi.org/10.1080/07380569.2016.1139912

Mikalef, K., Giannakos, M. N., Chorianopoulos, K., & Jaccheri, L. (2012). "Do not touch the paintings!" The benefits of interactivity on learning and future visits in a Museum. In *Lecture Notes in Computer Science* (pp. 553–561). https://doi.org/10.1007/978-3-642-33542-6_73

Mueller, F., Agamanolis, S., & Picard, R. W. (2003). Exertion interfaces. *Proceedings of the SIGCHI Conference on Human Factors in Computing Systems*, 561-568. https://doi.org/10.1145/642611.642709

Paivio, A. (2013). Imagery and verbal processes. In *Psychology Press eBooks*. https://doi.org/10.4324/9781315798868

Papert, S. A. (2020). *Mindstorms: Children, Computers, and Powerful Ideas*. Hachette UK.

Putz, L., Hofbauer, F., & Treiblmaier, H. (2020). Can gamification help to improve education? Findings from a longitudinal study. *Computers in Human Behavior, 110*, 106392. https://doi.org/10.1016/j.chb.2020.106392

Salen, K., & Zimmerman, E. (2003). *Rules of Play: Game Design Fundamentals*. Cambridge, MA, USA: MIT Press. https://ci.nii.ac.jp/ncid/BA63865022

Sketchfab. (n.d.-a). *Castle Village Scene – Download Free 3D Model by BlackSpire*. https://sketchfab.com/3d-models/castle-village-scene-675b3afce21a4acaa1140645a5206db3

Sketchfab. (n.d.-b). *DCEU Character Collage – Download Free 3D Model by DavidA3D*. https://sketchfab.com/3d-models/dceu-character-collage-07c7725906ae4b09a26ea12405dd1fd4

Sketchfab. (n.d.-c). *Library – Download Free 3D Model by Tridigon*. https://sketchfab.com/3d-models/library-2f385666f68a452091e646ea4f5a7b72

Sketchfab. (n.d.-d). *Low Poly Sci-Fi Lab – Download Free 3D Model by berkay.tari*. https://sketchfab.com/3d-models/low-poly-sci-fi-lab-e84fd3405b34421095c70b0e09cc3111

Sketchfab. (n.d.-e). *Treasure of the Sand King – Download Free 3D Model by GruntWorks*. https://sketchfab.com/3d-models/treasure-of-the-sand-king-af87c52c60f84defb0323fdae217dd07

Sketchfab. (n.d.-f). *Wood Bridge – Download Free 3D Model by boernie97*. https://sketchfab.com/3d-models/wood-bridge-fd55f51b15f44d8da3ba8957c6b879f5

Toker, S., & Baturay, M. H. (2016). Antecedents and consequences of game addiction. *Computers in Human Behavior, 55*, 668–679. https://doi.org/10.1016/j.chb.2015.10.002

Tosuncuoğlu, İ. (2019). Active language learning processes in Turkey. *Eurasian Journal of English Language and Literature, 1*(1), 23–29. https://dergipark.org.tr/en/download/article-file/856430

Tsekleves, E., Cosmas, J., & Aggoun, A. (2014). Benefits, barriers and guideline recommendations for the implementation of serious games in education for stakeholders and policy-makers. *British Journal of Educational Technology, 47*(1), 164–183. https://doi.org/10.1111/bjet.12223

Unity Real-Time Development Platform | 3D, 2D, VR & AR Engine. (n.d.). https://unity.com/

van Roessel, L., & van Mastrigt-Ide, J. (2011). Collaboration and team composition in applied game creation processes. *Digital Games Research Association Conference, 6*. https://www.digra.org/wp-content/uploads/digital-library/11301.53001.pdf

Vlachopoulos, D., & Makri, A. (2017). The effect of games and simulations on higher education: A systematic literature review. *International Journal of Educational Technology in Higher Education, 14*(1). https://doi.org/10.1186/s41239-017-0062-1

Willoughby, T. (2008). A short-term longitudinal study of Internet and computer game use by adolescent boys and girls: Prevalence, frequency of use, and psychosocial predictors. *Developmental Psychology, 44*(1), 195–204. https://doi.org/10.1037/0012-1649.44.1.195

Yang, J. C., Chien, K. H., & Liu, T. C. (2012). A digital game-based learning system for energy education: An energy conservation pet. *The Turkish Online Journal of Educational Technology, 11*(2), 27–37.

7 Future of Writing in Higher Education

A Review of Prospective Use of Artificial Intelligence and ChatGPT in Academia

Muhammad Imran, Norah Almusharraf, and Hala Dalbani

7.1 INTRODUCTION

Writing is one of the integral and crucial components of academic activities at educational institutions. Various writing tools and technologies, from the earliest forms of writing scripts to the contemporary word processor, have evolved to enhance writing in terms of accuracy, authenticity, and efficiency. These tools include predictive text, spell checking, syntax error detection, Grammarly, and many others (Imran & Almusharraf, 2023). Nevertheless, scholarly investigations in the Artificial Intelligence in Education (AIED) field have primarily focused on advancing AI-driven tools tailored to provide comprehensive support to learners. The overarching ambition has been to empower individuals to achieve greater self-sufficiency in their educational pursuits while simultaneously automating instructional functions traditionally carried out by human instructors (Holmes & Tuomi, 2022). AI's recently introduced chatbot, Chat Generating Pre-Training (ChatGPT), has emerged as a writing assistant that integrates AI tools and education to address the challenges learners encounter in the current era, particularly when it comes to academic writing tasks. Therefore, AI has significantly expanded its role by acting as a teacher's assistant, facilitating the adoption of online and blended learning methodologies, promoting a culture of virtual assistance, and meeting the growing demand for digital learning platforms worldwide (Imran & Almusharraf, 2023; Limna et al., 2023).

Nevertheless, the extensive use and integration of AI in academia raise significant concerns about the reliability of AI for educational tasks, its functions, acceptance in academic circles, and, most importantly, the ethical values connected with AI-supported tools' use for academic and research purposes. However, the trustworthiness and efficacy of AI systems are intricately associated with the data quality used for their training, underscoring the pivotal role of high-quality training datasets in ensuring the robustness and accuracy of these systems in various applications

DOI: 10.1201/9781003400691-7

(Imran & Almusharraf, 2023). This training raises concerns about biases and inaccuracies in AI-generated educational content and assessments. This necessitates a continuous effort to ensure accuracy and fairness. Furthermore, the functions of AI in education, while offering personalized learning and content creation, spark a debate about the optimal balance between human and AI expertise in teaching and learning. Acceptance within academic circles encounters resistance based on unfamiliarity and job displacement concerns, which require trust-building and evidence of AI's educational benefits. The ethical dimension looms significant, encompassing issues related to data privacy, transparency, and algorithmic decision-making. This makes it essential for institutions to establish ethical guidelines and frameworks governing AI use in education to ensure alignment with fairness, accountability, and transparency principles (van Slyke et al., 2023). Addressing these concerns is pivotal in responsibly harnessing AI's potential in the future of education.

Hence, learning with AI has gained significant attention in nearly all subjects since the COVID-19 pandemic began. The incorporation of AI tools and technologies within education involves the implementation of three distinct approaches to AI-driven tools for teaching and learning purposes. First, AI technology provides direct support to instructors both inside and outside the classroom. It offers real-time data analysis and insights into learners' progress, enabling more effective and targeted teaching approaches. This direct support from AI technologies empowers instructors to adapt their teaching strategies in response to the individual needs of the learners, fostering a more engaging and personalized learning environment. Second, AI is utilized to assist students who are actively engaged in the learning process and benefit from AI tools, including chatbots, online learning software, AI support for learners with disabilities, intelligent tutoring systems, learning network orchestrators, and machine writing evaluation and assessment. Lastly, AI technology serves a vital function in managing the educational and learning processes to support administrative tasks that are essential for the seamless operations of educational institutions. These tasks include, among other things, timetable management, faculty meeting plans, learning management systems, efficient class scheduling and resource allocation, and faculty and staff recruitment. The integration of AI into these administrative functions enhances the overall efficiency and effectiveness of educational institutions, contributing to a more streamlined and productive learning environment (Nemorin et al., 2023).

This chapter is dedicated to AI's latest chatbot, ChatGPT, and its role in education, particularly in writing programs. ChatGPT's rapid surge as a writing assistant and creative tool has prompted educators to consider future education policies. The swift rise of ChatGPT and other AI applications such as QuillBot, Grammarly, and Copy.ai has expedited the academic writing process and posed challenges to traditional educational tools and approaches (Karakose, 2023). The extensive integration of AI tools and applications raises questions about the future roles of instructors and teachers in writing classrooms, significantly impacting the traditional education system. The necessity lies in comprehending and equipping learners and educators on how to leverage the latest AI tools, from assessing and measuring a student's pace and performance to employing machine learning, deep learning, translation, and advanced analytics. These AI functions are increasingly deployed in advancing higher education settings, identifying gaps in learning and teaching, and enhancing educational competency (Nemorin et al., 2023).

ChatGPT, shortly after its launch, has received an enormous global response; however, discussions regarding its use in academic and research projects and its consequences have been somewhat slow to develop. Concerns and opportunities are linked to this chatbot, and since its introduction in November 2022, many researchers and practitioners have engaged in discussions regarding its advantages and disadvantages. This study holds significance in comprehending ChatGPT's role in facilitating writing tasks and its potential implications for Higher Education Institutions (HEIs) and research institutes. Additionally, this chapter reviews the opportunities, ethical concerns, and challenges presented by AI tools for writing purposes. Furthermore, AI tools can aid in overcoming writer's block and simplifying writing and editing tasks, enhancing overall efficiency. Consequently, it may help improve learners' performance by integrating teachers and AI technologies into the learning process for future educational programs (Sullivan et al., 2023; Young & Shishido; 2023; Imran & Almusharraf, 2023a). This chapter aims to address the following three questions:

a. What is the impact of the latest AI technologies on the educational process?
b. How can ChatGPT influence academic writing skills, prospects, and challenges?
c. What does the future hold for academic writing in the presence of advanced AI chatbots?

7.2 LITERATURE REVIEW

7.2.1 AI and Its Use in Education

The integration of AI in academia has been accepted and acknowledged globally over the past two decades (Imran & Almusharraf, 2023). AI is considered a driving force across all fields and has evolved into an essential component of the Industry 4.0 revolution, potentially sparking a 4.0 revolution within the education sector. Educators have initiated a review of their current policies and AI capabilities to optimize learning and explore new potential pathways (Karakose, 2023). Holmes and Tuomi (2022) critically examined the relationship between AI technologies and education. They further emphasize this connection through three distinct characteristics. Firstly, the report delves into both the teaching and application of AI in education, collectively referred to as AI and education. Secondly, it explores AI and education through the perspective of the Council of Europe's core values. Finally, it adopts a deliberately critical approach to AI and education rather than assuming the inherent benefits of AI in the educational context. This report offers a comprehensive perspective on AI technologies that empower both learners and instructors, with an emphasis on ensuring that future practices and developments genuinely serve the common good.

Similarly, Sullivan et al.'s (2023) study explores AI's integration in education, highlighting the substantial expansion of AI integration within various disciplines of educational technology. The focus has centered on the observation that both instructors and learners may not possess comprehensive familiarity with AI and its applications within HEIs. Policymakers should, therefore, invest additional efforts in educating academia about harnessing AI's potential for pedagogical and instructional purposes on a broader scale. Furthermore, it is crucial to engage in a discourse on the significance of AI's impact on learning, training, and teaching at the higher education level (Sullivan et al., 2023). This discussion can facilitate a more profound understanding among individuals within academia regarding AI's influence on education, including its opportunities and challenges. This chapter also emphasizes the importance of developing methodologies and techniques for the utilization of AI technologies within an AI-embedded learning and teaching platform. This ensures that both learners and instructors are well informed about the implications of AI in education.

7.2.2 ChatGPT

This chatbot is a product of cutting-edge OpenAI technology that has been trained to process extensive volumes of big data, employing deep learning techniques to deliver both reactive and analytical responses (Imran & Almusharraf, 2023). Kung et al. (2023) further discuss the core values and the underlying AI technology used in the development of ChatGPT programs. ChatGPT-3.5 is an openly accessible public tool created by OpenAI, leveraging generative pre-trained transformer (GPT) technology to respond to a wide range of text-based queries. ChatGPT exhibits a multitude of functions and is currently one of the most advanced chatbots available. It excels at executing numerous tasks, including composing letters, crafting creative pieces of poetry and fiction, comprehending and interpreting users' requests, and generating responses closely resembling natural human language (Kung et al., 2023). This study provides a comprehensive exploration of ChatGPT's functionality and its potential impact across various sectors, encompassing domains such as education and creative writing.

Furthermore, Lund and Wang (2023) conducted an analysis of the groundbreaking innovation of AI's chatbot, ChatGPT, and its possible effect on academia overall and, particularly, higher education. This study serves to enlighten educators and learners about the substantial capabilities and power that ChatGPT can have in advancing higher education. In a question-and-answer session, Lund and Wang (2023) explored its advantages and disadvantages, offering real-life examples of ChatGPT's application. However, their central focus is on the ethical considerations and responsibilities associated with the adoption of this new technology within academic settings. They also underscore concerns regarding security implications and data privacy when utilizing this chatbot. Consequently, there is a pressing need for a comprehensive awareness campaign to educate users and institutions about potential issues and to establish suitable measures for addressing potential security risks.

In other studies that examined the impact of AI on education, Taecharungroj (2023) conducted an evaluation of the effects of AI chatbots on three broad topics: technology, reactions, and news by posing the question, "What can ChatGPT do?" This study is highly recommended for educators and learners as it identifies five functional domains where this AI chatbot can exert both positive and negative influences on humans and technology. These domains encompass responsive writing, answering diverse questions, essay composition, code generation, and tasks related to prompt writing. Taecharungroj (2023) concludes his study by highlighting four pivotal issues in this age of AI: the challenges of ethical advancements, the emergence of a new technological landscape, the evaluation of the job market, and the pursuit of AI for everyday uses.

Regarding the reservations related to ChatGPT's integration into academic projects and scholarly writing endeavors, examining the multifaceted issues and challenges, McMurtrie (2022) poses a few critical questions for practitioners and learners regarding the incorporation of AI chatbots in education. These questions include: Is the college essay dead? Can AI-generated writing convincingly replace/imitate a first-year student's work? Are students using AI chatbots for unethical practices, such as cheating in writing exams or assignments? Similar questions and concerns have been raised by other scholars (Lund & Wang, 2023; Karakose, 2023) following the widespread adoption of ChatGPT within academic circles in its first month of launch. Experts and scholars specializing in writing programs and digital literacy acknowledge that AI chatbots are exerting a profound impact on everyday writing tasks, much like other technologies, such as computers and calculators, which have become integral to science, humanities, and engineering-related fields. Therefore, the pressing need of the moment is to initiate dialogues among teachers, scholars, and learners regarding how to harness and shape these AI technologies and tools to support the learning process, much like other educational aids, rather than fearing their integration.

7.2.3 Future of Academic Writing with ChatGPT

ChatGPT can have a profound impact on academic writing tasks at the higher education level in the future (Zhu et al., 2023). Nevertheless, through a case study conducted at Jonkoping University, Lopez and Qamber (2022) highlighted both the advantages and disadvantages connected to the use and extensive integration of chatbots and other AI technologies in education. This case study also offers guidance on the constructive application of AI tools while investigating user behaviors and opinions concerning the prospective use of chatbots, including ChatGPT, within the university context. The authors posit that AI chatbots are akin to 'chitchat' applications that are already widely utilized in various forms of communication. However, the underlying purpose of this advanced ChatGPT technology is to provide swift and immediate responses to common queries, aiming to enhance productivity among learners and users. Despite the myriad benefits and practicality that AI chatbots offer, there are also drawbacks to their utilization. For instance, learners may become overly reliant on automated solutions and shortcuts, chatbots may struggle with complex and personal inquiries, and they require regular updates and maintenance to ensure optimal performance (Lopez & Qamber, 2022).

Regarding the prospective application of technologies and AI-driven tools like ChatGPT and other chatbots in academia, Qasem (2023) and Imran and Almusharraf (2023) emphasized that AI technologies and ChatGPT represent remarkable innovations and their application. However, they also raised concerns about the potential impact on learners' ability to write and think independently. From the standpoint of future writing education, these technologies and software present significant challenges for developing young minds. Overreliance on AI applications and tools may diminish the capacity to think critically, write effectively, and communicate independently in the generations to come. Qasem (2023) further underscores the growing concern that the human ability to engage in written communication with others is at risk. Hence, it falls upon instructors to guide students and instill the importance of nurturing their writing skills. Furthermore, students should be encouraged to rely on their original work and provide statements confirming the authenticity of their produced content.

7.3 METHODOLOGY

A qualitative descriptive research design is employed in this study. According to Hennink et al. (2023, qualitative descriptive research is aimed at uncovering and exploring themes and ideas within ongoing processes. This research design is selected because the current study seeks to capture the opinions and perceptions of the sample population. Data are gathered from recently published literature, encompassing reports, essays, and research articles concerning the potential application of AI in academia, with a particular emphasis on ChatGPT. A total of 15 of the most pertinent items from the period between December 2022 and July 2023 have been chosen and subjected to qualitative analysis for this study. ChatGPT is also utilized as a primary source for generating sample writings to assess the chatbot's effectiveness in generating writing prompts (refer to images 1–4 in the discussion section) (Table 7.1).

The authors followed PRISMA flow chart guidelines for data collection and literature review (refer to Figure 7.1). To this end, they conducted searches exclusively for research articles and reviews indexed in Scopus, Web of Science, and Science Direct to gather relevant material. In the initial step, an initial search yielded a total of n = 495 articles, with the remaining count after eliminating duplicate entries (n = 76) and applying other inclusion eligibility criteria (n = 344), resulting in a total of 75. Subsequently, 55 articles were excluded during the screening process, along with five articles, blogs, reports, web essays, and other sources that did not align with the objectives of this study, leading to their exclusion. In the end, the qualitative synthesis comprised 15 journal articles.

TABLE 7.1
Selected Database Search Results with Customized Keywords

Database	Primary Keywords
Scopus	TITLE-ABS-KEYTITLE-ABS-KEY (((writing assistant) AND (chatgpt) AND (ai AND future AND of AND writing) OR(chatbots AND writing)))
WoS	TS = ((“ai and education” OR “chatbots” OR “ai applications” AND “future academic writing” AND “ai and scientific writing” OR “ai writings” OR “writing tools”))
Science Direct	ai and education AND chatgpt AND chatbots in academic OR ai technologies in education

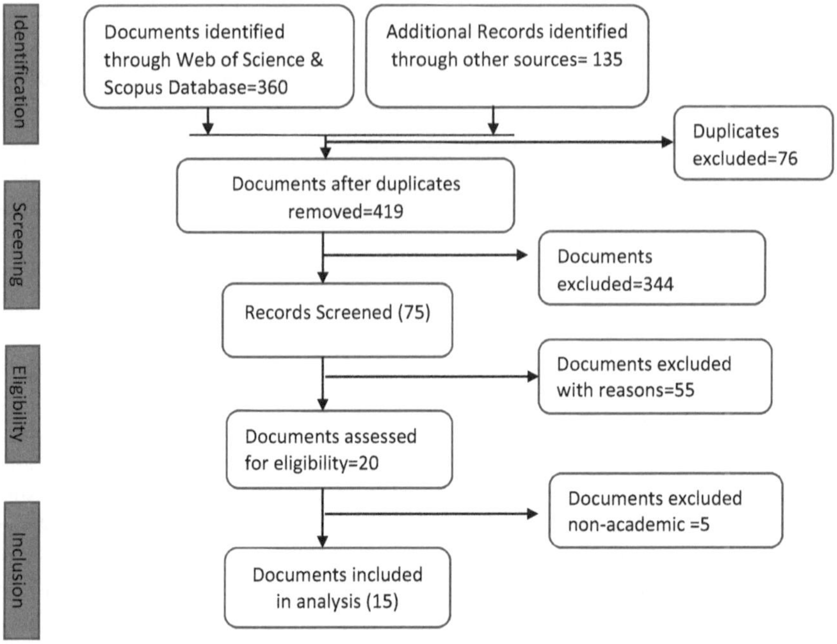

FIGURE 7.1 PRISMA flow chart.

FIGURE 7.2 Data structure according to Gioia's methodology.

For the purpose of qualitative analysis, Gioia's methodology (GM) (see Magnani & Gioia, 2023) is employed to analyze the collected data and identify key themes. This methodology represents a rigorous qualitative approach to data analysis, aiming to meet high standards. GM facilitates the categorization of the analyzed data into 1st-order analysis by developing comprehensive codes as 1st-order concepts. This methodology operates through a three-step process: first, organizing the relevant concepts; second, extracting prominent themes from the concepts assessed in the first step; and finally, further categorizing the refined themes within aggregated dimensions (Figure 7.2).

7.4 FINDINGS

7.4.1 ChatGPT as a Writing Tool

The findings demonstrate that the academic community has taken the integration of OpenAI ChatGPT quite seriously, resulting in the generation of numerous research articles, essays, blogs, and opinions across various platforms. All studies unanimously affirm that ChatGPT, along with other chatbots such as Bard and Tidio,

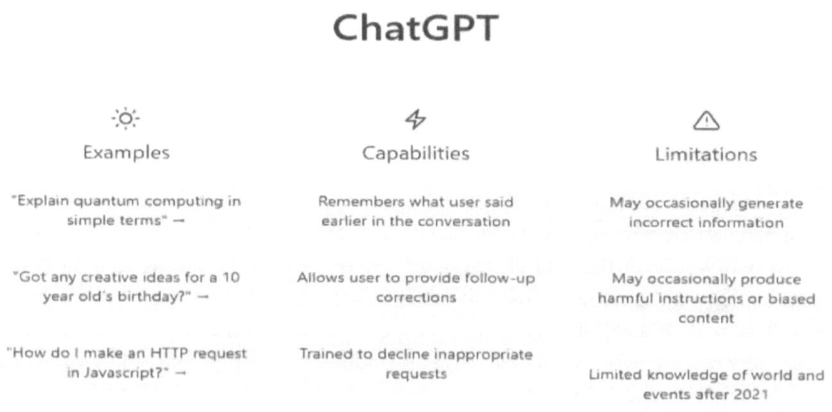

FIGURE 7.3 ChatGPT and its features.[1]

represents an interactive AI tool founded on Large Language Models (LLM) and GPT technologies. This chatbot is versatile and fine-tuned for multiple tasks, including chat interactions, conversational applications, acting as a writing assistant, and serving as a research engine (Imran & Almusharraf, 2023). Its reception has been overwhelming, primarily due to its remarkable features, especially its capability to generate human-like textual responses in various formats, be it short, lengthy, or in the form of letters. Within a few months of its launch, it harvested over 100 million users (Williams, 2023). ChatGPT can accomplish multiple tasks, along with writing, such as simplifying the explanation of quantum computing, providing creative ideas, and executing an HTTP request in JavaScript (see Figure 7.3). It stands apart from other chatbots due to its ability to remember prior user interactions within a conversation and employ contextual linking in responses, allowing users to offer follow-up corrections and decline inappropriate requests. Like other AI tools, this chatbot does have certain limitations. In its 3.5 version, information access is restricted up to 2021; however, the latest version, ChatGPT-4, has access to the most up-to-date information as well (Imran & Almusharraf, 2023).

Figure 7.3 illustrates the operational capacity of this technology. This chatbot excels in offering examples relevant to specific questions across various subjects. Nevertheless, as previously noted, complete reliance on this chatbot is discouraged due to its potential to generate misleading, biased, and potentially harmful information in response to specific queries. Consequently, ChatGPT serves as a valuable consultation tool; however, users are strongly advised to exercise caution and verify the information and other details before using them.

7.4.2 Use of ChatGPT for Writing

This systematic study shows that ChatGPT proves to be highly advantageous for a diverse range of writing tasks, primarily due to its proficiency in generating polished and sophisticated responses. This extends to its ability to comprehend and retain human instructions and input, resulting in the production of responses that closely resemble human language. The findings show that it serves as a highly valuable writing assistant for academic tasks through its advanced natural language processing capabilities (Taecharungroj, 2023). Operating on a machine learning model, ChatGPT utilizes an extensive text corpus to comprehend and produce language that closely emulates human linguistic patterns, making it proficient in various writing-related functions. Moreover, ChatGPT can aid researchers and students in several ways in academic writing. It further helps in brainstorming ideas, structuring content, and generating written drafts, thereby expediting the initial stages of the writing process (Xames & Shefa, 2023).

Additionally, it can provide clarity and coherence in articulating complex academic concepts, enhancing the overall quality of the written material. Furthermore, Zhu et al. (2023) believed that ChatGPT's capacity to expeditiously access and amalgamate information from a diverse array of sources can facilitate the incorporation of relevant research findings and citations. However, it is imperative for users to exercise critical judgment and oversight, ensuring that the content generated by ChatGPT aligns with academic standards, maintains originality, and adheres to ethical citation practices (Zhu et al., 2023). ChatGPT, being a great help in academic settings, streamlines and supports various aspects of the writing process, contributing to improved productivity and content quality while emphasizing the importance of responsible usage and academic integrity. Therefore, a few studies (van Slyke et al., 2023; Karakose, 2023; Alafnan et al., 2023) analyzed in this chapter have raised serious concerns over the unchecked use of AI chatbots, including ChatGPT. However, beyond academic writing, ChatGPT proved to be a valuable resource for virtual assistants in any subject or topic for explanation and elaboration like a tutor such as combing and solving various scientific problems and generating hypotheses for novice researchers (Imran & Almusharraf, 2023).

7.4.3 General Findings

By examining selected articles, this study has outlined the writing functions of ChatGPT and summarized them into the following key points:

- **ChatGPT as a Writing Tool:** According to Linma et al. (2023), Karakose (2023), and Talili et al. (2023), ChatGPT significantly reduces the time and effort needed to generate written content, exhibiting the ability to quickly produce human-like written responses.
- **Generation of New Ideas:** Several other studies (Xames & Shefa, 2023; Dashti et al., 2023; Zhu et al., 2023; Taecharungroj, 2023) have discussed ChatGPT's role in aiding researchers and learners in generating new ideas for their research tasks. It offers suggestions for topic selection, presents multiple choices for a single idea, and assists in crafting assignments covering various formats, including stems, multiple-choice questions, and both short and long answers.
- **Translation and Editing:** OpenAI's chatbot possesses versatile functions for translating text into multiple languages, providing valuable support to scholars and learners tackling the challenges of writing theses, projects, and articles in non-native languages, predominantly English (Alafnan et al., 2023; Salvagno et al., 2023).

- **Accurate and Error-Free Content**: Some studies (van Slyke et al., 2023; Kooli, 2023; Nemorin et al., 2023) contend that authors can generate more consistent and accurate content with ChatGPT's assistance. Leveraging GPT and natural language processing technologies, ChatGPT can produce precise and error-free responses to a query.
- **Collaborator and Facilitator**: Scholars (Young & Shishido, 2023; Qasem, 2023; Sullivan et al., 2023) concur that integrating ChatGPT can enhance collaborative efforts within the academic community, involving scholars, students, and educators. The utilization of AI for content generation fosters concurrent contributions from multiple individuals, thereby enhancing efficiency and effectiveness in collaborative scholarly projects.

The findings are further categorized in Table 7.2 according to the focus and findings of the selected studies.

TABLE 7.2
Selected Articles' Focus and Findings

Authors and Year	Focus	Findings
Xames and Shefa (2023), Karakose (2023), Alafnan et al. (2023), van Slyke et al. (2023)	Introduction, opportunities, and challenges in using ChatGPT	Its use will help learners, editors, and researchers work more efficiently; however, there are concerns about its authorship status, plagiarism, biased results, and nonexisting references. Therefore, new policies and guidelines immediately needed to be defined for using chatbots in academia.
Nemorin et al. (2023) and Taecharungroj (2023)	AI ecosystem	AI in education has created hype and is conceived as a gamechanger in future education. The ChatGPT innovation has the potential to impact humans and technologies.
Sullivan et al. (2023), Limna et al. (2023), Tlili et al. (2023)	Academic integrity and learning	There are apprehensions regarding academic integrity in consuming ChatGPT without any policies in academia. Most people are still unaware of its potential advantages and disadvantages in academia. Therefore, its safe, secure, and responsible use should be well-explained to the academic community.
Zhu et al. (2023), Salvagno et al. (2023), Kooli (2023)	Integration of ChatGPT in academia	It is a multipurpose tool with many innovative functions and utilities. Moreover, this tool can extract customized information, findings, and knowledge for educational uses. For more in-depth integration, future versions may be more supportive without limitations than current versions. However, a consensus about integrating AI tools into academic tasks is needed.
Qasem (2023), Dashti (2023), Young and Shishido (2023)	Future implications	This innovative chatbot is helpful in language tasks and will lead to more efficient tools for future language-related tasks. However, many things must be fixed before considering this chatbot as a fully reliable tool in the academic process. The future is bright with technology and innovative ideas; however, a particular focus should be laid on people's fears and concerns, such as writing tools might take writers' jobs and promote unethical and plagiarized content.

7.5 ANALYSIS AND DISCUSSION

7.5.1 Challenges and Opportunities

Out of the 15 articles analyzed, some studies (Xames & Shefa, 2023; Karakose, 2023; Zhu et al., 2023) advocated for the integration of AI and its applications in the domain of education. The majority of these studies shared the perspective that the AI chatbot ChatGPT has the potential to bring about a revolution in learning. Specifically, these studies addressed the main advantage for non-native speakers by aiding them in generating writing prompts accurately. Furthermore, some of these studies (Dashti et al., 2023; Taecharungroj, 2023) assert that employing AI tools for writing essays and assignments is not only acceptable but also equitable. This perspective stems from the recognition that non-native speakers often grapple with language-related challenges that impede their academic progress.

Few studies (Young & Shishido, 2023; Tlili et al., 2023; Limna et al., 2023) have delved into the role of AI systems in academia, shedding light on their advantages, challenges, and prospects for forthcoming research projects. This study contributes to the reader's understanding of the possible opportunities that learners and scholars can derive from the use of this AI chatbot, viewed through the lens of the constructivist theory of learning. It presents five key opportunities, each accompanied by its potential benefits, including innovative assessment methods, language editing, adaptive learning, feedback, and the automation of administrative tasks.

The analysis of opportunities and benefits of ChatGPT and its potential use has been summarized according to the GM approach (Figure 7.4).

Nevertheless, a few scholars (Alafnan et al., 2023; Qasem, 2023) have raised substantial concerns and challenges regarding the utilization of ChatGPT in education, particularly with respect to writing tasks. These studies view ChatGPT and other AI technologies as essential drivers of ongoing transformation in every field of life. It is anticipated that, by 2065, up to 65% of conventional learning and teaching tools may become obsolete. Consequently, it is argued that these technologies should be integrated into the learning process with human guidance to ensure they serve as tools rather than crutches. This approach would provide learners with increased opportunities to engage with the latest inventions to solve the challenges of the modern world. Despite ChatGPT's utility, several challenges have been identified, including limitations related to learning outcomes, the absence of real-time access to figurative

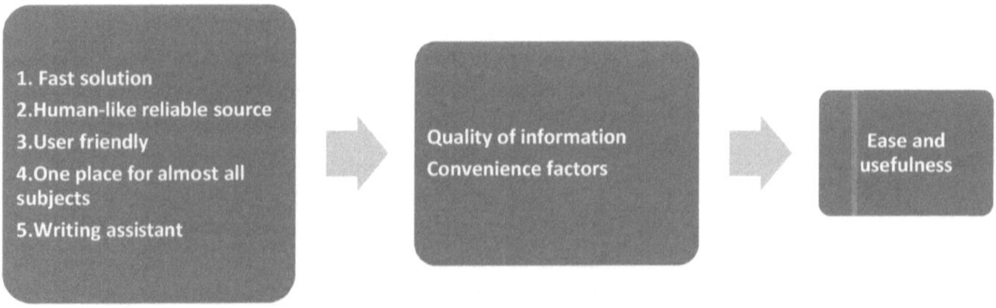

FIGURE 7.4 Potential benefits and opportunities of ChatGPT's use for writing.

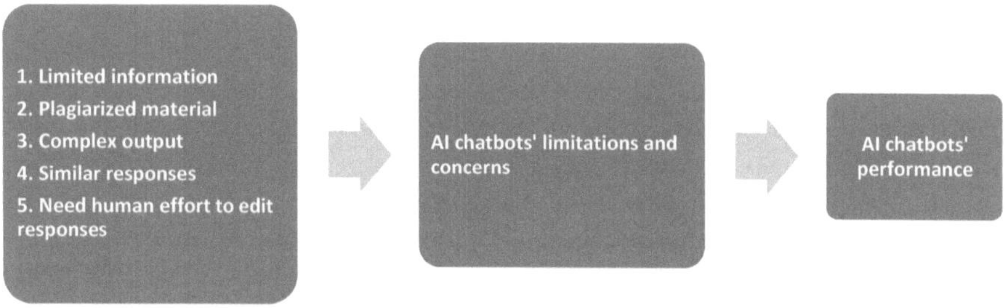

FIGURE 7.5 Concerns and challenges over the performance of ChatGPT.

data, concerns pertaining to academic integrity, issues surrounding misinformation, questions about source reliability, and potential difficulties in fostering independent learning skills.

Furthermore, studies like van Slyke et al. (2023) and Kooli (2023) have delved into the ethical challenges and potential misuse of AI technologies in the context of research and writing. These scholarly works emphasize the importance of considering the other side of the AI chatbot, ChatGPT. Numerous challenges emerge when applying this technology in academia, including the fairness of assessment and response generation during online and home-based exams. Similarly, during the initial stages of adopting AI technology, users are advised to exercise caution and refrain from unquestioningly relying on AI applications for official and academic tasks (Imran & Almusharraf, 2024). A significant concern pertains to the ethical implications, wherein users are urged to exercise vigilance when extracting data and results from ChatGPT because there might be unfair or biased outputs.

Additionally, the content generated may encounter citation issues, as ChatGPT has been reported to include fake citations (Polonsky and Rotman, 2023). Furthermore, challenges such as plagiarized content, poor contextual understanding within paragraphs, factual errors, and inconsistent text responses cast doubts on this chatbot's performance. Hence, the studies mentioned above recommend a thorough review and editing of all AI-generated responses before employing them for official or academic purposes.

The above discussion can be summarized according to GM's lens in Figure 7.5.

7.5.2 Future of Writing in Higher Education

Numerous studies (Taecharungroj, 2023; Sullivan et al., 2023; Salvagno et al., 2023) raise concerns about the transformative prospective influence of AI tools in shaping the trajectory of educational paradigms and practices. The capability of generating extensive texts, passages, essays, and letters on a wide array of subjects is a pivotal aspect of this technology. However, academic and scholarly writings are still evolving in their integration of AI applications and still have a long journey ahead. This study analyzes recent research publications to investigate the scholarly community's stance on the future of writing following the introduction of AI applications, notably ChatGPT.

Dashti et al. (2023) predicted ChatGPT as a writing assistant rather than a complete author capable of entirely replacing humans. AI chatbots, including ChatGPT, are crafted using advanced AI techniques, particularly Reinforcement Learning. They rely on human feedback to enhance their conversational capabilities and train their language models. However, this chatbot grapples with certain challenges in generating responses, such as crafting plausible-sounding yet nonsensical or erroneous answers. Additionally, it occasionally provides inaccurate references, particularly in scientific studies that do not exist in reality. In light of such AI tools, educators are prompted to reconsider their teaching and assessment methods in the classroom, while learners should be prepared for ethical use and tasked with assignments that present challenges beyond the capabilities of such chatbots (Dashti et al., 2023). In the context of writing courses and the impact of AI tools such as ChatGPT, it is imperative to emphasize that human involvement continues to be of paramount importance to ensure the quality and comprehensiveness of written content, underscoring the symbiotic relationship between AI and human expertise in academic settings.

Similarly, Alafnan et al. (2023) probe into the effectiveness and potential use of AI tools in prospective writing courses within the field of business studies. They conducted 30 tests using ChatGPT, from which they obtained various responses, including some misleading and plagiarized statements. Their study concluded that ChatGPT functions more like a research engine, and public opinions within the last two months have showcased a high degree of polarization regarding its use in academia and its effectiveness. Nevertheless, this chatbot is better suited to replace traditional search engines rather than being considered a substitute for human writers. For learners, ChatGPT serves as a valuable tool for obtaining more customized and precise research results compared to the vast volume of results produced by traditional search engines.

According to Kooli (2023), ChatGPT may not be particularly suitable for future writing tasks due to its tendency to generate plagiarized content. In an official statement, the Turnitin website disclosed that it had developed a tool with 97% accuracy for detecting AI-generated content by April 2023. Notably, even before the launch of Turnitin's new AI material detection tool, traditional plagiarism detection software was effective at identifying much of the content produced by ChatGPT. Alafnan et al. (2023) conducted tests on ChatGPT's responses and discovered a substantial similarity index in the writing prompts without any proper references. Furthermore, this study emphasizes that the prompts generated by chatbots exhibit other significant issues, such as organizational, structural, and syntactical problems (p. 13).

To ensure the secure use of AI tools for writing purposes, text-generating LLM, such as ChatGPT, are facing significant challenges related to transparency and credibility. Many journals and institutions have initiated policy revisions to monitor the ethical use of AI technologies in writing exams, articles, and other assignments. In his study, Qasem (2023) outlines strategies to maintain academic transparency and integrity in the era of ChatGPT. The future of writing in higher education is at a crossroads due to the emergence of advanced AI tools, software, and technologies. Nevertheless, these challenges can be addressed effectively through various approaches, including educating both staff and students about awareness and ethical considerations that can potentially damage their writing skills when they excessively rely on such writing tools. ChatGPT retains the context of user interactions and tailors responses accordingly, which means that similar answers are generated for the same questions posed by different users. Consequently, a limited number of students can benefit from these creative response tasks within a class.

In short, ChatGPT's performance is shady and raises questions for future writing tasks. Blindly relying on this chatbot is not advisable due to its inherent limitations, as highlighted by concerned users regarding the quality of content it generates. The emergence of AI technologies should serve as a compelling wake-up call for writing course instructors and educational institutions to reevaluate their conventional assessment methods, aiming to uphold integrity and credibility within the academic process. The findings from this chapter emphasize that the utilization of ChatGPT for scientific research-based writing introduces a range of potential concerns and challenges that researchers and scholars need to address. One significant concern pertains to the authenticity and accuracy of the generated content, as certain studies have underscored (van Slyke et al., 2023). While ChatGPT's responses can be impressive, they may not consistently meet the rigorous standards demanded for scientific research, potentially incorporating non-original and outdated material in scholarly work, which poses risks of false and incorrect citations, ultimately affecting academic integrity and credibility. Additionally, ethical considerations regarding authorship and intellectual property rights represent another challenge (Polonsky et al., 2023).

Moreover, the increasing use of AI systems like ChatGPT in content creation has introduced complexities related to determining appropriate authorship and citation practices, giving rise to concerns regarding the proper recognition of AI-generated content and tools. As a result, the processes of plagiarism detection and referencing may need to adapt to accommodate AI-generated content while ensuring its originality and accuracy. An additional challenge is the potential danger of extreme dependence on AI-generated content, which could potentially undermine the development of critical thinking and analytical skills among researchers, as noted by Nemorin et al. (2023). Finally, the potential for biases in AI-generated content, which may perpetuate existing biases in scientific literature, remains a significant concern. Therefore, the current study delves into the role of ChatGPT and its promise for scientific research. These issues underscore the importance of ethical guidelines, rigorous oversight, and critical evaluation in integrating AI technologies into the scholarly landscape.

The future of education and writing programs, particularly after the integration of the latest AI technologies in education, presents significant concerns regarding academic integrity, transparency, and fairness. It's important to emphasize that AI will never replace the human brain. However, the performance of ChatGPT and other AI tools serves as a wake-up call for the academic community to closely monitor the capabilities of chatbots and reconsider assessment methods to uphold integrity and address issues of intellectual dishonesty among learners and instructors. In response to this new AI advancement in chatbots, some counter technologies are also enhancing their capabilities to meet these challenges. Notably, Turnitin announced its efforts to incorporate the latest AI writing detection features into its products by mid-April 2023. According to Williams (2023), this addition will include the recognition of ChatGPT-generated content with 97% accuracy. The awareness campaign by Turnitin and other publishing houses like Springer and Nature and development in AI content detection highlights the evolving landscape of academic integrity in an era of advancing AI technologies (Williams, 2023).

However, in response to the challenges and complexities associated with this AI advancement, academia should actively seek and implement clear solutions to address the anticipated issues (Tlili et al., 2023). These solutions may include promoting public awareness about the safe use of AI technology and emphasizing the importance of viewing AI as a tool to aid in daily tasks rather than as a means to cheat. A few other studies by Zhu et al. (2023) and Limna et al. (2023) have emphasized to provide learners with a clear understanding of plagiarism policies, actively teaching students about proper citation and the ethical use of materials and tools. Institutions should establish and enforce strict ethics and policies to deter unfair use of materials or tools, emphasizing the importance of integrity in academic work. Additionally, encouraging a positive approach toward the use of AI tools can help students and scholars leverage the technology effectively while maintaining academic integrity. As AI continues to advance, it may become an integral part of the educational system (Karakose, 2023). Machines, including AI, will play an indispensable and vital role in helping learners and scholars generate ideas, execute plans, and present their research outcomes. This transition represents a shift in the educational landscape where AI authorship becomes increasingly prevalent and accepted.

7.5.3 Practical and Future Implications

Through this systematic review of the literature and its findings, the present study provides a comprehensive analysis of the role of ChatGPT as an educational aid, particularly for writing purposes across various subjects. However, as discussed in detail by Qasem (2023) and Dashti (2023), there is a need to develop clear policy documents and user guidelines for the utilization of ChatGPT and similar tools in academic tasks. More importantly, the use of these tools in scientific and research-based projects should prioritize transparent, accountable, and ethical knowledge production. Furthermore, considering the practical implications of ChatGPT in advancing academic progress, it is essential to address potential concerns and risks related to ethical issues, such as coauthorship involving chatbots in scientific writings. While authors may seek assistance or consultation from these tools, adding them as coauthors in writing manuscripts is a controversial issue, and many leading publishers and independent journals have established authorship policies warning against including chatbots as coauthors (Polonsky et al., 2023). This authorship matter remains a subject of debate and requires further attention from policymakers and the academic community to explore the future perspectives of AI's role as an author and facilitator.

Similarly, in the context of ChatGPT's role in academic writing production, it is essential to assess the authenticity of the content generated by GPT and LLM technologies. ChatGPT relies on information and datasets that are limited to existing internet data sources. This raises questions about the originality of its generated content. Is it a customized summary of findings or a rephrased version of already published material? The current chapter and its findings emphasize the need for further exploration of this aspect of chatbots to better understand their practical and future implications in ensuring balanced and accurate information, free from biases and limitations.

7.6 CONCLUSION

AI and its various tools, including ChatGPT, are integral components of the ever-evolving educational landscape, influencing learning methods, assessment strategies, and educational management. It is understandable that the rapid advancement of AI technologies in the education sector may induce apprehension. This study examined the utilization and concerns associated with the OpenAI tool ChatGPT. The key takeaway from this chapter is that AI, along with its various products like ChatGPT, holds immense importance and plays a significant role in advancing education. It is essential for individuals to refrain from engaging in overly polarized debates at this early stage. The adoption of AI tools and software like Grammarly, Turnitin, and QuillBot, etc., for daily tasks is already commonplace. Consequently, ChatGPT should not be perceived as a threat to writing programs. Instead, it serves as a potential substitute for conventional search engines, offering users more tailored search results in less time compared to sifting through billions of search results on platforms like Google. Academic institutions should embrace this emerging technology and take steps to educate teachers and learners about its limitations and challenges. While learners may find AI chatbots tempting for assignments and study-related tasks, they must remain mindful of the potential consequences, such as reduced engagement with coursework, leading to total reliance and dependence on AI tools. Similarly, educators should adapt their assessment methods and provide clear guidelines to promote academic integrity. Given that ChatGPT is a new launch and limited research has been conducted on its applications and challenges, more in-depth research projects in this domain are recommended to gain a better comprehensive understanding of the opportunities and challenges users may encounter in the long term.

NOTE

1 Source ChatGPT official website https://chat.openai.com/chat.

REFERENCES

Alafnan, M. A., Dishari, S., Jovic, M., & Lomidze, K. (2023). ChatGPT as an educational tool: Opportunities, challenges, and recommendations for communication, business writing, and composition courses. *Journal of Artificial Intelligence and Technology*, *3*(2), 60–68.

Dashti, M., Londono, J., Ghasemi, S., & Moghaddasi, N. (2023). How much can we rely on artificial intelligence chatbots such as the ChatGPT software program to assist with scientific writing?. The Journal of Prosthetic Dentistry. Online first. https://doi.org/10.1016/j.prosdent.2023.05.023

Moghaddasi, N. (2023). How much can we rely on artificial intelligence chatbots such as the ChatGPT software program to assist with scientific writing? *The Journal of Prosthetic Dentistry*. Online first copy.

Hennink, M., Hutter, I., & Bailey, A. (2020). *Qualitative research methods*. Sage Publications Ltd.

Holmes, W., & Tuomi, I. (2022). State of the art and practice in AI in education. *European Journal of Education, 57*(4), 542–570.

Imran, M., & Almusharraf, N. (2023). Analyzing the role of ChatGPT as a writing assistant at higher education level: A systematic review of the literature. *Contemporary Educational Technology, 15*(4),1–14

Imran, M., & Almusharraf, N. (2023a). Teaching innovation in university education: Case studies and main practices: edited by Jose Ramon Saura, Hershey. *The Social Science Journal,* 1–3. Online first. https://doi.org/10.1080/03623319.2023.2201973

Imran, M., & Almusharraf, N. (2024). Digital Learning Demand and Applicability of Quality 4.0 for Future Education: A Systematic Review. *International Journal of Engineering Pedagogy, 14*(4), 38–53.

Karakose, T. (2023). The utility of ChatGPT in educational research – Potential opportunities and pitfalls. *Educational Process: International Journal, 12*, 7–13.

Kooli, C. (2023). Chatbots in education and research: A critical examination of ethical implications and solutions. *Sustainability, 15*(7), 5614.

Kung, T. H., Cheatham, M., Medenilla, A., Sillos, C., De Leon, L., Elepaño, C., ... & Tseng, V. (2023). Performance of ChatGPT on USMLE: potential for AI-assisted medical education using large language models. *PLoS Digital Health, 2*(2), 1–12.

Limna, P., Kraiwanit, T., Jangjarat, K., Klayklung, P., & Chocksathaporn, P. (2023). The use of ChatGPT in the digital era: Perspectives on chatbot implementation. Journal of Applied Learning and Teaching, 6(1), 1–10

Lopez, T., & Qamber, M. (2022). The benefits and drawbacks of implementing chatbots in higher education : A case study for international students at Jönköping University (Dissertation). Retrieved from https://urn.kb.se/resolve?urn=urn:nbn:se:hj:diva–57482

Lund, B. D., & Wang, T. (2023). Chatting about ChatGPT: How may AI and GPT impact academia and libraries? Library Hi Tech News, *40*(3), 26–29.

Magnani, G., & Gioia, D. (2023). Using the Gioia methodology in international business and entrepreneurship research. *International Business Review, 32*(2), 102097.

McMurtrie, B. (2022). AI and the future of undergraduate writing. *The Chronicles of Higher Education.* Retrieved from https://www.chronicle.com/article/ai-and-the-future-of-undergraduate-writing

Nemorin, S., Vlachidis, A., Ayerakwa, H. M., & Andriotis, P. (2023). AI hyped? A horizon scan of discourse on artificial intelligence in education (AIED) and development. Learning, Media and Technology, *48*(1), 38–51.

Polonsky, M. J., & Rotman, J. D. (2023). Should artificial intelligent agents be your coauthor? Arguments in favour, informed by ChatGPT. *Australasian Marketing Journal, 31*(2), 91–96.

Qasem, F. (2023). ChatGPT in scientific and academic research: Future fears and reassurances. *Library Hi Tech News, 40*(3), 30–32.

Salvagno, M., Taccone, F. S., & Gerli, A. G. (2023). Can artificial intelligence help for scientific writing? *Critical Care, 27*(1), 1–5.

Sullivan, M., Kelly, A., & McLaughlan, P. (2023). ChatGPT in higher education: Considerations for academic integrity and student learning. Journal of Applied Learning and Teaching, *6*(1), 1–10.

Taecharungroj, V. (2023). "What can ChatGPT do?" Analyzing early reactions to the innovative AI Chatbot on Twitter. *Big Data and Cognitive Computing, 7*(1), 35.

Tlili, A., Shehata, B., Adarkwah, M. A., Bozkurt, A., Hickey, D. T., Huang, R., & Agyemang, B. (2023). What if the devil is my guardian angel: ChatGPT as a case study of using chatbots in education. *Smart Learning Environments, 10*(1), 15.

Van Slyke, C., Johnson, R. D., & Sarabadani, J. (2023). Generative artificial intelligence in information systems education: Challenges, consequences, and responses. *Communications of the Association for Information Systems*, 53(1), 14.

Williams, T. (2023). Turnitin announces AI detector with 97 per cent accuracy. Timeshighereducation.com. Accessed on February 25, 2023, from https://www.timeshighereducation.com/news/turnitin-announces-ai-detector-97-cent-accuracy

Xames, M. D., & Shefa, J. (2023). ChatGPT for research and publication: Opportunities and challenges. *Journal of Applied Learning and Teaching*, 6(1), 390–395.

Young, J. C., & Shishido, M. (2023). Investigating OpenAI's ChatGPT potentials in generating Chatbot's dialogue for English as a foreign language learning. *International Journal of Advanced Computer Science and Applications*, 14(6), 65–72.

8 An Exploratory Study on the Advantages and Disadvantages of OpenAI's ChatGPT on Higher Education

Sahib Khatoon, Muhammad Usman Thaheem, and Habibullah Pathan

8.1 INTRODUCTION

Artificial intelligence (AI) has revolutionized many professions and sectors in recent years, with astonishing gains. The World Economic Forum in Davos, Switzerland, concentrated on the development of AI, particularly the ChatGPT text generator (Atlas, 2023). There is potential for AI, like ChatGPT created by OpenAI, to supplement or completely replace human intelligence.

Educators can benefit from this technology's rapid feedback, personalized interactions, and administrative support by utilizing it to their advantage. That can improve their pedagogical practices. Besides, it gives students the freedom to study different topics on their own and learn at their speed. Individual learning demands may be addressed with the use of ChatGPT's individualized support, which will enhance learning overall. Due to ChatGPT's sophisticated language processing abilities, which are based on the GPT-4 model, it has attracted a lot of interest and adoption. ChatGPT has the potential to be a useful tool in the open education space by fostering autonomous learning, enhancing engagement, and providing individualized help (Baidoo-Anu and Owusu Ansah, 2023). Thus, education and the way students approach their studies could be revolutionized by AI tools like ChatGPT.

Furthermore, it's critical to remember that ChatGPT is still a relatively new technology when it comes to self-taught and self-learning, and more study is necessary to completely understand both its advantages and disadvantages. Seeing ChatGPT and other AI technologies develop over time and observe how they alter the paradigm of education in the 21st century would be fascinating (Firat, 2023).

In the current study, these benefits will be discussed in detail with an emphasis on how ChatGPT might help teachers create more efficient, customized, and engaging learning environments. Additionally, this study has looked at the drawbacks of ChatGPT in research, teaching, and learning contexts.

 DOI: 10.1201/9781003400691-8

8.2 RELATED LITERATURE

8.2.1 Artificial Intelligence in Education

With the advancement of computing and information processing technologies, AI has become increasingly prevalent in education. Artificial Intelligence in Education (AIEd) presents a range of new challenges and opportunities for educational practices (Ouyang and Jiao, 2021). AI has begun to reshape conventional teaching strategies and learning environments in the field of education. An overview of the various ways AI is being incorporated into educational practices is provided in this study.

A lot of AIEd has been created (intentionally or unintentionally) to replace instructors or to reduce them to a purely functional position, not to help them teach more effectively (Holmes et al., 2023). Although it may be helpful in situations when there aren't many teachers, it underestimates teachers' special talents and experiences as well as students' needs for social learning and direction. Instead of just automating the education of learners seated at computers, AI may help open teaching and learning opportunities that are otherwise difficult to acquire, that challenge conventional pedagogies, or that help teachers be more effective (Knox, 2020).

AIEd is largely concerned with applying AI principles to commonplace learning paradigms, and thus frequently replicates (or automates) pre-existing educational presumptions and practices (Holmes et al., 2023). Additionally, learners interact with AI, which is used to empower learning as learners take ownership of their education, and AI is utilized to represent knowledge models and direct cognitive learning while learners are service recipients. Increasing learner agency and personalization, allowing students to evaluate their education and provide feedback to AI systems, and encouraging the iterative development of learner-centred, data-driven, personalized learning have been the general trends in AIEd development (Ouyang and Jiao, 2021).

8.2.2 A Historical Snapshot of ChatGPT

ChatGPT was first introduced as a prototype and was made available to the public on November 30, 2022, and it represented a tremendous advancement in language modelling technology. This is an AI-powered conversational agent created by OpenAI that has drawn interest from engineers, social media users, company owners, authors, teachers, and students (Jürgen Rudolph et al., 2023). ChatGPT, which is based on the sophisticated GPT-3 language mode, surpassed the growth rates of well-known platforms like Facebook, Netflix, Instagram, and Twitter by gaining 1 million members in just five days. These services took a lot longer than Facebook (300 days), Netflix (1200 days), Instagram (75 days), and Twitter (720 days) to reach the same user milestone. The great demand and interest in ChatGPT's capabilities may be shown in the quick adoption. GPT-3, which has a huge model size of 175 billion parameters, is extremely good at tasks requiring natural language processing because it can produce text that closely mimics human English. The official release of GPT-4, which is a major improvement over its predecessor, was announced by OpenAI on March 13 with improved originality, dependability, and linguistic nuance and are the hallmarks of GPT-4 (King, 2023).

Additionally, ChatGPT is strengthened by the GPT-4 model, which enables it to hold multiple conversations at once, comprehend and respond to input in plain language, and provide customized and interactive help. Because of its multiple features, ChatGPT has the potential to be an important tool for open education (Dwivedi et al., 2023). Its versatility and utility make it interesting as a teaching tool. You might be asking how ChatGPT can be used in other aspects of higher education by students, teachers, and professionals. ChatGPT can alter how we teach and learn in higher education. It can be used for a variety of purposes, such as writing and brainstorming, business communications, and individualized learning (Atlas, 2023).

Researchers, teachers, and learners from a wide range of fields have shown a great deal of interest in the integration of ChatGPT in university-level teaching, learning, and research. The existing literature is summarized in this part, with an emphasis on both the advantages and drawbacks of using ChatGPT in academic and research settings.

8.2.3 Benefits of ChatGPT in Higher Education Teaching and Learning

The benefits of incorporating ChatGPT in higher education contexts have been emphasized by several recent studies, blogs, and media outlets highlighting its potential to engage students, simplify administrative processes, and generate dynamic learning experiences (Baidoo-Anu and Owusu Ansah, 2023; Jürgen Rudolph et al., 2023; Xiaoming Zhai, 2023). Approximately 650 million Google Search results as of February 23, 2023, reflect the media and academic community's extraordinary interest. ChatGPT's use in this field is a by-product of AI innovation, even if it's unlikely that it was intended to be used for academic purposes (Benuyenah, 2023). The exceptional capabilities of ChatGPT, which mimic human interaction, outperform many current tools (Illingworth, 2023). ChatGPT can be used for several things, including exam preparation, language translation, creating custom code, and even managing difficult jobs like summarizing scientific writing (Neumann et al., 2023).

There are five main benefits to using ChatGPT in education: it may be used to do learning assessments, enhance pedagogical practices, offer virtual one-on-one coaching, construct outlines, and generate ideas (Sok and Heng, 2022). The study by Hoai Thu et al. (2023) emphasizes the growing significance of online learning systems in contemporary education, with ChatGPT promising to enhance the efficacy and quality of teaching and learning. According to this study, a more customized learning environment could be created by adding ChatGPT to the Vietnamese online learning system. Despite the challenges, they could be overcome with focused efforts and continued technological improvement.

Besides, one of the studies investigated how educators use ChatGPT in their classrooms and the underlying elements that determine whether they use it or not. Based on the results it was evident that ChatGPT is a useful tool for teachers and learners, helping them with tasks like information retrieval, idea production, and language learning (Tuhuteru et al., 2023). Such studies made this idea clear that ChatGPT has much utilization in the educational context.

According to a study, assessments are the cornerstone of excellent instruction. ChatGPT provides academic staff and instructors with a flexible and innovative way to create learning assessments that include instantaneous feedback and reporting. Additionally, since many teachers spend a lot of time creating quizzes, monthly exams, and examinations, there is an opportunity for educators to lessen the strain of assessment by obtaining assistance from ChatGPT (Sok and Heng, 2022; Atlas, 2023b; Firaina and Sulisworo, 2023; Tuhuteru et al., 2023).

Moreover, the results of a study showed that teachers may create visual aids – like worksheets or slides – that explicitly state the learning objectives and success criteria for a session using ChatGPT. It's crucial to remember that these instructional tools have a strong likelihood of motivating students to study and engage in a classroom setting (Tlili et al., 2023).

Similarly, in one of the studies, it is discussed how AI will impact learning outcomes, assessment and evaluation processes, learning materials and approaches, and educational procedures. The authors contend that integrating AI and ChatGPT into education has the potential to improve educational quality oversight, promote the dissemination of educational resources, and increase student learning effectiveness. Moreover, education should prioritize encouraging students to be creative and critical thinkers so they can effectively address and resolve a variety of issues in the environment, resources, economy, politics, and other spheres of future existence. The authors argue that ChatGPT has the potential to be a significant developmental channel for future education and a vital catalyst for educational improvements (Shidiq, 2023; Xiaoming Zhai, 2023).

Likewise, ChatGPT has a lot of innovation potential; it can offer fresh suggestions for structuring a lecture or writing a task, and activities for teaching or learning but it needs proper understanding of those techniques which can help a teacher or a learner to take much benefit by using it (Neumann et al., 2023; Biswas, 2023).

8.2.4 Concerns About the Integration of ChatGPT in Higher Education

Every technology has its advantages, disadvantages, affordances, and hazards; none of them are neutral. Teaching and learning approaches have undergone radical modifications because of the adoption of AI technology in higher education, such as ChatGPT. Although these technologies have many advantages, it is important to consider any potential drawbacks to deploying them in a balanced and knowledgeable manner (Tlili et al., 2023).

One of the studies discusses serious concerns about the potential effects of ChatGPT on students' ability to acquire critical thinking skills. The ease of AI-generated solutions can deter students from doing the in-depth study and problem-solving necessary for real knowledge. Students rely too heavily on ChatGPT, they could adopt a passive learning style in which they just look for quick solutions rather than understanding the underlying ideas (Biswas, 2023b).

Students would utilize it improperly to copy and paste the material and they do not practice certain skills which they have to sharpen by practice (Benuyenah, 2023). According to Benuyenah, students could not understand how and when they are permitted to utilize ChatGPT; if there are some rules governing its use designed by teachers this issue can be addressed.

Besides, a problem with ChatGPT's ability to understand complex contexts in academic subjects has been noted in the literature. This shortcoming may cause inappropriate or erroneous responses, confounding students and impeding their ability to learn (Firat, 2023). The usefulness of deploying AI in educational contexts may be hampered by the inability to recognize the complexities of a particular situation, which could result in inaccurate information (ThankGod et al., 2023).

Noam Chomsky referred to ChatGPT as "*high-tech plagiarism*" (Bozkurt et al., 2023). There are issues of academic integrity for which researchers must be careful. According to Atlas (2023), Higher education may benefit greatly from the usage of ChatGPT and other language models, which can be applied in a number of settings, such as writing support, language acquisition, research, and administration. Consequently, one could argue that ChatGPT has the potential to be a useful resource for research and instruction.

Research scholars/students are using ChatGPT for paraphrasing and making their research presentation better, but some research students are using it for copy and paste which creates the problem of plagiarism; the academic community's strength is a big reassurance because companies like Turnitin have already created ChatGPT and AI detection technologies to cope with potential malfeasance.

Besides, the potential concern of AI-generated feedback in comparison to the empathy and understanding offered by human instructors has been raised as a problem by researchers (Firat, 2023; Bozkurt et al., 2023; Xiaoming Zhai, 2023; Illingworth, 2023; Benuyenah, 2023; Firaina and Sulisworo, 2023; Atlas, 2023a). Effective learning experiences must include human teachers' constructive criticism, encouragement, and individualized assistance. Furthermore, threats do exist, but institutions can deter them, if not right away (Benuyenah, 2023). Likewise, Zulkepli et al. (2018) emphasize the value of face-to-face communication in learning. The likelihood that students will communicate with real professors and peers may decrease if AI techniques like ChatGPT are overused.

While ChatGPT and related AI technologies have the potential to revolutionize higher education, it is important to recognize and solve any potential drawbacks they may have. By being aware of these issues, educators, administrators, and politicians can make well-informed choices on how to incorporate AI tools into the educational environment.

8.3 METHODOLOGY

The purpose of this study was to explore the benefits and challenges vis-à-vis ChatGPT usage for teaching and learning in higher education. The study followed a qualitative research approach to examine and learn more about the specific perspectives on the experiences of the students who use ChatGPT for learning and researching. The reason to follow a qualitative approach in this study is to know thorough and extensive experiences of the ChatGPT users; according to Creswell and Creswell (2017), a semi-structured interview guide is a schematic arrangement of interrogations that the interviewer is supposed to investigate to get the comprehensive data. Thus, this study conducted semi-structured interviews and collected saturated data from the respondents.

8.3.1 Sample

In this study, convenient sampling was utilized to find and choose those engineering students and teachers who are proficient in using ChatGPT for their teaching, learning, and research purposes. The convenient sampling is used because of the availability of the respondents who are ChatGPT users. Besides, the number is limited also because of the availability of respondents. The sample size is limited since this study is qualitative, allowing for the collection of more in-depth information from the participants. To do this, engineering final-year students (n = 05) were interviewed who were using ChatGPT for their learning and researching. Moreover (n = 05) teachers were also interviewed who use ChatGPT for their teaching purposes. The respondents were from the Mehran University of Engineering and Technology Jamshoro (MUET), a local public engineering university.

8.3.2 Research Tools and Techniques

Since this research is entirely qualitative, the sampling instrument was used to collect the data needed for analysis: a pre-planned interview to find out the students' opinions about ChatGPT for learning and researching their subjects. To compare student responses to in-depth, semi-structured interview questions previously created by the researcher, a face-to-face interview style was employed. Semi-structured interviews are typically "*organised around a set of predetermined open-ended questions, with other questions emerging from the dialogue between the interviewer and interviewee(s)*" (Chu and Chang, 2017). The participants were first made aware of the goal of the study before the interview session began, and their identities were kept anonymous. Student participants S1, S2, S3, S4, S5 were listed, similarly teacher participants were identified as T1, T2, T3, T4, T5 by keeping their identities anonymous.

8.3.3 Data Handling

This study follows the Braun and Clarke thematic analysis; thus, the interviewees' replies were captured on audio tapes and typed out for documentation and more analysis. Each interviewee's input was categorized using data coding, which resulted in the creation of thematic codes. To find patterns and create codes, the participant-collected data underwent thorough analysis and organization. This process made it simple to combine and arrange the data from both the students and the teachers, which improved the qualitative analysis. "*The primary aspect of qualitative data analysis involves the categorization of objects, individuals, and occurrences, along with the attributes that define them*" (Braun and Clarke, 2006).

8.4 FINDINGS AND DISCUSSION

To answer the predetermined study questions, the information gleaned from the interviews was gathered, arranged, categorized, and analysed. To examine the interview data and provide a more accurate and up-to-date grasp of the relevant subject matter, an inductive analytical approach was used. This required a careful evaluation of respondent inputs and viewpoints, allowing the identification of nuanced meanings concealed within each interviewee's unique ChatGPT usage experiences to identify the advantages and disadvantages of ChatGPT in higher education. Following that, these meanings were interpreted and relevant quotes from the participants were added to the report to give the data interpretation and conclusions credibility and validity. This strategy strengthens the dependability of the inferred results and improves the transparency of the data analysis process. This effort would make data interpretation more transparent and reliable and validate the conclusions that were produced (Tong et al., 2007).

Students and teachers highlighted various advantages of ChatGPT usage in their teaching, learning, and researching context along with the disadvantages traced out from their utterances.

Table 8.1 describes teachers' responses and themes taken out from their responses' vis-à-vis ChatGPT's advantages:

TABLE 8.1
Teachers' Responses

Themes	Verbatim
1. Possibility of dynamic explanations	**T1:** "*I've been using technology in my lectures for the past five years as an assistant professor of computer science. I use ChatGPT daily for designing my lectures, ChatGPT offers a **dynamic aspect** to lectures by enabling in-the-moment clarification of challenging subjects.*"
2. Simplify the ideas	**T3:** "*I used ChatGPT to clarify those abstract ideas which I felt difficult to simplify but with ChatGPT's assistance **I simplified the ideas**, and students found it useful for understanding the subject matter.*"
3. Creating interesting content	**T5:** "*In my experience teaching, ChatGPT has been a wonderful gem. It enables me to create **content that appeals to the interests** and learning preferences of my students. I could develop interesting case studies, dialogues, and scenarios that make learning both interesting and riveting.*"
4. Tailored responses	**T4:** "*I usually use it to make my lessons according to the level of my students to provide them **tailored explanations and responses**, catering to each student's unique needs.*" ***T2:*** "*It enabled me to **personalize the explanations** and accommodate the facilitation according to the need.*"
5. Instant feedback	T1: "*ChatGPT has completely changed how I respond to student inquiries because I get possible answers already and can give **instant feedback**.*"
6. Enhance critical thinking	**T3:** "*I believe, ChatGPT fosters **critical thinking**, students are not merely passive users of information when they engage with the content it produces. For their intellectual development, this tool acts as a catalyst.*"
7. Recreate lessons	T5: "*This AI tool works like my assistant, it **makes my lessons** and helps me in making creative activities according to my lesson.*"

The teacher participants had overwhelmingly good impressions of using ChatGPT in their classroom settings and highlighted its important role in enhancing their teaching material arrangements. The themes in Table 8.1, which are drawn from the data they gave, show that ChatGPT is a blessing for them to utilize and a source of help for generating and updating their teaching materials. All the teachers who were interviewed expressed satisfaction and recommended using this to their advantage when customizing their lesson plans and activities. The data showed positive feedback for ChatGPT's practicality and utility, and responses suggested that it was a reliable tool for instruction and task and technique improvement. They acknowledged ChatGPT's contribution to enhancing their teaching activities and saw it as a flexible tool for developing personalized learning resources.

Table 8.2 describes students' responses and themes taken out from their responses vis-à-vis ChatGPT's advantages.

While student participants were interviewed their facial expressions were also noticed which showed they were very excited about using the ChatGPT for their learning and research purposes. Moreover, they considered ChatGPT as a tool for

TABLE 8.2
Students' Responses

Themes	Verbatim
1. Enhanced probing skills	*S1: "I feel ChatGPT encourages me to explore topics from various aspects by **prompting creative questions**."* *S2: "ChatGPT trained me about **posing the right and creative question** to get the desired answer."*
2. Enhanced curiosity	*S5: "ChatGPT's quick responses enticed my **curiosity and my drive to learn** more was stoked."*
3. User-friendly	*S4: "ChatGPT is a very **user-friendly** and accessible platform there is no need of training required to use it."*
4. Enhanced brainstorming	*S3: "Madam I feel ChatGPT **simplifies the brainstorming** process."*
5. Flexible access	*S5: "I think its **flexibility of access** allows us to access it whenever we want, whether it's during the day, late at night, or over the weekend allowing for a more seamless and continuous learning experience."*
6. Help silent learners	*S5: "I used to be **silent in class and never ask questions**, but it helped me to chat with it and learn which I cannot do in class."*
7. Comprehensive resources	*S3: "I do not open so many **books to make my notes and gather material and arrange** it ChatGPT does it for me."*
8. Enhance creative writing	*S2: "I participated in a **poetry** competition and my poem won the prize which I wrote with the help of ChatGPT."*
9. Research assistance	*S1: "I was writing my research proposal it **helped me in format and research compilation**."*
10. Language assistance	*S4: "My English composition is not that good ChatGPT **helps me in paraphrasing, summarizing, and sentence structures**."*

preliminary investigation in doing research and value, its capacity to deliver concise summaries, paraphrasing ability, and produce preliminary study directions. According to these respondents, ChatGPT makes them more able to delve into more in-depth discussions, clarify questions, discuss difficult ideas, and request sources or references. They appreciated ChatGPT's ability to assist them in studying their areas of interest in greater depth and see it as a good resource for doing so.

Table 8.3 describes teachers' responses and themes taken out from their responses' vis-à-vis ChatGPT's disadvantages.

While there are many advantages to using ChatGPT in the classroom, there are also several issues that need to be resolved.

Table 8.4 describes students' responses and themes taken out from their responses vis-à-vis ChatGPT's disadvantages.

While using ChatGPT and other AI technologies can be helpful for students, several potential disadvantages should be considered as well. As respondents shared their opinions, many claimed to rely excessively on ChatGPT for their academic or research endeavours. Students admitted that they do not utilize their skills, which compromises their learning and prevents them from gaining the in-depth knowledge. Additionally, they said that every piece of information that was given to them was faulty in some way and contained incorrect citations. Therefore, before using, such information must be validated.

TABLE 8.3
Teachers' Responses

Themes	Verbatim
1. Time-sensitive content	**T1**: "*I always keep myself updated with the new trends, but ChatGPT is providing me **limited material which it is trained for till 2021** which makes me perplexed in understanding the trendy topics.*"
2. Improper content	**T4:** "*I believe that teachers need to stay aware of contemporary contents because sometimes the **information provided by ChatGPT is not accurate**, so teachers shouldn't rely on it and instead, it should be double-checked, verified, and filtered by them.*"
3. Lack of interaction (discussions/hands-on activities)	**T5:** "*Hmm, ChatGPT can deliver knowledge, but it **may not encourage students to participate in active learning,** students become computer-bound only.*"
4. Over-dependency and lack of critical thinking	**T4:** "*I believe we only **rely** nowadays on such devices and tools but avoid thinking, creating, and sharpening our **critical thinking skills**.*"
5. Time restrictions	**T3:** "*I think being the teacher I have to plan my lessons according to my learner and learning context and ChatGPT gives me huge material sometimes inaccurate which **wastes my time** in filtering it.*"

TABLE 8.4
Students' Responses

Themes	Verbatim
1. **Inaccurate information**	**S3:** "*When I used it for searching material related to any research topic, I felt it was providing me a wrong citation, because when I checked other databases for checking the reliability of that material it was wrong.*"
2. **Plagiarism**	**S2:** "*I use ChatGPT but use my mind and skills also whereas I have noticed my friends just copy and paste the material and use this unethically.*"
	S4: "*We must not copy and paste the content, but rather conduct our research, but usually we unethically plagiarize content and try to own it to complete our research papers and assignments.*"
3. **Compromised learning**	**S5:** "*While AI technologies can offer quick solutions to every problem, they might not motivate students to research a subject in-depth, which could result in a* ***superficial understanding of complicated topics***."

Teachers and student respondents appreciated and directed and mentioned possible challenges and recommended that these benefits be availed but challenges be considered also to achieve better outcomes.

8.5 CONCLUSION

According to the findings and discussions, several benefits and drawbacks of using ChatGPT in research, teaching, and learning purposes must be carefully considered. Positively, ChatGPT makes it quick and easy to get information, assisting academics, students, and researchers in their work. Its ability to produce content helps with the development of instructional materials, discussion facilitation, and idea generation. Additionally, because ChatGPT is accessible, it is a useful tool for students with a variety of backgrounds and abilities, promoting a more inclusive educational environment. Nevertheless, these benefits do have some built-in issues, though. Users must take caution and double-check information from credible sources due to the possibility of errors and incorrect information. By encouraging a propensity to take information at face value, excessive dependence on ChatGPT may inhibit deep understanding and critical thinking. The integrity of academic work must also be protected, which calls for addressing ethical issues like plagiarism and appropriate technology use.

In conclusion, ChatGPT has incredible potential as a tool for teaching, learning, and research. Users must be cautious about the potential hazards even though it offers substantial advantages in terms of accessibility, information creation, and efficiency. Users can maximize ChatGPT's benefits while minimizing its drawbacks by adopting a balanced strategy that combines its advantages with conventional approaches, encouraging a more enriched and thorough educational experience.

8.6 PRACTICAL IMPLICATIONS

The study emphasizes the growing significance of ChatGPT and other AI tools in higher education. ChatGPT can be utilized by educational institutions to create AI-powered academic support resources, aiding students and educators in creating effective instructional materials and discussions. This could lessen the administrative effort for faculty members and benefit students in their academic endeavours.

Academic practitioners, such as students and researchers, should receive training on how to critically assess the data collected via ChatGPT. This suggests that the information acquired is correct and trustworthy. Moreover, teachers should support a well-rounded approach that blends ChatGPT's benefits with traditional teaching, learning, and research techniques. This would promote a more comprehensive and richer educational experience while assisting in reducing the possible negative effects of an over-reliance on AI tools.

Furthermore, faculty and students should be given priority in educational institutions when it comes to ethical consideration in research. This entails resolving plagiarism-related issues and encouraging technological use that is appropriate. It is important to create and disseminate explicit norms and guidelines for academic integrity and the use of AI tools to all relevant parties. According to the study, educators should put more of an emphasis on encouraging critical thinking and information analysis because excessive usage of ChatGPT can impair critical thinking and comprehension.

8.7 RECOMMENDATIONS

8.7.1 For Policymakers

Academia is hierarchical where policymakers, teachers, and students are involved in educational activities; thus, policymakers should encourage the design and set of precise rules for using AI tools like ChatGPT in academic programmes. These guidelines should place a strong emphasis on a well-rounded strategy that improves learning without dispensing with crucial interactive teaching techniques. Moreover, teachers are given access to training programmes that will give them the right skills and information to use AI tools for research and teaching successfully. When adopting AI tools, protecting student data should be a top priority with set standards. To protect sensitive information, ensure compliance with privacy law, and put data security measures in place, policymakers can ensure that all students, regardless of background, have equitable access to AI technologies and can utilize them to their advantage vis-à-vis learning.

8.7.2 For Teachers

According to the findings of the current study, it is suggested to educators/teachers/practitioners/facilitators that instead of using ChatGPT to replace conventional teaching techniques, teachers must think about integrating technology as a supplement. Teachers should take care to prevent it from obscuring meaningful classroom interactions so that it doesn't overshadow discussions and teaching materials. Encourage learners to evaluate AI-generated content critically to place emphasis on the development of critical thinking abilities. Besides, teachers must encourage students to use trustworthy sources to confirm information from ChatGPT before using it in their coursework projects or research works. Teachers should create practical exercises that demand students to apply the knowledge they gain through ChatGPT to real-life situations, encouraging active learning and problem-solving skills by involving students in such scenarios. Moreover, teachers should make sure that students are aware of the ethical consequences of utilizing AI tools for learning by addressing the ethical feature by talking about AI ethics, plagiarism, and responsible technology use.

8.7.3 For Students/Learners

The main stakeholders of the educational system are students who should realize the limits of ChatGPT in giving thorough comprehension before using it as a starting point for study and learning. They must always perform a careful analysis of the data produced by ChatGPT and cross-reference it with trustworthy sources to guarantee correctness and take help from teachers, peers, or parents. By individually researching subjects outside of what AI tools can offer, they can hone their research and critical thinking skills. They must indulge in group activities and conversations to promote a deeper comprehension of subjects. When they are employing AI-generated content in assignments and projects, they must be mindful of ethical aspects including correct citation and avoiding plagiarism.

8.7.4 For Researchers

Researchers can be students or teachers so to support their initial research and idea development, they could make use of AI technologies like ChatGPT. To validate and go deeper into themes, they must do a thorough study using conventional techniques, nevertheless. They must be aware that AI-generated content could be lacking in context and nuance, so they must be sure their interpretations are accurate. Researchers have to work together with subject matter experts to improve and confirm the data generated by AI tools. They need to try to balance any potential biases in AI-generated material by consulting a variety of sources.

REFERENCES

Atlas, S. (2023). ChatGPT for higher education and professional development: A guide to conversational AI. University of Rhode Island DigitalCommons. https://digitalcommons.uri.edu/cba_facpubs/548

Baidoo-Anu, D., & Ansah, L. O. (2023). Education in the era of generative artificial intelligence (AI): Understanding the potential benefits of ChatGPT in promoting teaching and learning. *Journal of AI*, *7*(1), 52–62.

Benuyenah, V. (2023). Commentary: ChatGPT use in higher education assessment: Prospects and epistemic threats. *Journal of Research in Innovative Teaching & Learning*, *16*(1), 134–135.

Biswas, S. (2023). ChatGPT and the future of medical writing. *Radiology*, *307*(2), e223312.

Bozkurt, A., Xiao, J., Lambert, S., Pazurek, A., Crompton, H., Koseoglu, S., ... & Jandrić, P. (2023). Speculative futures on ChatGPT and generative artificial intelligence (AI): A collective reflection from the educational landscape. *Asian Journal of Distance Education*, *18*(1), 53–130.

Braun, V., & Clarke, V. (2006). Using thematic analysis in psychology. *Qualitative Research in Psychology*, *3*(2), 77–101.

Creswell, J. W., & Creswell, J. D. (2017). Research design: Qualitative, quantitative, and mixed methods approaches. Sage publications.

Dwivedi, Y. K., Kshetri, N., Hughes, L., Slade, E. L., Jeyaraj, A., Kar, A. K., ... & Wright, R. (2023). "So what if ChatGPT wrote it?" Multidisciplinary perspectives on opportunities, challenges and implications of generative conversational AI for research, practice and policy. *International Journal of Information Management*, 71, 102642.

Firaina, R., & Sulisworo, D. (2023). Exploring the usage of ChatGPT in higher education: Frequency and impact on productivity. *Buletin Edukasi Indonesia*, *2*(01), 39–46.

Firat, M. (2023). How chat GPT can transform autodidactic experiences and open education. *Department of Distance Education, Open Education Faculty, Anadolu Unive.*

Thu, C. H., Bang, H. C., & Cao, L. (2023). Integrating ChatGPT into Online Education System in Vietnam: Opportunities and Challenges.

Holmes, W., Bialik, M., & Fadel, C. (2023). *Artificial Intelligence in Education*. Globethics Publications. https://doi.org/10.58863/20.500.12424/4276068.

Illingworth, S. (2023, June 20). ChatGPT is the push higher education needs to rethink assessment. The Conversation. https://theconversation.com/chatgpt-is-the-push-higher-education-needs-to-rethink-assessment-200314

Rudolph, J., Tan, S., & Tan, S. (2023). ChatGPT: Bullshit spewer or the end of traditional assessments in higher education?. *Journal of Applied Learning and Teaching*, *6*(1), 342–363.

King, M. R., & Chat GPT. (2023). A conversation on artificial intelligence, chatbots, and plagiarism in higher education. *Cellular and Molecular Bioengineering*, *16*(1), 1–2.

Knox, J. (2020). Artificial intelligence and education in China. *Learning, Media and Technology*, *45*(3), 298–311.

Neumann, M., Rauschenberger, M., & Schön, E. M. (2023). "We Need To Talk About ChatGPT": The Future of AI and Higher Education. In 2023 IEEE/ACM 5th International Workshop on Software Engineering Education for the Next Generation (SEENG) (pp. 29–32). IEEE.

Ouyang, F., & Jiao, P. (2021). Artificial intelligence in education: The three paradigms. *Computers and Education: Artificial Intelligence*, *2*, 100020.

Shidiq, M. (2023). The use of artificial intelligence-based chat-GPT and its challenges for the world of education; from the viewpoint of the development of creative writing skills. In *Proceeding of International Conference on Education, Society and Humanity 1*(1), 353–357. https://ejournal.unuja.ac.id/index.php/icesh.

Sok, S., & Heng, K. (2023). ChatGPT for education and research: A review of benefits and risks. *Available at SSRN 4378735*.

ThankGod Chinonso, E. (2023). The impact of chatgpt on privacy and data protection laws. The Impact of ChatGPT on Privacy and Data Protection Laws (April 16, 2023).

Tlili, A., Shehata, B., Adarkwah, M. A., Bozkurt, A., Hickey, D. T., Huang, R., & Agyemang, B. (2023). What if the devil is my guardian angel: ChatGPT as a case study of using chatbots in education. *Smart Learning Environments*, *10*(1), 15.

Tong, A., Sainsbury, P., & Craig, J. (2007). Consolidated criteria for reporting qualitative research (COREQ): A 32-item checklist for interviews and focus groups. *International Journal for Quality in Health Care*, *19*(6), 349–357.

Tuhuteru, L., Pratiwi, E. Y. R., Suryowidiyanti, T., Mahendika, D., & Abdullah, D. (2023). Strategies For Primary School Students Understanding Of Character Education Through The Active Role Of Teachers. *Journal on Education*, *5*(4), 13569 13577.

Zhai, X. (2023). Chatgpt and AI: The game changer for education. *Available at SSRN*. https://ssrn.com/abstract=4389098

Zhai, X. (2023). ChatGPT for next generation science learning. XRDS: Crossroads, The ACM Magazine for Students, *29*(3), 42–46.

Zulkepli, N., Tajuddin, S. N. A. A., Atan, A., & Khaja, F. N. M. (2018). A study on autonomous use of technology for language learning among ESL learners at tertiary level. *International Journal of Academic Research in Business and Social Sciences*, *8*(11), 1093–1107.

9 Quality of Distance Education in the Era of Industry 4.0 and Society 5.0

An Exploration of the Effective Frameworks and Models to Design a Criteria

Zafarullah Sahito, Noor Ahmed, and Anna-Marie Pelser

9.1 INTRODUCTION

Education is a process that is explained as schools' work (Mowen, 2015) for the betterment of society through a remarkable curriculum around the globe. Little is known about the intellectual customs and traditions that have shaped education (Mowen, 2015) and training to increase awareness among individuals in society, fostering healthy and peaceful life through thinking and acting (Dewey, 1938). While strongly believing in democracy and education (Creighton, 1916); critical thinking (Dewey, 1933); reflective thinking about educative process (Dewey, 1938); and the relationship between education and experience (Dewey, 1929a), which support the exploration of nature in a real sense and have a strong relationship with the external world, the mind and knowledge, skills and their disposition. Education explores truth and responsibility through some scientific processes in society, which can work as a bank to make deposits of knowledge (Freire, 1972, 1993; Soler-Gallart, 2018[2023]) for the benefit of the individuals and society, because it is a social process to prepare the members for future living (Dewey, 1929a). Here, education focuses on the social interest of the individuals, which must be maintained and not distracted from learning new things in a well-organized manner (Lieberman, 2013), because education is learning and learning is social activity that can be cultivated through cognitive and emotional activities (OECD, 2016). Education is considered as a garden and the teachers are its gardeners, who can take a high level of care of their students (Gopnik, 2016)

 DOI: 10.1201/9781003400691-9

and bring up their flowers and plants that are very much sensitive but capable enough to be great human resources in the future.

In the 21st century, the definition of education has become multifaceted, multidirectional and multidimensional, extending beyond the classroom to encompass what is going inside the room. Nowadays, individuals acquire their education through many ways and resources, i.e. they learn from their own observations and experiences (May and Aikman, 2003; Rogoff et al., 2016) inside and outside of the classroom and the school (Crowley et al., 2017; Ullah et al., 2012). The learning is through different styles, ways and systems directly connected with the main three types of education, i.e. Formal, Informal and Non-formal (Curtis et al., 2013; Eshach, 2007). Whereas each system has its characteristics, strengths, weaknesses, advantages and disadvantages, they directly and indirectly support the individuals to obtain education and learn new things (Crowley, 2017; Sharma et al., 2016; Ullah et al., 2012) scientifically in every condition of life, status, position, etc. But the majority of theorists did distinguish between formal and informal education (Strauss, 1984). A formal education system that takes place in a complex institutionalized framework in a chronological and hierarchical order starts from primary to university level organizations, guided by a governmental instruction(s) with certain compulsory age limits (Tudor, 2013), while non-formal and informal education systems vary from formal education due to lack of governmental institutionalized framework, fixed place of learning, freedom, time limit, etc. Non-formal education constitutes the middle ground of formal and informal systems with organized and systematic sort of balance to carry out everything with a clear purpose, i.e. tutoring, assessing, etc. (Curtis et al., 2013), finally and occasionally visits are to be arranged in non-formal education (Eshach, 2007; Tudor, 2013). The students' motivation is responsible for intrinsic engagement in the non-formal system of education (Eshach, 2007; Tudor, 2013). The difference between all three types of education systems depends on the exemplary cases that support the open education system (Marín et al., 2020) to appreciate the individuals to learn as per their freedom, interest, wish and will.

9.1.1 Industry 4.0

The Fourth Industrial Revolution (4IR) is also known as Industry 4.0 (Bai et al., 2020). In the 21st century, Industry 4.0 was widely used in the scientific literature (Colombo et al., 2014) that conceptualizes the rapid change to technologies, industries and societal patterns and processes and their relationships due to increase in interconnectivity and smart automation. Afterwards in 2015, Industry 4.0 was popularized by Klaus Schwab, who declared that the changes seen due to technologies are more than the improvements to product efficiency, but is a significant shift in industrialization and capitalization. Actually, it is a phase of industrial change in which the technologies like artificial intelligence, gene editing, advanced robotics shape the lines between the physical, digital and biological worlds for betterment of the world and humans. This paradigm shift is directly concerned with operations of global production and supply network through technology, especially modern smart technology at a large scale by machine-to-machine communication (M2M) and the internet of things (Schwab, 2018). This paradigm shifts of technology integration result in increased automation, improvement in communication, and self-monitoring system through efficient usage of smart machines that can examine and establish the issues with scientific solutions without any human intervention. Therefore, it is also called the era of embedded connectivity, representing the social, political and economic shift from the digital age and the widespread technological use throughout society changing the ways humans experience the world around them (Yun et al., 2019). The role of Industry 4.0 in the development of distance education system is very important, focusing on the process of the production supported by the technology to make systems smooth and easy. In the validation of the process, data are taken from different sources especially from learners and teachers about their learning analytics to ensure the feasibility of the technology and its related systems to complete the tasks in time for fruitful and scientific results to achieve the personal, professional and organizational goals.

9.1.2 Society 5.0

Society 5.0 means a society that relies on the well-being and happiness of human beings, which ensures that no human being is left behind as expressed in the United Nation's SDGs. Society 5.0 is a human-centred society making a balance between economic progress and the resolution of social problems through a scientific and socialized system, highly connected with cyberspace and the real world (Ellitan, 2020). Society 5.0 is a quiet revolutionary concept that was started in Japan and then flourished throughout the world; assurances to transform society and improve the quality of life of all human beings are provided by focusing on the central role of human beings through innovation and technological transformation. The prime aim of Society 5.0 is to develop the quality of public's lives, through using available potentialities acquired by Industry 4.0 (Ferreira and Serpa, 2018).

Society 5.0 began as a Japanese government programme, with a wish to establish a better, improved and healthier super-smart and prosperous human-centred society, and with the backing of the technological modernizations and transformations. The unique goal of the programme was to recover the Japanese economy for future sustainability. Society 5.0 was founded as part of the programme of Japanese Prime Minister Shinzo Abe to revive the Japanese economy called "Abenomics" consisting of "three arrows", a symbol of the monetary acceleration, fiscal stimulus and growth strategy. The first two arrows show the economy of Japan to stay floating and the third arrow shows the socio-economic development and sustainability. The first growth strategy was made by the government of Japan in 2013 and revised each year till 2017 when the concept of Society 5.0 was incorporated.

In 2050, Japan's 40% population will be over 65 years old, which proves that it will be the oldest society (Granrath, 2019). There is a directional relationship between age and the usage of technology among Japanese people to be happy to increase their well-being, as they use technology, i.e. artificial intelligence, robotics, Big Data, autonomous trucks and drone deliveries, on priority basis that will have to be integrated to deal with the problem of today's society. This new concept of society having happy employees aims to promote the idea that industries will be successful if they have happy and fulfilled people at work, by creating all possible required optimal conditions to improve the day-to-day routines concerned with well-being and happiness of their employees to make them more satisfied and motivated for a productive society (Ferreira and Serpa, 2018; Ellitan, 2020). Economic growth and development and addressing social problems can be achieved by the exhibition and proper utilization of technologies to service humanity in order to reduce greenhouse gases, increase the production and minimize the loss of food and foodstuff by moderating the costs associated with society through sustainable industrialization, redistribution of wealth and rectification of regional inequalities among all individuals (Oliveira, 2019).

9.1.3 Link between Industry 4.0 and Society 5.0

The relationship between Industry 4.0 and Society 5.0 can be explained in many ways, i.e. Society 5.0 wants to develop a smart society through Industry 4.0 especially through the solution of social and technological issues, challenges and problems. The solution is especially focused on humanization and the expansion of technological, organizational, and economic transformations of industrial production grounded in innovative scientific and technical developments aimed at fostering economic development and sustainability; industrial sector and its development are significantly important for the social structure of society (Chang et al., 2021).

Industry 4.0 and Society 5.0 are both strongly connected with technology, i.e. artificial intelligence, cyber systems, big data, internet of things, robotics, augmented reality and the cloud technology (Pereira et al., 2020). Society 5.0 consists of the entire society with industry which means Industry 4.0 is also a part of it and works for the betterment of all stakeholders of society (Kansha and Ishizuka, 2019). Furthermore, Society 5.0 addresses social issues faced by humans concerned with lowering costs through capable manufacturing methods and techniques (Deguchi et al., 2020) to solve the health, poverty, prosperity, easy access to food and water, smart agricultural practices, gender equality and social justice issues on a priority basis scientifically; these are also the need and requirement of SDGs. Industry 4.0 and Society 5.0 are directly connected with the factors of sustainability, i.e. energy, economic growth, smart working models, high and effective production, smart waste management and smart infrastructure systems, given in the SDGs of 7, 8, 9, 11, 12 and 13 (Karamustafa et al., 2022). Entrepreneurial growth and development converts the knowledge society into a smart society having knowledge-dependent social groupings exercising social and interpersonal relationships (Ritzer, 2015). Students in Industry 4.0 and Society 5.0 are closely related and interact technologically with technological gadgets and accessories to receive knowledge and learn new skills to use and advance for obtaining any job (Collins and Halverson, 2018). Industry 4.0 and Society 5.0 provide opportunities to all individuals to follow their passion and goals through rich social learning environment to reach the peak of their success. Industry 4.0 and Society 5.0 handle everyone and everything carefully through educational institutions by providing their services as the educational institutions prepare goods and materials for a smart society.

9.2 OBJECTIVES AND THE RESEARCH QUESTIONS

The objectives of the chapter are (a) To understand the concepts of Industry 4.0 and Society 5.0; (b) To understand the concepts of quality in distance education (E-learning) and quality assurance of distance education (E-learning); and (c) To explore the effective frameworks and models of quality assurance of distance education (E-learning) to design criteria. While the research questions are: (a) What is Industry 4.0 and Society 5.0? (b) What is Distance Education, Quality Education, Quality Assurance, and Criteria of Quality? (c) How are effective frameworks and models of Quality Assurance of Distance Education to be designed and used to maintain Quality Education?

9.3 DISTANCE EDUCATION (E-LEARNING)

Distance education (E-learning) is known as distance learning, a system of education in which the students and teachers are not available physically at the same place, i.e. in class, school and institution but learning take places by their interaction through technology (Plass and Kaplan, 2016). The correspondence of courses can be done through emails, WhatsApp groups, Facebook chatting, Zoom meetings, Google classrooms, etc., as per needs and requirements of the students, teachers and institutions (Aristovnik et al., 2020; Deflem, 2019; Deflem and Silva, 2021; Hood, 2020; Jelińska and Paradowski, 2021; El-Kasim, 2020; Raymond-Hayling, 2020). It is also known as online education, when each and every process can be done through virtual forms (Burns, 2012; Garrison, 2016; Garrison, 2016; Hood, 2020; Oblinger et al., 2005; Sazmand Asfaranjan et al., 2013). The distance learning programme can be an entirely distance learning one or a mixture of distance learning and traditional classroom instructions known as a hybrid or a blended programme (Garrison, 2016; Vaughan, 2010). Massive open online courses (MOOCs) offer large-scale collaborative contribution and open access through the World Wide Web and other network technologies that have become the latest educational methods and approaches in the field of distance education and learning (Plass and Kaplan, 2016). Distance education is also known with different names, i.e. distributed learning, e-learning, m-learning, online learning, virtual learning or virtual classroom learning, blended learning (Kumar et al., 2021), deep learning, etc., where e-learning is an educational tool used for an interactive process with multiple learning modes and various levels for all types of learners to learn efficiently and effectively (Bower, 2011; Veletsianos, 2020; Veletsianos et al., 2021). Distance learning environments are exciting places and platforms to learn new things collaboratively with others through the interactive approaches to maintain self-discipline and self-directed learning (Bower, 2011; Garrison, 2016; Nhan, 2020; Nguyen, 2015; Stanton, 2001).

E-Learning context has been elevated and is guided by several studies, which conclude that there is a dearth of literature available about e-Learning and the Web 4.0 to support the teaching-learning process (Babori et al., 2022). E-Learning has been instrumental in the transformation of higher education institutions, in the situation of pandemic associated with remote teaching supported by information and communication technology (ICT) tools and techniques and e-learning solutions, i.e.

3.0, 4.0, and 5.0, that were conceptualized and connected with the framework of the e-learning 5.0 system to prioritize for future work (Kierzyk and Pólkowski, 2021). Remote learning or e-learning had become the norm during COVID-19 to avoid the spread of diseases and to continue the teaching-learning process because the sudden outbreak disrupted the education systems worldwide and forced the students, teachers and educators to switch to online learning (Sahito et al., 2022). This switching process was acknowledged by some students and teachers and they made themselves ready for online learning (Sharin, 2021). E-learning in higher education has been exponentially increased during the last couple of years due to its unescapable benefits in critical situations, such as the COVID-19 pandemic, etc., where exams were conducted through e-learning platforms without the physical availability of students and instructors, which facilitates the concerned stakeholders to select the right decisions to continue the process of education (Muzaffar et al., 2021). E-learning focuses on the development of web-based learning that supports how e-learning can be initiated and what characteristics, specifications and shortcomings can be overcome in future through educators' engagements at different levels of experiences through different levels of education, communication skills and managing technical issues for better performance (Sun and Ochiai, 2022). Digital adaptation also known as e-learning is needed in today's emerging landscape for organizations depending on innovations to transform them based on new technologies to increase the organizational ability (Rosário, 2022). E-government is a type of e-learning that works from a distance in which most cases remained successful due to efficient systems of e-government that were examined through the success model of DeLone and McLean, the Technology Acceptance Model (TAM). But some experts also proposed new hybrid models, a combination of several models available, especially the integrated IS model that remained successful in the context of e-government (Stefanovic et al., 2021).

9.4 QUALITY ASSURANCE OF DISTANCE EDUCATION (E-LEARNING)

The fast growth of computer networks brings a huge modification in the processing capability of computers to store more files and data for the betterment of education, research, etc., specially to teach students from any place and time virtually. The advancement in the field of ICT allows to bring about any change in the teaching-learning methodologies through suitable and authentic use of new learning materials depending on the student-centric environment of e-learning, and the effectiveness potential of distance education for educational organizations. With the growing stream of e-learning, quality assurance has undoubtedly become an area of strategic plan and value to be counted among top ranking organizations through ensuring the learning outputs of acceptable and confirmable quality (Marciniak, 2018) by developing quality assurance criteria in the field of e-learning (Kishabale and Hassan, 2018). A taxonomy of online quality criteria comprehended the course content, resources, alignment, instructional methodologies, feedback, accessibility and information design to support the students, institutions and faculty (Mahdiuon et al., 2017) to achieve the desired outcomes. Mhlanga (2013) examined the quality assurance criteria of online quality indicators, i.e. course design, evaluation, assessment (Nayyar et al., 2020) and technology, through research to improve the quality of e-learning and found them fit to evaluate and maintain quality. E-Learning programmes have been found to be aligned and crucial with the components of quality learning, i.e. methodologies, learner profiles, thematical material, educational technology and evaluation approaches (Nayyar et al., 2020). For quality maintenance a set of components for e-learning has been proposed consisting of objectives, thematic nature of material, learning activities, evaluation methodologies and the virtual classroom environment (Marciniak, 2018).

Quality assurance in distance or e-learning is the hottest problem of the institutions of the world (Avota, 2018; Batsila, 2019; Li et al., 2022; Taftaf and Williams, 2020; Scheunemann et al., 2018; Stambekova et al., 2022; Roxå and Mårtensson, 2008; Vlachopoulos and Makri, 2019, Zawacki-Richter and Qayyum, 2019), where special attention is directly dedicated to the distance or e-learning with especial focus on the virtual character and the physical distance between students and teachers (Vlachopoulos and Makri, 2019). The content, instructional techniques and learning outcomes are regularly assessed (Mukhopadhyay et al., 2020) to maintain quality (Pandey and Indrakanti, 2017) because there are no one-size-fits-all model as the prime input to maintain quality. All quality criteria are used to measure the quality connected with the situations, goals and objectives to develop the quality standards to assist the concerned firms to develop the quality mechanism to match the unique needs of all stakeholders. Because, previously the quality was judged by the quality of content, pedagogy and learning outcomes, whereas the quality maintenance process focuses on the enhancement of the educational experiences and learning (Vlachopoulos and Makri, 2019).

The quality criteria of distance education or e-learning have been described in various forms and perspectives, i.e. pedagogical quality (Al-Harthi and Ani, 2023; Al Hosni and AlAli, 2022; World Health Organization, 2015; Benner et al., 2008; Holsombach-Ebner, 2013), technological quality and mechanisms (World Health Organization, 2015; Al-Sharhan and Al-Hunaiyyan, 2012; Holsombach-Ebner, 2013; Nsiah and Oti-Boadi, 2015), quality assessment, content quality (Bates et al., 2016), organizations' quality and mechanisms (Nsiah and Oti-Boadi, 2015; Roxå and Mårtensson, 2008). In this regard, Harman et al. (2000) define quality assurance as a systematic management and assessment procedure adopted in order to monitor performance against objectives and to ensure the achievement of quality outputs and quality improvements, which were also cited by many researchers, especially Belawati and Zuhairi (2007). Farrell-Evans and Warren (2014) advocated and explained the quality assurance after conducting a study on ICT in 53 countries of the African continent, i.e. the system that encloses the programme design, its approval, review, management and delivery to support students' learning and communication through demonstration and assessment (Mukhopadhyay et al., 2020). Furthermore, Norman and Redfern (1993) suggested some prime aspects of quality assurance in distance or e-learning, i.e. products, processes, production and delivery system, and general philosophy, which means that the courses, materials and graduates; functions of distance education, i.e. registering students, tutoring and assessment; course development, etc.; and individual roles, responsibilities, mission and accountability, respectively, were also cited by Tripathi and Jeevan (2009).

9.5 METHODOLOGY

Desk review research method was used for this chapter, which helps to explore the data from the existing literature (Sileyew, 2019) available online (Hewson et al., 2015; Kara, 2015) and offline (Bolander and Locher, 2014). The secondary data were used to gather information on the particular phenomena to understand (Clark et al., 2021) it with different lenses of the authors in different directions, which provide solid arguments to elaborate and explain the topic in detail (Kraus et al., 2020). It is focused on a necessary process to provide baseline information on any topic of study for better understanding, using an advanced level of data and its interpretation (Hennink and Bonnie, 2022) and focusing on the latest trends in the body of knowledge and research on the advanced areas of the knowledge directly and indirectly (Sovacool et al., 2018) connected with society, communities, markets, infrastructure, landscapes, industries, situations, trends, behaviours and the identification of sources and solutions (Kozinets et al., 2014) to reach a scientific conclusion (Grover and Kumar Kar, 2017; Merriam and Tisdell, 2015).

9.6 DISCUSSION

The criteria (Appendix I) can be used as an evaluation tool or the checklist as a benchmark or rubric to assess the quality of distance education. The criteria have five columns, i.e. serial number, criterion, indicators with statements, four levels of performance and evidences. The criteria have been designed to assess the quality of the distance education system based on the eight criteria or indicators, i.e. (a) Curriculum and course quality; (b) Content quality; (c) Pedagogical quality; (d) Learner support system; (e) Technological support; (f) Assessment mechanism; (g) Quality assurance and evaluation; and (h) Organizational management, administration and leadership.

Curriculum and Course Quality possesses 10 statements, i.e. (a) Logical presentation; (b) Adoptable course structure; (c) Links between course and library research databases and Learning Management System (LMS); (d) Licence of course; (e) Adherence of course with institution; (f) Alignment of course format with learning tools; (g) Flexibility and change between first and current course; (h) Standards for e-learning delivery; (i) Incorporation of the national curricular standards; and (j) Testing of materials and courses. ***The Content Quality possesses*** 16 statements, i.e. (a) Learning outcomes; (b) Linkage between learning outcomes and the applicable learning framework; (c) Addition of text and media aid in subject; (d) Student involvement and engagement; (e) Approved monitoring and evaluation criteria; (f) Specified level of course; (g) Verification of course content; (h) Satisfaction of content standards; (i) Trustworthiness of content references; (j) Reliability and validity of contents with plagiarism-free material; (k) International or national alignment of contents; (l) Standards of content writers and designers; (m) Quality of editing and review; (n) Piloted materials for students and teachers; (o) Quality assurance of final version of content; (p) Checking of typographical errors, etc. Furthermore, the ***Pedagogical Quality possesses*** 14 statements, i.e. (a) Learners' participation; (b) Relationship between the learning activities, resources and outcomes; (c) Relationship between pedagogical resources and desired learning goals; (d) Use of active learning tactics for students' engagement; (e) Learning activities and active learning; (f) Access to external programmes and software; (g) Media incorporation in course contents; (h) User-friendly LMS; (i) Instructional exercises and the students' encouragement; (j) Tutoring-based system to support students; (k) Ongoing professional development opportunities; (l) Efficient use of audiovisual aids; (m) Orientation and pre-training for instructors to use; (n) Active participation of instructors.

Learner Support System possesses 15 statements, i.e. (a) Clearly defined instructions about learning objectives; (b) Tutors necessary training and certification; (c) Orientation and introduction of course's components; (d) Students' advance feedback; (e) Guidelines for navigating the course website; (f) Instructions to navigate the course site; (g) Guidelines for learners to interact with a blended learning course; (h) Opening of course materials as a separate Word doc; (i) Sufficient indication of materials' usage; (j) Availability of FAQs; (k) Following the proper selection procedure; (l) Mentoring, coaching and support services availability; (m) Availability of registration and examination procedures; (n) Availability of essential component of teacher education programmes; (o) Availability of multiple assessment styles and formats. ***Technological Support possesses*** 10 statements, i.e. (a) Infrastructure and learner access to technical tools; (b) Selection of technologies for improvement of learning outcomes; (c) Efficient use of LMS; (d) Active interaction between learners and technology; (e) Availability of independent browser; (f) Relationship between learning objectives, material and selection of technology; (g) Specific use of technology as teacher-learner centred; (h) Efficient use of technology by teacher-learners; (i) Availability of simple instructions for technical operations; and (j) Availability of guidelines for the utilization and the management of databases.

Assessment Mechanism possesses eight statements, i.e. (a) Advance information about course evaluation; (b) Availability of marking guides for papers and exams; (c) Formative and summative assessment measures; (d) Assessment tools for self-evaluation and feedback; (e) Assessment tools to assess the mastery of learning outcomes; (f) Measurement of students' readiness through diagnostic evaluation; (g) Identification of learners' issues through formative assessment; (h) Alternative types of teaching and assessment. ***Quality Assurance and Evaluation possesses five statements***, i.e. (a) Insurance of the quality assurance methods like face-to-face course; (b) Availability of plans and timelines for reviewing online courses; (c) Submission of learners' comments on the blended learning experience; (d) Availability of quality assurance and control mechanisms; and (e) Standards and quality of materials. ***Organizational Management, Administration and Leadership possesses 14 statements***, i.e. (a) Employment and engagement of basic databases or an Educational Management Information System (EMIS); (b) Regular updating of the database/EMIS; (c) Easy access to relevant information; (d) Availability of additional means; (e) Controlled material distribution and delivery; (f) Accountability of distribution of learning materials; (g) Readiness of all materials and equipment; (h) Availability of related policies with values of distant education programmes; (i) Availability of institutional vision statement; (j) Availability of policies and processes for training, support, materials, resources and technology; (k) Promotion of quality learning culture; (l) Availability of learning objectives for foundation provision; (m) Problems confrontation and resolution system; (n) Readiness and dedication of top management for distance programme.

The Quality Criteria have been decided with the help of studies conducted by Avota (2018), Bandeira and Cardoso (2021), Cifuentes (2021), Jung and Latchem (2012), Jung et al. (2013), Mahlangu (2018), Mahlangu (2021), McCabe and González-Flores (2017), Melton (2004), Ramdass and Ṋemavhola (2018), Shattuck (2023), Stavredes and Herder (2013), Welch and Reed (2005), Hope (2014), Zuhairi et al. (2020).

9.7 CONCLUSION

Education provides opportunity, sense and wisdom to explore the truth with the sense of duties and responsibilities through scientific processes in any society for the maximum common benefits of the individuals through social processes to prepare the members for future living. The term education has been defined as a multifaceted, multidirectional and multidimensional way to learn through different styles, ways and systems, directly connected with the main three types of education, i.e. Informal, Formal and Non-formal. Non-formal and informal education systems vary from formal education due to lack of governmental institutionalized framework, fixed place of learning, freedom, time limit, etc. The difference between all three types of education systems depends on the exemplary cases that support the open education system. Industry 4.0 is the paradigm shift in this direction, connected and concerned with operations of global production and supply network through technology, especially smart technology, at a large scale by machine-to-machine communication. Society 5.0 wants to develop the quality of public's lives using available potentialities acquired by Industry 4.0. Society 5.0 has been described by researchers as a human-centred society in which regional oriented solutions are found for their sustainability. Industry 4.0 and Society 5.0 are strongly connected in the relation ships of technology, human welfare and sustainability where all members will be able to use technology on equality, equity and justice basis to work for their mutual benefit and the future sustainability.

For achieving sustainability, the members of the society must be educated to work as experts and professionals, which can be achieved only through a strong and updated education system. Having a strong and updated education system for every member of the society is impossible due to financial constraints especially for hugely populated nations, which can be achieved through distance schooling system to offer services at a lower cost to everyone who wants to get education as per his or her will, wish and needed time as both campus-based and distance education stand to benefit greatly through integration of the systems. Learners in the Industry 4.0 and Society 5.0 eras are technologically cultured and classy to increase their interest in learning fluctuations (Nayyar and Kumar, 2020) as a result of advances and changes in the job. Educational institutions must accept technology developments in the new Society 5.0 to digitalize their learning environment and atmosphere for a rapidly changing world to utilize the internet to access information and learning tools through distance educational platform.

The future of the distance (e-learning) education is quite bright, clear and attractive because of the advancement of the technology and its proper usage, i.e. the

Internet of Things (IoT) and cyber-physical system (CPS) that support the industrial applications and automation, which is about to bring an incredible change for the betterment of the people and industry in the era of Society 5.0 and Industry 4.0. The prime aspect is the enhancement of the technology that will be a great, easy and affordable use of modern technology, possible and feasible by a great role played by educational leaders and managers in their educational organizations and systems. If the educational leaders and managers have a positive, feasible and implementable vision, mission and plans to upgrade the system of technology to meet the modern needs and requirements of the education system then it will be quite possible to have efficient and effective systems of education especially the distance education system purely based on the quality, excellence and suitability to achieve the MDGs.

REFERENCES

Al-Harthi, Aisha Salim Ali, and Wajeha Thabit Al Ani. (2023). "Learner readiness for MOOCs in Omani higher education institutions: disparities between projections and reality." *Education and Information Technologies, 28*(1): 303–319.

Al Hosni, Asma K., and Said F. AlAli. (2023). Original paper MOOCs in Omani Higher Education Institutions: Use and popularity. *Education, Language and Sociology Research, 3*(1): 1–24.

Al-Sharhan, Salah, and Ahmed Al-Hunaiyyan. (2012). "Towards an effective integrated e-learning system: Implementation, quality assurance and competency models." In Seventh *International Conference on Digital Information Management (ICDIM 2012)*, pp. 274–279. IEEE.

Aristovnik, Aleksander, Damijana Keržič, Dejan Ravšelj, Nina Tomaževič, and Lan Umek. (2020). "Impacts of the COVID-19 pandemic on life of higher education students: A global perspective." *Sustainability, 12*(20): 8438.

Avota, Baiba. (2018). "Assessment of distance learning quality criteria for continuing education of medical practitioners." In *Proceedings of the International Scientific Conference* on Society, Integration, Education, vol. 5, pp. 15–24.

Babori, Abdelghani, Noureddine Falih, and Hicham Ouchitachen. (2022). "A framework for adaptive E-learning systems based on self-regulated learning strategies." In International *Conference on Digital Technologies and Applications*, pp. 380–387. Cham: Springer International Publishing.

Bai, Chunguang, Patrick Dallasega, Guido Orzes, and Joseph Sarkis. (2020). "Industry 4.0 technologies assessment: A sustainability perspective." *International Journal of Production Economics, 229*: 107776.

Bandeira, Fernando, and João Casqueira Cardoso. (2021). "Quality practices in higher education distance learning: The perspective of practitioners." In *Handbook of Research on Determining the Reliability of Online Assessment and Distance Learning*, pp. 1–23. IGI Global.

Bates, Meg S., Lena Phalen, and Cheryl Moran. (2016). "Online professional development: A primer." *Phi Delta Kappan, 97*(5): 70–73.

Batsila, Anastasia. (2019). "Inter-learner communication and collaborative learning as quality criteria of distance vocational education and training." *European Journal of Open, Distance and E-Learning (EURODL), 22*(2): 98–112.

Belawati, Tian, and Amin Zuhairi. (2007). "The practice of a quality assurance system in open and distance learning: A case study at Universitas Terbuka Indonesia (The Indonesia Open University)." *The International Review of Research in Open and Distributed Learning, 8*(1): 1–15.

Benner, Patricia, Molly Sutphen, Victoria Leonard-Kahn, and Lisa Day. (2008). "Formation and everyday ethical comportment." *American Journal of Critical Care, 17*(5): 473–476.

Bolander, Brook, and Miriam A. Locher. (2014). "Doing sociolinguistic research on computer-mediated data: A review of four methodological issues." *Discourse, Context & Media, 3*: 14–26.

Bower, Matt. (2011). "Synchronous collaboration competencies in web-conferencing environments-their impact on the learning process." *Distance Education, 32*(1): 63–83.

Burns, Rebecca. (2012). "The pedagogy of ESL in elementary teacher education." *Linguistics and Education, 23*(4): 412–413.

Chang, Victor, Yuanyuan Kate Xu, Jingqi Zhang, and Qianwen Xu. (2021). "Research on intelligent manufacturing development approach for China's local valve industry." *Smart and Sustainable Built Environment, 10*(2): 293–321.

Cifuentes, Lauren. (2021). *A Guide to Administering Distance Learning*, vol. 2. Brill.

Clark, Tom, Liam Foster, Alan Bryman, and Luke Sloan. (2021). *Bryman's Social Research Methods*. Oxford University Press.

Collins, Allan, and Richard Halverson. (2018). *Rethinking Education in the Age of Technology: The Digital Revolution and Schooling in* America. Teachers College Press.

Colombo, Armando W., Stamatis Karnouskos, and Thomas Bangemann. (2014). "Towards the next generation of industrial cyber-physical systems." In *Industrial Cloud-Based Cyber-Physical Systems: The IMC-AESOP Approach*, pp. 1–22. Cham: Springer International Publishing.

Creighton, J. E. (1916). "Democracy and education," *The Philosophical Review*, *25*(5), 735–741.

Crowley, Kevin. (2017). "Child development: A practical introduction." *Child Development*, 1–320.

Curtis, David J., Mark Howden, Fran Curtis, Ian McColm, Juliet Scrine, Thor Blomfield, Ian Reeve, and Tara Ryan. (2013). "Drama and environment: Joining forces to engage children and young people in environmental education." *Australian Journal of Environmental Education, 29*(2): 182–201.

Deflem, Mathieu, (ed.) (2019). The Handbook of Social Control. Wiley.

Deflem, Mathieu, and Derek MD Silva, (Eds.) (2021). *Media and Law: Between Free Speech and Censorship*. Emerald Publishing Limited.

Deguchi, Atsushi, Chiaki Hirai, Hideyuki Matsuoka, Taku Nakano, Kohei Oshima, Mitsuharu Tai, and S. Tani. (2020). "What is Society 5.0",. Society, *5*, 1–24.

Dewey, John. (1929a). "Experience and philosophic method."

Dewey, John. (1933). "Why have progressive schools?" *Current History, 38*(4): 441–448.

Dewey, John. (1938). "What is social study?" *The Modern Language Journal, 23*(3): 219–219.

Diana Oblinger, E. D. U. C. A. U. S. E., James Oblinger, Greg Roberts, Ben McNeely, Carie Windham, Joel Hartman, Patsy Moskal, Chuck Dziuban, and Robert Kvavik. (2005). *Educating the Net Generation*, vol. 272. Brockport Bookshelf, Book.

El-Kasim, Mukhtar. (2020). "Nigerian public relations practitioners' use of social media: Validation of an online interactivity model applying structural equation modeling." *Media Watch, 11*(2): 337–347.

Ellitan, Lena. (2020). "Competing in the era of industrial revolution 4.0 and society 5.0." *Jurnal Maksipreneur: Manajemen, Koperasi, dan Entrepreneurship, 10*(1): 1–12.

Eshach, Haim. (2007). "Bridging in-school and out-of-school learning: Formal, non-formal, and informal education." *Journal of Science Education and Technology, 16*: 171–190.

Farrell-Evans, M., and W. Warren. (2014). "Food safety assurance systems: quality assurance and good laboratory practice," 293–300. https://www.doi.org/10.1016/B978-0-12-378612-8.00374-7

Ferreira, Carlos Miguel, and Sandro Serpa. (2018). "Society 5.0 and social development." *Management and Organizational Studies, 5*(4): 26–31.

Freire, Paulo. (1972). Pedagogy of the Oppressed 1968. Trans. Myra Bergman Ramos, pp. 273–290. New York: Herder.

Freire, Paulo. (1993). *Pedagogy of the Oppressed* (20th-Anniversary ed.). New York Continuum.

Garrison, D. Randy. (2016). *E-Learning in the 21st Century: A Community of Inquiry Framework for Research and Practice*. Taylor & Francis.

Gopnik, Alison. (2016). *The Gardener and the Carpenter: What the New Science of Child Development Tells Us About the Relationship between Parents and Children*. Macmillan.

Granrath, Lorenz. (2019). "Large scale optimization is needed for industry 4.0 and society 5.0." *Optimization in Large Scale Problems: Industry 4.0 and Society 5.0 Applications*, pp. 3–6.

Grover, Purva, and Arpan Kumar Kar. (2017). "Big data analytics: A review on theoretical contributions and tools used in literature." *Global Journal of Flexible Systems Management,* 18: 203–229.

Harman, Grant Stewart, Vincent Lynn Meek, G. Harman, and L. V. Meek. (2000). *Repositioning Quality Assurance and Accreditation in Australian Higher Education*. AusInfo.

Hennink, Monique, and Bonnie N. Kaiser. (2022). "Sample sizes for saturation in qualitative research: A systematic review of empirical tests." *Social Science & Medicine,* 292: 114523.

Hewson, Claire, Carl Vogel, and Dianna Laurent. (2015). *Internet Research Methods*. SAGE.

Holsombach-Ebner, Cinda. (2013). "Quality assurance in large scale online course production." *Online Journal of Distance Learning Administration, 16*(2): 1–22.

Hood, N. (2020). *Research Use and Professional Identity*, pp. 1–5. Encyclopedia of Teacher Education.

Hope, Andrea. (2014). "*Quality Assurance in Distance Education and E-Learning: Challenges and Solutions from Asia*," pp. 86–88.

Jelińska, Magdalena, and Michal B. Paradowski. (2021). "Teachers' engagement in and coping with emergency remote instruction during COVID-19-induced school closures: A multinational contextual perspective." *Online Learning, 25*(1): 303–328.

Jung, Insung, and Colin Latchem. (2012). *Quality Assurance and Accreditation in Distance Education: Models, Policies and Research*, Routledge.

Jung, Insung, Tat Men Wong, and Tian Belawati, eds. (2013). *Quality Assurance in Distance Education and E-Learning: Challenges and Solutions from Asia*. SAGE Publications India.

Kansha, Yasuki, and Masanori Ishizuka. (2019). "Design of energy harvesting wireless sensors using magnetic phase transition." *Energy,* 180: 1001–1007.

Kara, Helen. (2015). *Creative Research Methods in the Social Sciences: A Practical Guide*. Policy Press.

Karamustafa, Esin Yücel, Burcu Arsan, and Kağan Beşoğul. (2022). "How do the circular economy and industry 4.0 contribute to the achievement of sustainable development goals?" *Bilgi Sosyal Bilimler Dergisi, 24*(2): 294–323.

Kierzyk, Tadeusz, and Zdzisław Pólkowski. (2021). "Transformation of higher educational institutions from distance learning to the e-learning 5.0: An analysis." In *Data Analytics and Management: Proceedings of ICDAM*, pp. 923–934. Springer Singapore.

Kishabale, Bashir, and Sharifah Hassan. (2018). "A Predictive study on instructional design quality, learner satisfaction and continuance learning intention with e-learning courses: data screening and preliminary analysis." *Interdisciplinary Journal of Education, 1*(2): 122–137.

Kozinets, Robert V., Pierre-Yann Dolbec, and Amanda Earley. (2014). "Netnographic analysis: Understanding culture through social media data." In *Sage Handbook of Qualitative Data Qnalysis*, pp. 262–275.

Kraus, Sascha, Matthias Breier, and Sonia Dasí-Rodríguez. (2020). The art of crafting a systematic literature review in entrepreneurship research. *International Entrepreneurship and Management Journal, 16*: 1023–1042.

Kumar, Adarsh, Rajalakshmi Krishnamurthi, Surabhi Bhatia, Keshav Kaushik, Neelu Jyoti Ahuja, Anand Nayyar, and Mehedi Masud. (2021). "Blended learning tools and practices: a comprehensive survey." *IEEE Access, 9*(2021): 85151–85197.

Li, Richard Chen, Simon KS Cheung, Peter HF Ng, Leung-Pun Wong, and Fu Lee Wang, (Eds.) (2022). Proceedings *on the Blended Learning: Engaging Students in the New Normal Era: 15th International Conference, ICBL 2022*, Hong Kong, China, July 19–22, 2022, vol. 13357, Springer Nature.

Lieberman, Matthew D. (2013). *Social: Why Our Brains Are Wired to Connect.* OUP Oxford.

Mahdiuon, Rouhollah, Davoud Masoumi, and Maghsood Farasatkhah. (2017). "Quality improvement in virtual higher education: A grounded theory approach." *Turkish Online Journal of Distance Education, 18*(1): 111–131.

Mahlangu, Vimbi Petrus. (2018). "Quality assurance and accreditation in open and distance learning." In Koksal Buyuk, Serpil Kocdar, and Aras Bozkurt (Eds.), *Administrative Leadership in Open and Distance Learning Programs*, pp. 146–169. Hershey, PA: IGI Global.

Mahlangu, Vimbi Petrus. (2021). "Quality assurance and accreditation in open and distance learning." In *Research Anthology on Preparing School Administrators to Lead Quality Education Programs*, pp. 1079–1097. IGI Global.

Marciniak, Renata. (2018). "Quality assurance for online higher education programmes: Design and validation of an integrative assessment model applicable to Spanish universities." *International Review of Research in Open and Distributed Learning, 19*(2): 126–154.

Marín, Victoria I., Olaf Zawacki-Richter, and Svenja Bedenlier. (2020). "Open educational resources in German higher education-an international perspective." In *EDEN Conference Proceedings*, no. 1, pp. 85–94.

May, Stephen, and Sheila Aikman. (2003). "Indigenous education: Addressing current issues and developments." *Comparative Education, 39*(2): 139–145.

McCabe, Margaret Foley, and Patricia Gonzalez-Flores. (2017). *Essentials of Online Teaching: A Standards-Based Guide*. Taylor & Francis.

Melton, Reginald F. (2004). *Planning and Developing Open and Distance Learning: A Framework for Quality*. Routledge.

Merriam, Sharan B., and Elizabeth J. Tisdell. (2015). *Qualitative Research: A Guide to Design and Implementation*. John Wiley & Sons.

Mhlanga, Ephraim. (2013). "*Quality Assurance in Higher Education in Southern Africa: Challenges and Opportunities*," Oxford: Peter Lang.

Mowen, Thomas J. (2015). "Parental involvement in school and the role of school security measures." *Education and Urban Society, 47*(7): 830–848.

Mukhopadhyay, Moutan, Saurabh Pal, Anand Nayyar, Pijush Kanti Dutta Pramanik, Niloy Dasgupta, and Prasenjit Choudhury. (2020). "Facial emotion detection to assess learner's state of mind in an online learning system." In *Proceedings of the 2020 5th International Conference on Intelligent Information Technology*, pp. 107–115.

Muzaffar, Abdul Wahab, Muhammad Tahir, Muhammad Waseem Anwar, Qaiser Chaudry, Shamaila Rasheed Mir, and Yawar Rasheed. (2021). "A systematic review of online exams solutions in e-learning: Techniques, tools, and global adoption." *IEEE Access,* 9: 32689–32712.

Nayyar, Anand, and Akshi Kumar, (Eds.) (2020). *A Roadmap to Industry 4.0: Smart Production, Sharp Business and Sustainable Development*. Berlin: Springer.

Nayyar, A. N. A. N. D., R. U. D. R. A. Rameshwar, and A. R. U. N. Solanki. (2020). "Internet of Things (IoT) and the digital business environment: a standpoint inclusive cyber space, cybercrimes, and cybersecurity." *The Evolution of Business in the Cyber Age, 10*: 9780429276484–6.

Nguyen, Tuan. (2015). "The effectiveness of online learning: Beyond no significant difference and future horizons." *MERLOT Journal of Online Learning and Teaching, 11*(2): 309–319.

Nhan, Nguyen Thanh. (2020). "Theoretical framework for education distance."

Norman, Ian J., and Sally J. Redfern. (1993). "Quality assurance in nursing care: from inspection to opportunity." *International Journal of Health Care Quality Assurance, 6*(5): 5–13.

Nsiah, Gabriel Kofi Boahen, and Mable Oti-Boadi. (2015). "Assessing the effectiveness of distance education within the context of traditional classroom." *Creative Education, 6*(08): 707–710.

OECD. (2016). "The three dimensions of learning: Instructional leadership, by principals' training in instructional leadership, in lower secondary education: Average scores on the instructional leadership scale for principals who did or did not attend training or a course in instructional leadership". In *School Leadership for Learning: Insights from TALIS 2013*, TALIS. Paris: OECD Publishing.

Oliveira, Lídia, (Ed.) (2019). *Managing Screen Time in an Online Society.* IGI Global.

Pandey, Umesh Chandra, and Verlaxmi Indrakanti, (Eds.) (2017). *Open and Distance Learning Initiatives for Sustainable Development.* IGI Global.

Pereira, Andreia G., Tânia M. Lima, and Fernando Charrua Santos. (2020). "Industry 4.0 and Society 5.0: opportunities and threats." *International Journal of Recent Technology and Engineering, 8*(5): 3305–3308.

Plass, Jan L., and Ulas Kaplan. (2016). "Emotional design in digital media for learning." In *Emotions, Technology, Design, and Learning*, pp. 131–161, Academic Press.

Ramdass, Kemlall, and Fulufhelo Ṋemavhola. (2018). "Quality practices: An open distance learning perspective." *Turkish Online Journal of Distance Education, 19*(1): 234–246.

Raymond-Hayling, Olivia. (2020). "What lies in the year ahead for medical education? A medical student's perspective during the COVID-19 pandemic." *Medical Education Online, 25*(1): 1781749.

Ritzer, George. (2015). *Essentials of Sociology*, SAGE Publications.

Rogoff, Barbara, Maureen Callanan, Kris D. Gutiérrez, and Frederick Erickson. (2016). "The organization of informal learning." *Review of Research in Education, 40*(1): 356–401.

Rosário, Albérico Travassos. (2022). "The organizational impact of digital adoption: A literature review." In *Handbook of Research on Smart Management for Digital Transformation*, 27–52.

Roxå, Torgny, and Katarina Mårtensson. (2008). "Strategic educational development: A national Swedish initiative to support change in higher education." *Higher Education Research & Development, 27*(2): 155–168.

Sahito, Zafarullah, Sayeda Sapna Shah, and Anna-Marie Pelser. (2022). "Online teaching during COVID-19: exploration of challenges and their coping strategies faced by university teachers in Pakistan." In *Frontiers in Education*, vol. 7, p. 880335. Frontiers Media SA, 1–12.

SazmandAsfaranjan, Yasha, Farzad Shirzad, Fatemeh Baradari, Meysam Salimi, and Mehrdad Salehi. (2013). "Alleviating the senses of isolation and alienation in the virtual world: Socialization in distance education." *Procedia-Social and Behavioral Sciences, 93*: 332–337.

Scheunemann, Silvia, Anarosa Brandão, and Daniela Brauner. (2018). "Towards defining quality criteria for digital educational resources in distance learning." In 2018 *IEEE World Engineering Education Conference (EDUNINE)*, pp. 1–4. IEEE.

Schwab, K. (2018). "The Fourth Industrial Revolution (Industry 4.0) a social innovation perspective." *Tạp Chí Nghiên Cứu Dân Tộc, 7*(23): 12–21.

Sharin, Ana Nurdini. (2021). "E-learning during COVID-19: a review of literature." *Jurnal Pengajian Media Malaysia, 23*(1): 15–28.

Sharma, Sue Ann, Wendy C. Kasten, and Lynn A. Smolen. (2016). *Online Focus Groups: Investigating Diversity Training Issues in Literacy Teacher Education*. SAGE Publications Ltd.

Shattuck, Kay, (Ed.) (2023). *Assuring Quality in Online Education: Practices and Processes at the Teaching, Resource, and Program Levels*. Taylor & Francis.

Sileyew, Kassu Jilcha. (2019). Research design and methodology. *Cyberspace*, 1–12.

Soler-Gallart, Marta. (2018). "Pedagogy of the oppressed: Paulo Freire. Bloomsbury Academic, New York and London, 232 pp. Translated by Myra Bergman Ramos. With a new introduction by Donald Macedo and an afterword by Ira Shor. ISBN 978-1-5013-1414-8 (hbk), ISBN 978-1-5013-1413-1 (pbk), ISBN 978-1-5013-1416-2 (ePDF), ISBN 978-1-5013-1415-5 (ePUB)." (2023): 1–3.

Sovacool, Benjamin K., Jonn Axsen, and Steve Sorrell. (2018). "Promoting novelty, rigor, and style in energy social science: Towards codes of practice for appropriate methods and research design." *Energy Research & Social Science, 45*: 12–42.

Stambekova, Assel, Saule Zhakipbekova, Kunslu Tussubekova, Bagdat Mazhinov, and Mariya Shmidt. (2022). "Education for the disabled in accordance with the quality of inclusive education in the distance education process." *World Journal on Educational Technology: Current Issues, 14*(1): 316–328.

Stanton, Sue. (2001). "Going the distance: Developing shared web-based learning programmes." *Occupational Therapy International, 8*(2): 96–106.

Stavredes, Tina, and Tiffany Herder. (2013). *A Guide to Online Course Design: Strategies for Student Success*. John Wiley & Sons.

Stefanovic, Sandra, Urlica Dragoescu, Alina Andreea, and L. Kamberi. (2021). "The Use of E-learning Management Systems to Support English Language Learning at Engineering Study Programs: Quality and Learners' Satisfaction."

Strauss, Claudia. (1984). "Beyond" formal" versus" informal" education: Uses of psychological theory in anthropological research." *Ethos, 12*(3): 195–222.

Sun, Yuwei, and Hideya Ochiai. (2022). "Homogeneous learning: Self-attention decentralized deep learning." *IEEE Access, 10*: 7695–7703.

Taftaf, Randa, and Christy Williams. (2020). "Supporting refugee distance education: A review of the literature." *American Journal of Distance Education, 34*(1): 5–18.

Tripathi, Manorama, and V. K. J. Jeevan. (2009). "Quality assurance in distance learning libraries." *Quality Assurance in Education, 17*(1): 45–60.

Tudor, Sofia Loredana. (2013). "Formal-non-formal-informal in education." *Procedia-Social and Behavioral Sciences, 76*: 821–826.

Ullah, Shahnoor M., Andrew Bodrogi, Octav Cristea, Marjorie Johnson, and Vivian C. McAlister. (2012). "Learning surgically oriented anatomy in a student-run extracurricular club: An education through recreation initiative." *Anatomical Sciences Education, 5*(3): 165–170.

Vaughan, Norman D. (2010). "A blended community of inquiry approach: Linking student engagement and course redesign." *The Internet and Higher Education, 13*(1–2): 60–65.

Veletsianos, George. (2020). "How should we respond to the life-altering crises that education is facing?" *Distance Education, 41*(4): 604–607.

Veletsianos, George, Royce Kimmons, Ross Larsen, and Jacob Rogers. (2021). "Temporal flexibility, gender, and online learning completion." *Distance Education, 42*(1): 22–36.

Vlachopoulos, Dimitrios, and Agoritsa Makri. (2019). "Online communication and interaction in distance higher education: A framework study of good practice." *International Review of Education, 65*(4): 605–632.

Welch, Tessa, and Yvonne Reed. (2005). "Designing and delivering distance education." *Designing and Delivering Distance Education: Quality Criteria and Case Studies from South Africa. NADEOSA, Pretoria, South Africa.*

World Health Organization. (2015). "eLearning for undergraduate health professional education: A systematic review informing a radical transformation of health workforce development," World Health Organization, 1–6.
Yun, Jinhyo Joseph, MinHwa Lee, KyungBae Park, and Xiaofei Zhao. (2019). "Open innovation and serial entrepreneurs." *Sustainability, 11*, no. 18: 5055.
Zawacki-Richter, Olaf, and Adnan Qayyum. (2019). *Open and Distance Education in Asia, Africa and the Middle East: National Perspectives in a Digital* Age. Springer Nature.
Zuhairi, Aminudin, Maria Rowena Del Rosario Raymundo, and Kamran Mir. (2020). "Implementing quality assurance system for open and distance learning in three Asian open universities: Philippines, Indonesia and Pakistan." *Asian Association of Open Universities Journal 15*(3): 297–320.

APPENDIX I: CRITERIA FOR QUALITY ASSURANCE IN DISTANCE EDUCATION

These criteria can be used as an evaluation tool or the checklist as a benchmark or rubric to assess the quality of distance education. The criteria have five columns, i.e. serial number, criterion, indicators with statements, four levels of performance and evidence. There are eight criteria with statements to mark at an appropriate place found suitable for you and your system: **1** is **Unsatisfactory**; **2** is **Satisfactory**; **3** is **Good**; **4** is **Excellent**.

Criterion 1 – Curriculum and Course Quality						
S#	**Indicator(s) with Statement(s)**	**Performance Levels**				**Evidence**
		1	**2**	**3**	**4**	
01	The presentation is logical (e.g. sequential, nicely paced).					
02	The course structure is adaptable, allowing for simple updating of content in units, activities, assignments and learning materials.					
03	Links to library research databases are included in the course framework, as are links to LMS and associated websites.					
04	Where possible, the course is appropriately licensed.					
05	The course adheres to institutional rules or is consistent in terms of formatting, templates and so on.					
06	The course format allows for an appropriate combination of learning tools and teacher help.					
07	Following the first course, more changes are made.					
08	Standards for e-learning delivery are followed.					
09	The national curricular standards are followed.					
10	Materials and courses have been tested in the field.					

Criterion 2 – Content Quality						
S#	Indicator(s) with Statement(s)	Performance Levels				Evidence
		1	2	3	4	
11	The learning outcomes are clearly defined compared to academic/workplace standards.					
12	The level of learning outcomes is linked with the applicable learning framework.					
13	The detailed text and media aid in the comprehension of the subject topic.					
14	The material encourages student involvement as well as interaction with the teacher/tutor.					
15	The monitoring and evaluation criteria for the amount of study hours are satisfied as per approved policies.					
16	The course level is specified.					
17	The course content is verified/piloted/tested, up to date, and applicable to market/society demands.					
18	The content satisfies the standards for institutional and transfer credit.					
19	The references are trustworthy (i.e. correct information, live links), relevant, and updated.					
20	The contents/notes/reading materials are appropriately credited, referenced, and plagiarism-free.					
21	Content complies with international or national content requirements.					
22	Writers and designers are taught to a certain standard.					
23	Qualitative editing and review, language and design are used to validate the content.					
24	Students and teachers pilot materials, and adjustments are made based on their comments.					
25	Final revisions include quality assurance.					
26	Copy editing was performed on the material to ensure accuracy, typographical errors, format, style, substance and functioning links.					

Criterion 3 – Pedagogical Quality						
S#	Indicator(s) with Statement(s)	Performance Levels				Evidence
		1	2	3	4	
27	Learners participate in a wide range of learning activities.					
28	The learning activities and educational resources are related to the learning outcomes.					
29	Pedagogical resources help students attain the desired learning goals.					
30	Active learning tactics are used to engage students and encourage them to accomplish the intended learning goals.					
31	Learning activities promote active learning by providing chances for contact (student-student, student-tutor and with materials/technology).					
32	There is access to external programmes/ software.					
33	The course incorporates media with care (e.g. brief movies and a range of interactive elements).					
34	The LMS is user-friendly (as few as possible clicks and external websites or resources to access).					
35	There are instructional exercises meant to encourage pupils to think critically.					
36	Tutoring system is in place to help struggling learners.					
37	Instructors, tutors, mentors, coaches, course designers, and student support workers all get ongoing professional development.					
38	Audio, video, and instructional texts are sampled to ensure quality, then amended and confirmed.					
39	Instructors receive training in the distant education media that they will be using.					
40	Instructors have participated in the distant education medium in which they will be teaching as students.					

Criterion 4 – Learner Support System						
S#	Indicator(s) with Statement(s)	Performance Levels				Evidence
		1	2	3	4	
41	The course provides clearly defined instructions for learners to follow in order to complete the course requirements (e.g. activities, assignments) as well as the course learning objectives.					
42	Tutors have received necessary training and certification to administer the blended course.					
43	Students are given an introduction to the course's online components as well as registration options.					
44	There are certain guidelines for how and when students should anticipate feedback.					
45	There are guidelines for navigating the course website.					
46	There are instructions on how to navigate the course site.					
47	There are guidelines for how learners should interact with a blended learning course, as well as what percentage of the course should be online and face-to-face.					
48	Materials should be opened in the course site rather than as popup windows, etc., if possible (i.e. separate Word docs).					
49	There is sufficient indication of where materials should be used online and in person.					
50	There is a help option to assist learners with FAQs and other resources.					
51	A proper selection procedure is followed.					
52	Distance learners benefit from mentoring, coaching and support services.					
53	Registration and examination procedures are straightforward and available in a variety of formats (electronic and paper based).					
54	Distance education is an essential component of teacher education programmes and is not seen as "inferior" or auxiliary.					
55	Multiple assessment styles and formats (performance-based, Web-based, portfolio-based and so on) encourage a shift away from sit-down examinations that focus on declarative knowledge.					

Criterion 5 – Technological Support						
S#	Indicator(s) with Statement(s)	Performance Levels				Evidence
		1	2	3	4	
56	In terms of infrastructure and learner access, the selection of technical tools has been carefully studied.					
57	The chosen technologies facilitate the execution of tasks that improve learning outcomes.					
58	The LMS allows users to download resources for offline use.					
59	Learners can interact and cooperate thanks to technology.					
60	The course website is device and browser independent (i.e. operational on mobile devices, multiple browsers).					
61	Learning objectives and material are created first, and then technology is chosen.					
62	The technology used is the most effective way of providing knowledge and training to that specific teacher-learner audience.					
63	Teacher-learners understand how to utilize the technology chosen (radio, CD players, laptops, MP3 players, etc.).					
64	Teacher-learners are instructed on how to learn with and through technology (beyond simple technical operations).					
65	Students are taught how to utilize and manage databases/EMIS.					

Criterion 6 – Assessment Mechanism						
S#	Indicator(s) with Statement(s)	Performance Levels				Evidence
		1	2	3	4	
66	Learners are informed about how they will be evaluated during the course.					
67	There are marking guides for papers and exams.					
68	Formative and summative assessments are examples of assessment measures.					
69	Assessment tools can help with self-evaluation and feedback.					
70	Assessment tools are used to assess mastery of learning outcomes.					

(*Continued*)

Criterion 6 (*Continued*) – Assessment Mechanism						
S#	Indicator(s) with Statement(s)	Performance Levels				Evidence
		1	2	3	4	
71	A student's "readiness" to be a distant learner is determined via a diagnostic evaluation.					
72	Formative assessment identifies learner issues and provides support or remediation.					
73	Alternative types of teaching and summative assessment accommodate learning styles and learner aptitudes while also providing a valid method of evaluation of the learner's entire range of skills.					

Criterion 7 – Quality Assurance and Evaluation						
S#	Indicator(s) with Statement(s)	Performance Levels				Evidence
		1	2	3	4	
74	The course has been/is being conducted to the same quality assurance methods as a face-to-face course.					
75	Plans and timelines for reviewing online courses are in place to guarantee that the provider keeps up with changes in technology and content.					
76	Learners can submit comments on the blended learning experience, including the course content, through course assessment.					
77	Quality assurance and control mechanisms are in place and being applied.					
78	The materials are aesthetically appealing, and they adhere to excellent design and layout standards.					

Criterion 8 – Organizational Management, Administration and Leadership						
S#	**Indicator(s) with Statement(s)**	**Performance Levels**				**Evidence**
		1	**2**	**3**	**4**	
79	It is necessary to employ basic databases or an EMIS.					
80	The database/EMIS information is regularly updated.					
81	Those in need have access to relevant information.					
82	Additional means (email, SMS, messaging, automated phoning, and so on) are used to disseminate information from EMIS to learners.					
83	Material distribution and delivery are closely controlled and overseen.					
84	Someone or something is in charge of distributing learning materials and is held accountable.					
85	All materials, equipment, and so on are ready and available or delivered to learners prior to the start of distance learning courses.					
86	The stated policies of educational institutions emphasize the value of distant education programmes.					
87	The institution creates a vision statement and learning objectives, which serve as the framework for the distant learning programmes.					
88	Policies are backed up by processes (training, support, materials, resources, and technology) to ensure the quality of distant learning programmes.					
89	All levels of education promote a culture of continuous improvement in the effectiveness and efficiency of all aspects of distance learning.					
90	Vision statement and learning objectives are developed by institution and provide foundation for distance learning programme.					
91	Problems are not ignored or buried, but rather confronted and resolved.					
92	Top management is active in and dedicated to the remote learning programme in general, as well as excellent distance education processes in particular.					

Thanks for Completely Filling up the Form

Demographic Information

Name	
Programme	
Level	
Organization	
Country	
Email	
Address	
Cell/Telephone	

10 Applying Artificial Intelligence to Remote Students Learning Assessment in Higher Education

Toward a New Approach Based on Information Extraction Model

G. Faouzi, N. Amrous, N. El Faddouli, and M. Khabouze

10.1 INTRODUCTION

In the era of the coronavirus pandemic, higher education is undergoing a radical transformation. Most universities around the world have suspended or cancelled on-campus activities such as lectures, workshops, sports and conferences due to the outbreak of the virus. Following the logic of the exception, Morocco's Ministry of Education has announced the closure of all educational establishments and has adopted e-learning as a pedagogical tool in place of face-to-face teaching (Faouzi et al., 2022). This radical change of the educational system mainly involves the switch from face-to-face to online teaching, and the integration of information and communication technologies (ICTs) to ensure pedagogical continuity.

Higher education institutions worldwide face many challenges in their teaching activities, such as educational assessment, which is the process of acquiring information that can be used to evaluate students and make educational decisions. It has also posed many challenges and difficulties due to the large number of students. In addition, online assessment, which provides computerized, comprehensive and formative feedback that can support the learning process and help the learner to self-assess and improve learning outcomes while preparing for summative assessment, is being adopted throughout the pandemic period; to assess students and make decisions about their performance, a variety of methods and approaches are adopted, including

 DOI: 10.1201/9781003400691-10

a combination of techniques such as traditional assessment methods, like quizzes and exams, or authentic ones (such as e-portfolios, online journals, group work, etc.).

In this study, the challenges faced by higher education institutions when implementing distance assessment were investigated, using Mohammed V University in Rabat as a case study. For that, a questionnaire was prepared and data were collected by integrating questions relating to distance assessment into the survey.

Faced with the different challenges encountered, especially technical problems and internet outages, students find themselves in a state of uncertainty regarding the exams; the aim of this contribution is to analyze the different approaches to distance assessment adopted by Mohammed V University, and to propose a better model of student assessment based on IEM that determines the keywords that should appear in students' responses, which can contribute to self-assessment as it is considered an effective means of increasing the quality and effectiveness of assessment and building self-confidence among students.

For this purpose, this chapter firstly presents the related works carried out around online assessment. Secondly, methods used are cited, the third section is about results founded, the fourth section presents the proposed model, section five concerns discussion and finally the conclusion.

10.2 LITERATURE REVIEW

Technology has made an impact on almost every aspect of life today, and education is not an exception. The use of virtual classrooms has become a necessity to ensure distance learning. In spite of that Martin-Somer (2024) mentioned in his paper that students and teachers prefer face-to-face learning due to greater interaction, but this can be enhanced in online teaching by using interactive applications that help to maintain attention and the feedback of teachers and students, who should be motivated to promote distance learning (Mendoza et al., 2023).

The Covid-19 pandemic is an added factor that can exacerbate the challenges faced by students. Many studies have analyzed the impact of this pandemic on students' learning. For instance, Hantem (2020) in his paper exposed the conditions including tools during the period of quarantine due to the Covid-19 pandemic. This study revealed that most of the students were very worried about the pandemic and were afraid of failing their exams. Moreover, it sheds light on the teacher's efforts through the design of videos, knowing that a large portion have not been prepared. The major difficulties observed in distance learning in Moroccan higher education can be summed up in the fact that students are not familiar with the use of ICTs, as they do not always have the means to attend online classes: internet connection, smart phones, IT skills, etc. Indeed the students are bombarded with the tasks that need to be completed.

To counteract these problems, it is proposed to strengthen communication between teachers and students, use other platforms, avoid overloading the students and review the pace of work to be completed (Joahi et al., 2020).

In the context of Moroccan higher education and precisely in Mohammed V University in Rabat, Elmendili and Saaidi (2020) present an overview of distance education practices in the period of quarantine; the inventory includes various systems put in place for distance university education, such as Google Meet, Google classroom, Microsoft Teams, Social networks, etc. The difficulties encountered are those related to contact with teachers, technical problems, difficulties in accessing the different platforms, etc.

The studies conducted revealed a general overview about the shift from face-to-face education to online education; hence there is the necessity of discussing the case of online assessment, tools used, opportunities and challenges faced by both students and teachers. Many studies have examined remote assessment in the era of Covid-19 pandemic, approaches adopted, advantages and disadvantages of those practices and students' perceptions of assessment quality. In Senal and Senel's (2021) article, the authors exposed the important role of valid and reliable assessment results that are necessary to attend the educational goals.

Senel et al. confirm that among the approaches adopted in Turkey during this period there are online tests, assignments and projects. Results indicated that performance-based tools are mostly used and participants reported positively about the quality of online assessment; however, they reported negatively about just two items: the insufficiency of the use of rapid assessment and feedback. In Guangul et al.'s (2020) paper, the authors mentioned a study conducted where the infrastructure challenges, coverage of learning outcomes and commitment of students to submit assessments were particularly highlighted. On the other hand, it is proposed to adopt various assessment approaches and methods to remedy the above-mentioned challenges. In choosing the type of assessment, it is recommended to take into consideration some criteria in order to ensure the online assessment's objectives like validity, reliability and clarity. In this study there are two types of method assessment: question-based assessment (maximum 3 hours) and project-based assessment (from one to four weeks). In Saudi Arabia, the online assessment also faced many challenges such as lack of infrastructure, achievement of the learning outcomes and students' commitment to submitting assessments; therefore, staff need to be trained to use alternative assessment methods, which should be a part of the regular teaching plan rather than used only in emergency situations (Almossa , 2021).

In the faculty of medicine in Jordan, e-exams were adopted as a mode of assessment during the Covid-19 pandemic, and despite its challenges illustrated by inappropriate exam questions, efforts/time needed for exam preparation, etc., there was the need to use different exam forms, limit the exam duration and apply one-way navigation mode to reduce those problems (Elsalem et al., 2021). To support teachers, students and universities in online assessment process who weren't fully prepared for the pandemic, Gracia-Penalvo et al. (2020) proposed these recommendations: assessment of subjects continuously, use technologies already available at the university and those usually used, individual oral tests may be conducted and the casuistry of the students must be known and taken into account. This chapter considers two initial categories of online assessment: synchronous assessments that contain academic thesis dissertations and working on groups, and asynchronous ones with assessment of knowledge where self-assessment questionnaire tools were preferred and activities assessed based on defining clear rubrics or scales of assessment to be shared with students. In the United Kingdom, the typical online assessment methods used were online quizzes, continuous feedback, multiple-choice questions and automated assessment for essays, in this context (Tuah and Naing, 2021). It is proposed to consider the other alternative assessment options before examination (writing reports, for example).

All those works focused only on the analysis of general situation, tools used, challenges related to lack of infrastructures, technical problems, internet, etc., and proposed some recommendations.

Throughout the crisis, the students were afraid to sit for their exams due to many challenges; in our research we proposed a model to support learners and build student's self-confidence, which can be an added solution to the above-mentioned recommendations.

10.3 METHODOLOGY

This study is conducted to explore students' answers to a survey concerning their engagement with online learning and assessment during Covid-19, in order to highlight challenges that affected students' educational life and remote assessment.

The quantitative approach was used in this research study, employing the questionnaire instrument. Participants were asked to answer questions related to the research topic, enabling us to gain a comprehensive picture of the factors influencing student satisfaction with online learning and assessment.

Students were selected from Mohammed V University in Rabat, including students from the following: Faculty of Educational Sciences, Faculty of Dental Medicine, Faculty of Medicine and Pharmacy, Faculties of Legal and Social Economics Souissi, Faculties of Legal and Social Economics of Sale, Faculties of Legal and Social Economics Agdal, Higher Normal School for Technical Education, Faculty of Sciences of Rabat, National Higher School of Computer Science and Systems Analysis, Normal Higher School of Rabat, Higher School of Technology of Sale, Faculty of Letters and Human Sciences of Rabat, Faculty of Sciences, Mohammadia School of Engineers, Scientific institute, University Institute of Scientific Research, Institute of African Studies and Arabization Studies and Research Institute. The students were asked about evaluation methods used, quality of courses, level of satisfaction, problems encountered and the ability to carry out work requests (projects, etc.).

Table 10.1 shows the participants' details in order to describe the target students.

10.4 RESULTS

The adaptation of e-learning makes it possible to better organize educational activities and gives students more opportunities to learn on their own. Since online assessment is a pedagogical activity and part of adaptive education, which refers to a digital testing method where students can take exams, tests or assessments through a device connected to the internet, it must be discussed.

According to Mario Pezzino, online assessment, particularly multiple-choice question (MCQ)-based tests, provide valuable practice, especially when accessed multiple times, and helps reinforce classroom learning. They can also be used as a formal assessment within a module, if the quiz randomly selects questions from a set of questions. It can also provide students with immediate feedback (Pezzino, 2018).

TABLE 10.1
Participants' Details

Student's Sex	Age of the Participants	Level of Studies	Field of Studies
Women 74%	18 to 28 years old: 78%	Doctorate (12%)	Economic, Legal and Management Sciences (68%)
Men 26%	28 to 38 years old: 18%	Masters (76%)	Health Sciences (8%)
	Over 38 years old: 4%	Bachelor's Degree (6%)	Science and Technology (6%)
		Engineering (4%)	Educational, Human and Social Sciences (6%)
		DEUG (2%)	Finance (2%)
			Computer Science (2%)
			Engineering (2%)
			Strategic Human Resources Management (2%)
			Other Specialties (Architecture, etc.) (4%)

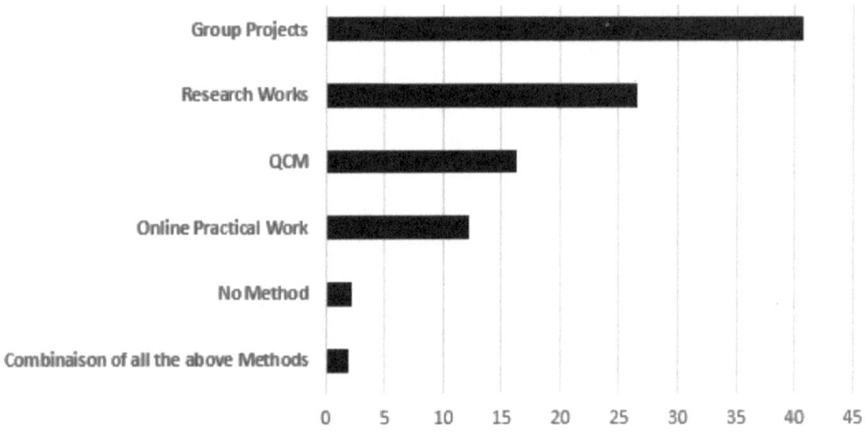

FIGURE 10.1 Methods used for remote assessment in Mohammed V University during Covid-19.

A description of the online assessment situation during quarantine in Mohammed V University follows:

The participants mentioned several assessment methods that they used during this remote teaching experience. These assessment methods are from the most mentioned method to the least mentioned method as shown in Figure 10.1.

- Group project: This was the most mentioned method in the survey where the students are grouped to realize projects in different subjects.
- Research works : Several teachers used this method to assess their students by assigning research topics to students.
- MCQs

- Online practical work
- No method: response written as "Others", who are not concerned with taking the exams.

The duration of exams as shown in Figure 10.2 differs from 1 hour to more than 2 hours, and there is a category "no exam" for doctorate students.

As far as the ability to put acquired theoretical knowledge into practice is concerned, the majority of students were able to put their theoretical skills into practice, demonstrating the success of the assessment methods used, as illustrated in Figure 10.3.

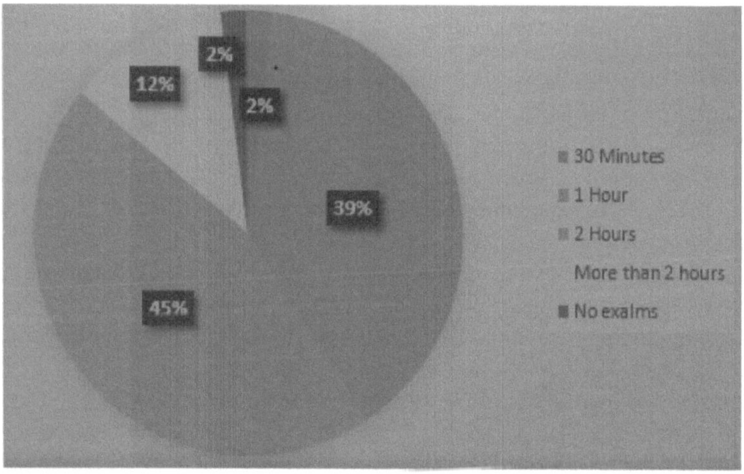

FIGURE 10.2 Duration of online exams during covid-19.

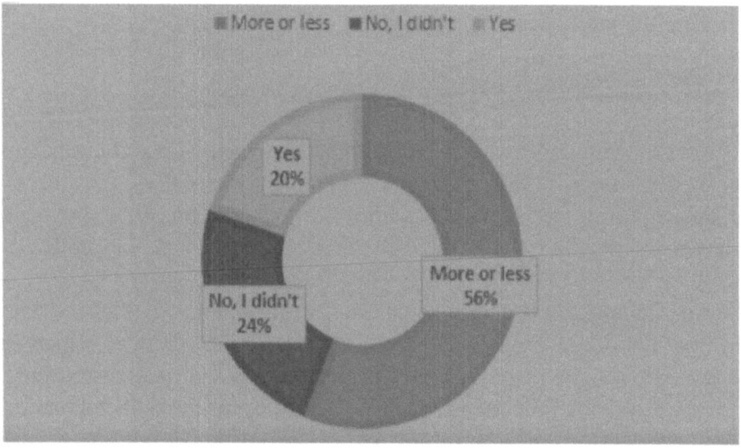

FIGURE 10.3 Implementation of theorical skills in practice.

TABLE 10.2
Comparing Online and Face-to-Face Assessment

Online Assessment Methods	Face-to-Face Assessment Methods
Individual work sent by email (46%)	Case studies (28%)
Contribution to focus groups (30%)	Team works (22%)
Automated tests (12%)	Oral exams (12%)
Oral or written exams in the presence of a supervisor or by videoconference (10%)	Practical works (14%)
	Projects (22%)
MCQs (2%)	Other methods (2%)

Regarding interactions, 72% of the students have the opportunity to participate in the course during distance learning, 88% of students are able to realize the projects requested by their educators, 16% of students are very satisfied about the online assessment, 52% are satisfied and 32% are not satisfied.

In general, the last questions show the success of distance assessment during Covid-19, despite some problems that are as follows: technical problems, accessibility issues, variable involvement of teachers, not receiving the exam link, internet issues and not receiving materials.

Another point concerns the methods used in face-to-face and distance learning, with the percentage of students' preference, as seen in Table 10.2.

Finally, the students formulated the following proposals to overcome this situation:

- Improve the quality of distance learning.
- Take advantage of the experience gained during the health crisis to integrate the option of hybrid teaching into higher education, while maintaining face-to-face assessment to ensure fairness.
- Train teachers to use the resources and tools available on distance learning platforms.
- Develop the level of distance learning.

10.5 THE PROPOSED MODEL (IEM)

The proposed model comprises three essential stages, the first is the acquisition of documents published by teachers, then the comprehension of those documents starting with the import of Python modules until the construction of the dictionary and finally database query by students through the interface to check the validity of their answers.

The desire to make students responsible for the quality of their education is reflected in a number of proposed models, such as self-assessment, which enables each student to highlight his or her educational strengths and weaknesses, but before letting students evaluate themselves, course objectives must be set before assessment begins, and the learners and learning rhythms should be clarified by teachers, as mentioned in Table 10.3:

TABLE 10.3
Fundamental Concepts Considered in the Evaluation Model

Objectives	Learners Need	School Level	Learning Rhythms
What should be memorized and appear in students answers?	– The learners goal – Needs due to the lack of resources or technical problems	Master Bachelor's Degree Engineering DEUG[a]	Ability to understand Problems that increase the pace of learning for students

[a] DEUG (Diploma of General University Studies).

The second task is to define the roles of teachers and students:
For teachers:

- Providing resources;
- Preparing the exams;
- Delimiting answers and the main keywords;
- Extracting keywords.

For students:

- Answering the exams;
- Then verify the compatibility between answers and the corrections provided by the teachers by using the system proposed.

The model limits the lexical field of keywords that must appear in the students' answers; after the questions have been prepared by teachers, and the exams distributed in the e-database, the students are led to give answers on the field. After answering the questions, the system confirms the validity of the question (true/false) using the corpus extracted from the keywords in the course materials, as shown in Figure 10.4.

In our study, we opted for the textual IEM proposed by Tang et al. (1996) as shown in Figure 10.5, which is comprised of the following elements:

- A document has two structures: a geometric (or layout) structure and a logical structure;
- Processing a document involves two phases: document analysis and document comprehension;
- Document analysis consists in extracting the geometric structure of a document;
- Document comprehension consists in transforming the geometric structure of a document into its logical structure; once the logical structure has been established, its meaning can be decoded by artificial intelligence or any other techniques.

The model for document processing is described (Tang et al., 1996) as follows:

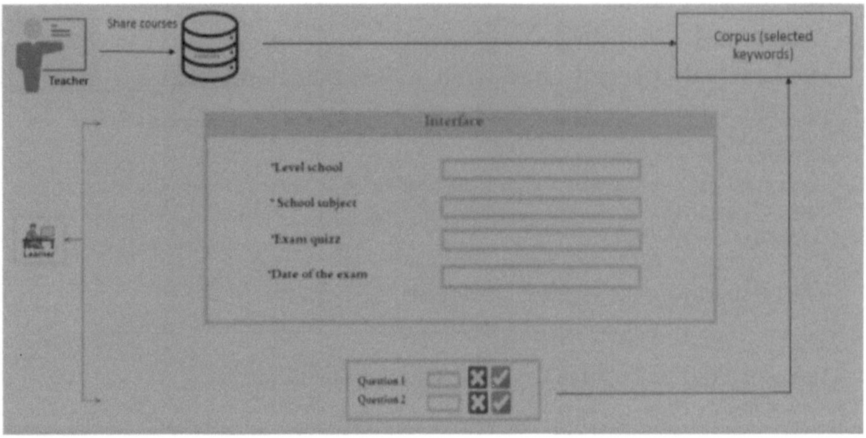

FIGURE 10.4 Students evaluation process.

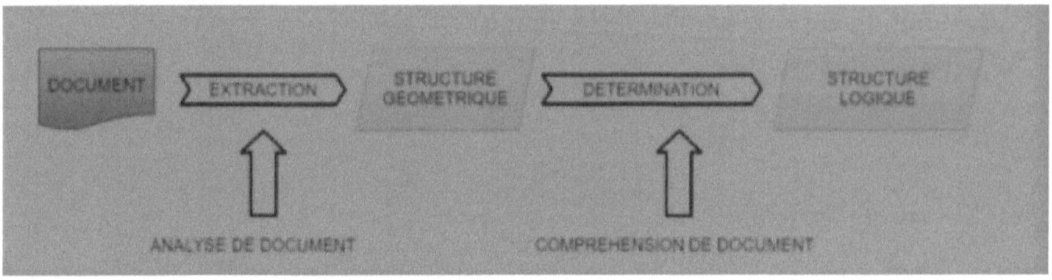

FIGURE 10.5 Shematic diagram of the basic document processing model proposed by Tang et.al.

Definition 1.

A document Ω is defined by a quintuplet

$$\Omega = (T, \Phi, \delta, \alpha, \beta) \quad (1.1)$$

$$\text{such that } T = \{\Theta^1, \Theta^2, ..., \Theta^i, ..., \Theta^m\}, \quad (1.2)$$

where $\Theta i = \{\Theta^i{}_j\}^*$

$$\text{and } \Phi = \{\alpha^1, \alpha^r\} \quad (1.3)$$

$\delta = T \times \Phi \rightarrow 2^T$

$\alpha = \{\alpha^1, \alpha^2, ..., \alpha^p\} \subseteq T$

$\beta = \{\beta^1, \beta^2, ..., \beta^q\} \subseteq T$,

T: is a finite set of document objects, these objects are sets of blocks Θ^i ($i = 1, 2, ..., m$);

$\{\Theta^i{}_j\}*$: denotes repetition of subdivisions;

Φ: is a finite set of linking factors;

α^l and α^r: represent bonds and bond repeats respectively;

δ: denotes a finite set of logical linking functions indicating a logical linking between document objects;

α: is a finite set of root objects;

β: is a finite set of final objects.

Definition 2.

Document processing consists in constructing the quintuplet represented by equations 1.1 to 1.3.

Document analysis consists in extracting the elements T, Θ^i and Θ^i_j from equation 1.2, in other words, extracting the geometric structure of Ω.

Document comprehension involves determining Φ, δ, α and β from equation 1.3 by considering the logical structure of Ω.

The document structure mentioned above (Pitou, 2017) according to the international standard ISO/CEI 8613-1:1994 is the successive division and subdivision of a document's content into progressively smaller areas called objects. An object that cant' be subdivided into sub-objects is called a base object. It is a combination of logical structure, which refers to the semantic properties of the document, and geometric structure, which signifies the geometric characteristics of a document.

Document processing involves dividing the document into homogeneous regions or blocks (Figure 10.6; text, graphics, tables, etc.) and then into lines of text and characters within these known as segmentation. This process extracts the document's geometric structure, which can be represented as the document's logical structure (Kreich et al., 2017).

As introduced before, the artificial intelligence technology can be used in the document comprehension phase by introducing algorithms that will help us to highlight

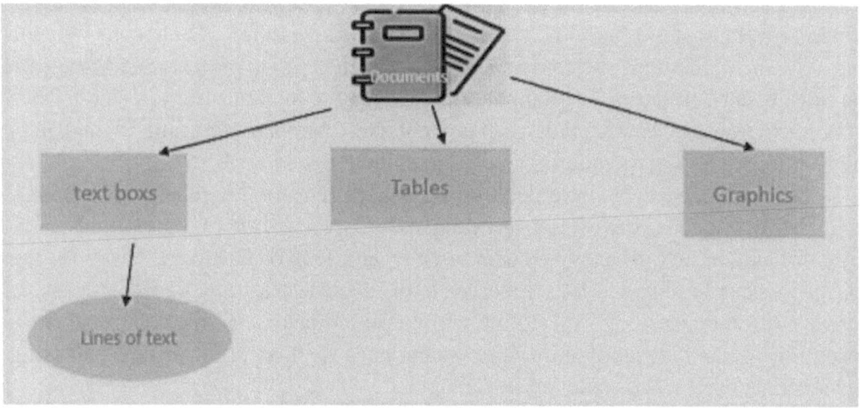

FIGURE 10.6 Document structure.

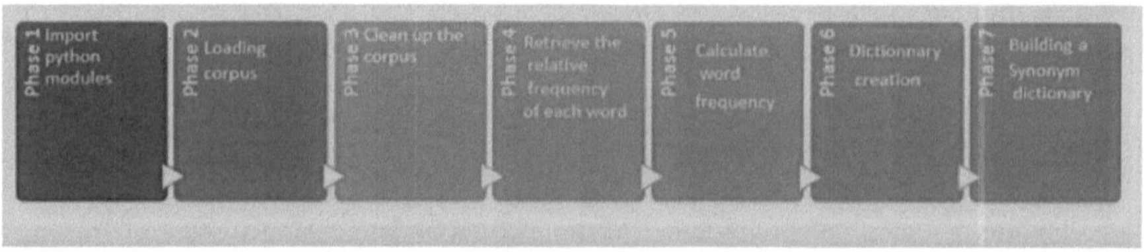

FIGURE 10.7 Document comprehension phase.

the keywords. To realize the work, the following procedure was adopted as shown in Figure 10.7.

- To begin with, several Python modules need to be imported such as: os (for file management), pandas (array management), numpy (matrix and number management), string (character management), unicodedata (management of special characters), etc.;
- Loading a corpus;
- Clean up the corpus by deleting unwanted characters;
- Retrieve the relative frequency of each word;
- Use word frequency to define a list of words that are too rare or too common;
- Create your own dictionary of important words;
- Build a synonym dictionary to group together all related terms.

10.6 DISCUSSION

The pandemic has affected all sectors, including education, because it is changing the way we teach. Since the start of the pandemic, the paradigm and way of teaching and learning have been changed, and this calls for urgent intervention to ensure pedagogical continuity, on the one hand, and on the other, distance learning is attracting more and more interest due to the strong evolution of learning environments.

In this research, we sought to examine distance assessment and students' opinions of the experience by asking research questions. Students who responded to the questionnaire were positive about the quality of distance assessment. The results indicate that assessment methods such as research assignments, group projects and MCQs are the most widely used tools. A comparison between the methods used in face-to-face and online assessment was made with percentages (obtained according to the study) in order to choose the method best suited to student assessment.

In fact, several articles have dealt with distance learning in general (AL-Nasa'h et al., 2021), which addresses two key points: the identification of discriminant indicators of online self-efficacy, general anxiety and fear of Covid-19, based on the ability to classify students into three levels of distance learning, and how learners perceive distance learning satisfaction with these indicators. In this context we proposed this model to discuss online assessment because it is part of distance learning and students experience some anxiety about it.

This model offers the opportunity for students to evaluate themselves by providing the answers to the elements that are mentioned; moreover, this mode of assessment allows students to complete the task when they wish, while reducing technical and network constraints and problems. It also helps to build self-confidence and ensures equal opportunity.

10.7 CONCLUSION

This research study was carried out to assess the challenges of distance assessment in Moroccan higher education, taking the example of Mohammed V University in Rabat, and to propose a model based on self-assessment to remedy, in particular, the technical problems encountered in online examinations, which can be used to assess students' skills and abilities acquired through distance learning.

This study contributes to current research on online assessment with the following findings. Firstly, it can be a method of avoiding technical problems encountered when accessing an online exam. Secondly, it can be used to identify the learner's weaknesses and to seek to boost self-confidence.

In the future works, we suggest to apply the proposed model to a specific class of students and a specific topic, while specifying text-mining algorithms to extract keywords and compare the best algorithm taking into consideration: accuracy, score, average, etc.

REFERENCES

Samar Yakoob Almossa (2021). "University student's perspectives toward learning and assessment during COVID-19." *Educ Inf Technol (Dordr)*. Volume 26, 7163–7181, https://doi.org/10.1007/s10639-021-10554-8

Mais AL-Nasa'h, Luae Al-Tarawneh, Ferial M. Abu Awwad and Ikhlas Ahmad (2021). "Estimating students' online learning satisfaction during COVID-19: A discriminant analysis". *Heliyon*. Volume 7, Issue 12, https://doi.org/10.1016/j.heliyon.2021.e08544

Soumaya Elmendili, Sara Saaidi. (2020). "Les pratiques de l'enseignement à distance dans l'université marocaine à l'ère du coronavirus: Cas de l'université Mohammed V de Rabat". *The Journal of Quality in Education (JoQiE)*. Volume 10.

Lina Elsalem, Nosayba Al-Azzam, Ahmed A. Jum'ah, Nail Oeidat. (2021); "Remote E-exams during Covid-19 pandemic: A cross-sectional study of student's perferences and academic dishonesty in faculties of medical sciences". *Annals of Medicine and Surgey*. Volume 62, 326–333. https://doi.org/10.1016/j.amsu.2021.01.054.

Ghita Faouzi, Naila Amrous and Nour-eddine El Faddouli. (2022). "Factors Influencing Students' Satisfaction in Learning with Virtual Platforms: Case of Higher Education in Morocco," *Digital Technologies and Application*. pp. 1–2. https://doi.org/10.1007/978-3-031-01942-5_37

Fransisco José Gracia-Penalvo, Alfredo Corell, Victor Abella-Garcia, and Mario Grande-de-Padro (2020). "Recommendations for Mandatory Online Assessment in Higher Education During the COVID-19 Pandemic". https://doi.org/10.1007/978-981-15-7869-4_6.

Fiseha M. Guangul, Adeel H. Suhail, Muhammad I. Khalit, Basim A. Khidhi. (2020). "Challenges of remote assessment in higher education in the context of COVID-19: A case study of Middle East College". *Educational Assessment, Evaluation and Accountability*. Volume 32, 519–535.

Aziz Hantem. (2020). "Les conditions de l'enseignement à distance pendant le confinement du à COVID19: Cas de l'enseignement supérieur au Maroc", pp. 4–7.

Anuradha Joahi, Amrit Virk, Shaista, Rajiv Mahajan and Tejinder Singh. (2020). "Online Assessment; Concept and Applications". *Journal of Research in Medical Education and Ethics*, 49–59. https://doi.org/10.5958/2231-6728.2020.00015.3

Joachim Kreich, Achim Luhn, Gerd Maderlechner, "An experimental environment for model based document analysis". In Proceedings of International Conference on Document Analysis and Recognition.

Miguel Martin-Somer, Cintia Casado, Gema Gomez-Pozuelo. (2024). "Utilizing interactive applications as educational tools in higher education: Perspectives from teachers and students, and an analysis of academic outcomes". *Education for Chemical Engineers*, Volume 46, 1–9.

Norman B. Mendoza, Zi Yan, Ronnel B. King. (2023). "Supporting Students' intrinsic motivation for online learning tasks: The effect of need-supportive task instructions on motivation, self-assessment, and task performance". *Computers & Education*. https://doi.org/10.1016/j.compedu.2022.104663

Mario Pezzino (2018). "Online assessment, adaptive feedback and the importance of visual learning for students. The advantages, with a few caveats, of using MapleTA". *International Review of Economics Education*, https://doi.org/10.1016/j.iree.2018.03.002

Cynthia Pitou (2017). "Extraction d'informations textuelles au sein de documents numérisés: cas des factures".

Selma Senal, Huseyin Can Senel. (2021). "Remote Assessment in Higher Education during Covid-19 Pandemic". Volume 8, Issue 2, 181–199, https://doi.org/10.21449/ijate.820140

Yuan Y. Tang, Seong Whan Lee, & Ching Y. Suen. (1996). Automatic document processing: "A survey." *Pattern Recognition*, Volume 29, Issue 12, 1931–1952.

NikAni Afiqah Tuah, Lin Naing. (2021). "Is Online Assessment in Higher Education Institutions during COVID-19 Pandemic Reliable?".

11 Zoom on the Boom

A Perspective of South East Asian ESL Teachers after Covid-19 Situations

Muhammad Mooneeb Ali

11.1 INTRODUCTION

The outbreak of COVID-19 pandemic along with its variation like Delta and then Omicron has impacted humans badly. Due to this life-threatening virus, there was a complete lockdown and resultantly a radical shift was observed towards online learning and teaching. In the Western world the use of information and communication technologies (ICTs) and other technological tools in education are a trend but it is rare in South East Asia (Zhang, 2021). Therefore, this study becomes significant to investigate the perspectives of English as a Second Language (ESL) teachers of the three major countries present in South Asia regarding technology use in English language classrooms.

11.1.1 INDIAN, PAKISTANI AND SRI LANKAN ENGLISH CLASSROOMS

The need of English has meaningfully amplified its position worldwide. English is accepted and understood by majority of the nations for trading, social connectivity, diplomacy, learning, teaching etc. (Ilyosovna, 2020). Further, in education, apart from English as native language there is a huge demand of ESL teaching and learning. Therefore, for immigration, admissions, jobs acquisition and many other opportunities English language learning is required (Tan & Yunus, 2023).

In South-East Asia, English enjoys high prestige in India, Pakistan and Sri Lanka both socially and academically (Gargesh, 2019). The classrooms in all the three countries have some common aspects like focusing on conventional curriculum that lacks latest techniques. Classrooms are teacher-centered while grammar-translation method is used (Kalyanpur et al., 2022). Rote learning and memorization are encouraged. Teachers are unfamiliar with modern teaching techniques, tools and apps (Gamage & Dilani, 2022) (Figures 11.1–11.3).

Furthermore, the learners are higher in ratio as compared to the teachers, which creates ineffective teaching. Though the demand for English in formal contexts is on the higher side, the overall ambiance is not encouraging for learners for learning and speaking English (Quyyum, 2021). Another aspect is the mother tongue influence that creates issues in English language learning.

DOI: 10.1201/9781003400691-11

FIGURE 11.1 Traditional Indian classroom.

FIGURE 11.2 Traditional Pakistani classroom.

11.1.2 Zoom Application as MALL Tool

Technology has added a new strip in the field of education which is known as MALL (Mobile-assisted language learning). MALL is described as the exploitation of mobile devices for language learning (Metruk, 2022). MALL defines the process and advantages of mobile phone usage for ESL learning. The utility of MALL has become indispensable in classrooms. A traditional ESL class is time and place

FIGURE 11.3 Traditional Sri Lankan classroom.

bound, whereas a MALL classroom is always flexible in place as well as time. On the other hand, any portable gadget that helps in language learning procedures is called as MALL (Sianturi & Hung, 2022). MALL features many latest applications that are enormously popular among students. The latest invasion of Web tools 2.0 has given way to purposefully developed apps that can run equally on mobile phone as well as computers (Wulansari et al., 2022). Since COVID-19 pandemic, the use of apps like Google Classrooms, Microsoft Teams and Zoom are on the rise. Among them Zoom is the most popular app used precisely in English language teaching and learning (Silalahi et al., 2022). Zoom app is developed in a way that it has features of audio, video, screen sharing, messaging and presentation. Though Zoom originated in 2011, it gained popularity only from 2020 with the outbreak of the pandemic.

11.1.3 Research Questions

1. What are the perceptions of South East Asian teachers regarding Zoom incorporation as a tool for MALL in Indian, Pakistani and Sri Lankan ESL classrooms?
2. Do the teachers find Zoom a beneficial MALL tool in their teaching scenarios after COVID-19?

11.1.4 Significance of the Study

This research will open new corridors of innovative learning ways through MALL apps. It will try to take opinions of teachers in all these three countries so that educational environments can be effectively transformed. This study will have implication for learners and other teachers as it will encourage both of them to use Zoom as a supportive tool to improve their teaching and learning. For the policymakers in India, Pakistan and Sri Lanka this would be one of the baseline investigations that explores the value of Zoom as MALL tool in the formal language learning process after COVID-19. This study would be significant to diminish the influence of traditional learning that has become ineffective in modern times. This will result in the change in the concept of ESL teaching-learning settings in South-East Asian countries. For the policymakers this study will be important to consider the opinions of teachers who are an important pillar in education.

11.2 RECENT LITERATURE

11.2.1 Mobile-Assisted Language Learning

The effectivity of MALL in ESL setting is indisputable. Karakaya and Bozkurt (2022) shared that modern classrooms are MALL classrooms using latest methods of learning. MALL has various features like portability, ubiquity, laxity in learning, informal settings, anywhere and anytime learning and teaching and rapid connectivity (Kessler et al., 2023). Earlier methods were teacher-dominant in which the learners were only focusing upon memorizing and rote learning. MALL classrooms are learner-dominant. Besides, the learners are active during learning affairs. MALL provides motivation and encouragement in learning and teaching due to attractive and novel styles. Abdollahpour and Asadzadeh (2023) are of the view that MALL gives ownership to the teachers and learners. Further, the atmosphere created by MALL has a sense of comfort that allows to collaborate and communicate freely. This view is supported by Lee and Xiong (2023) who shared that MALL classrooms are flexible and relaxed where learning is done informally as well. Finally, the tools utilized by MALL are its most pivotal aspect. Modern applications like Microsoft Teams, Google Classrooms, WhatsApp and Zoom have revolutionized the educational scenarios. Various studies have proved that MALL has been found significant in ESL teaching as well as learning. Lei et al., (2022) did a study on the role of MALL by assessing the attitudes of 39 English as a Foreign Language (EFL) learners towards vocabulary. The observation was of one year and in that span of time various MALL tools are used. Data analysis was done by latent curve modeling and it was observed that positive correlation was found between MALL and learning attitude towards vocabulary. Assapari and Hidayati (2023) executed research on exploring the readiness of EFL learners towards MALL. Forty-two learners were part of the study who were studying in non-English programs. Two projects were developed for the class, which were exploited through MALL. Data were collected through Google Forms and results were analyzed by content analysis. It was explored that learners showed positivity in using MALL. Learners also shared that MALL improved their speaking skills. Similarly, Ramadina et al., (2023) studied the experiences of instructors who were teaching young learners through MALL in a distance learning program. Data collection was done by photo-voice collection of learners and also from interview. Teachers explained that learners benefited a great deal from MALL and found it an attractive and motivating tool. Moreover, teachers also shared that MALL helped to provide comfort, innovation and impact in their teaching. These researches validated that as MALL is contrary to traditional teaching styles and it is also captivating for the learners.

11.2.2 Previous Studies on Zoom as MALL Tool

There are some researches that stressed on the vitality of Zoom app. Cheung (2023) did a case study of an ESL teacher in Hong Kong during COVID-19 about teaching by Zoom. Data were collected from interviews and class-recording. It was concluded that Zoom is an influential and effective tool for language learning. It is a modern pedagogy that attracts learners prolifically. Risma (2021) performed an exploration to investigate the attitudes of Algerian learners regarding the use of Zoom app. Through questionnaire they found out that Zoom has been much appreciated by the learners and it was beneficial and easy to use during their learning processes. Likewise, Souhila (2021) did an investigation and explored the efficacy of Zoom in the development of language and communication of learners. It was found that Zoom aids learners to upgrade their language learning skills, which enhances their communication. Further, Ayoub (2019) researched to find out the effect of Zoom app on the motivation of English language learners. Through a qualitative analysis they concluded that the learners felt motivated and energetic while using Zoom for learning process. They also suggested that Zoom app is interesting and appreciated by the English language learners. Finally, a study conducted by Barbosa and Barbosa (2019) used technology acceptance model (TAM) to view learners' perception regarding the acceptance of Zoom app. Conducted on EFL learners they explored how they experience Zoom app for learning during COVID-19 pandemic.

11.2.3 Theoretical Framework

11.2.3.1 Technology Acceptance Model

The canvas of exploring the behaviors of individuals and organizations towards technology usage is getting broader. Among various models, TAM has been found to be pertinent and effective to explore the individual's behavior towards technology. Initiated by Davis (1985) this model is derived from the theory of reasoned action, which is recognized and accepted as the base on behavior studies; the emphasis of this model is to explain that the behaviors adopted by individuals are the outcomes of psychological-social views. Theory of Reasoned Action looks at the individual norms, behavioral intentions and attitudes. TAM varies from TRA as it predicts the acceptance of a phenomena and explores the changes needed to make it a comprehensive system (Chang et al., 2012). TAM is the most popular model for examining behaviors towards technology. This model aims to provide a structure for exploring intentions, beliefs and attitudes (Davis et al., 1989). There are two reasons for the attitudes to any aspect: perceived usefulness (PU) and the perceived ease of use (PEOU). PU is the trust level of individual towards any object or system that elevates performance.

Moreover, the PEOU is the perception about the use of anything, i.e., in which an individual feels that the usage would be easy and without effort. PEOU is pivotal as it impacts on PU, i.e., if using a gadget, system or object is easy, consequently the performance will be elevated. Both of these opinions are the outer aspects of a person's response and are interconnected with mind and thus become part of behavior.

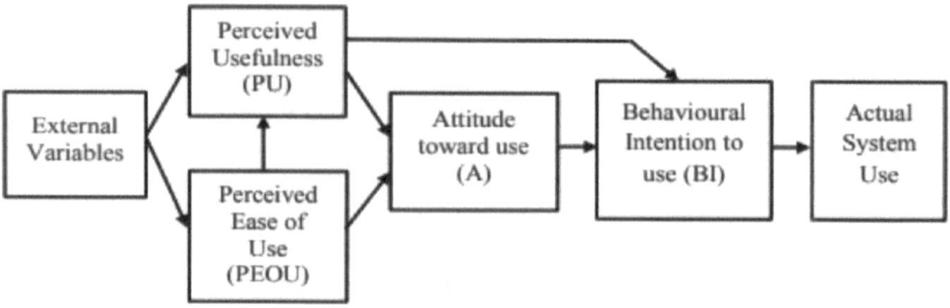

TAM has been used by various researchers as technological innovations are rapidly increasing. Al-Adwan et al., (2023) conducted research to explore the views of university learners towards metaverse technology. Using TAM as model, qualitative method was used in which questionnaire was utilized to collect data from 574 university learners. The outcomes of this study concluded that metaverse technology is appreciated by learners and they found it innovative, interesting and effective. Moving on, Shahid et al., (2023) did research on relations between TAM and language learning through application. A total of 100 learners were part of the study and it was found that learners' intentions and behavior were positive and they were ready to adopt it. Moreover, the study concluded a positive relation between TAM and language learning app. Further, Fan and Wang (2023) performed an exploration to view the factors that create an impact on the learners' behavior towards Web 2.0 tools. Data were collected through structured modeling method from 834 Chinese learners. TAM's major aspects, i.e., PEO and PEOU, were explored and it was found that the learners' intentions to use latest Web 2.0 tools are encouraging and they think it as a motivating and enjoyable aspect in their learning. Moreover, all the learners shared that latest tools are appreciated in learning language. Nadhifa and Kher (2022) investigated the behavior of 63 learners towards social media usage. Having TAM as their model the researchers explored PEO and PEOU of learners. Using a descriptive method to collect data through a questionnaire, it was concluded from the eyes of TAM that PU and PEOU learners both were positive and they found social media as an interesting and fun-learning platform for English language learning. All these studies validate the utility of MALL and Zoom.

These studies also validate the efficacy of exploring any new tool through the eye of TAM precisely in English language learning. However, there is hardly any research that has been conducted with an aim to explore the opinions of English language teachers from three major countries of the world regarding MALL and any of its tool that can be used for English language learning under the eye of TAM. This study will attempt to cover this research gap to explore the behaviors of Indian, Pakistani and Sri Lankan teachers regarding Zoom app.

11.3 RESEARCH METHODOLOGY

Research design is a significant step to choose as it provides a way to pick tools and methods that can give replies to the research problems under discussion. This research intends to explore the teacher's views regarding the use of MALL after

COVID-19. So, for the said reason qualitative research design was applied for this study. Qualitative method is significant in exploring the experiences, views and opinions of the participants. Rehman and Alharthi (2016) stated that it also aids in finding the efficacy of any said phenomenon.

There were 30 teachers (10 each from India, Pakistan and Sri Lanka) involved in this interview process. All of them were regularly using Zoom app in their teaching since COVID-19 pandemic. All teachers were picked from public sector universities of India, Pakistan and Sri Lanka and were teaching regularly to post-graduate classes. All the teachers willingly gave the interviews after providing consent. These teachers were chosen using convenient sampling method from all the three countries. To maintain consistency in the demographical aspects of the teachers some variables were kept constant. These variables were age, experience, designation, Android phone usage duration and gender and subject. All the teachers were between 40 and 50 years. Further those teachers who had more than eight years of ESL teaching experience were chosen and all were assistant professors. All of them had been using Android phone since 2012. No female was part of the study. This was done for maintaining the reliability and validity of research. In relevance with the objectives of this study semi-structured interviews were used for data collection. Semi-structured interviews are the most viable and authenticated methods to collected data.

11.3.1 Results of Interviews

These interviews were recorded from all the 30 teachers while considering TAM and its two contents. The questionnaire was constructed while keeping these two aspects:

1. perceived usefulness
2. perceived ease of use

Duration of interview for each teacher was 30–40 min. Ten open-ended questions were asked in the interviews. Prior permission was taken from the teachers for interview recording. Moreover, transcription of the interview was done later to extract themes. The analysis collectively includes the opinions of the teachers from all three countries. However, it is important to mention here that from 1 to 10 teachers were from India, from 11 to 20 Pakistan and from 21 to 30 Sri Lanka. While quoting the very words of teachers, the number has been given to make it explicit and comprehensible for the readers. These themes were analyzed using content analysis approach. The major themes taken from teachers' recordings are

1. Develops interest of the learners
2. Provides motivation to learn
3. Helpful in ESL learning situations
4. Problems while applying Zoom in local ESL setting
5. Zoom app a comfort or a trouble
6. The variety provided by Zoom is a distraction or attraction
7. Faced technical outrage while using Zoom

8. Advantages of Zoom app
9. Zoom on computers or on mobile phones
10. Anytime and anywhere learning

11.3.1.1 Theme One Analysis

The first theme explains the opinion of the teachers regarding their own and learners' interest towards Zoom. The novelty and the freshness by Zoom help the learners to indulge themselves in learning methods. The built-in features based on modernity and innovation are indeed a new learning style for ESL learning contexts. This is one of the reasons for making it a popular app. All the teachers opinioned that Zoom is an app that adds variety and variation in a complex ESL situation.

Teacher 6 stated,

> Surely it increases the curiosity and interest of the learners by giving them new channel of learning. The attraction of Zoom is actually online learning with live classes that adds a new concept which increases the interest of ESL learners.

Zoom can develop a scintillating environment and this can urge the learners to get them involved in learning process. Through the camera the teacher is monitoring the learners, which also makes them attentive during learning. The learners can also be assessed online using Zoom app. The use of audio video facility, the sharing of screen and other features also develop interest of learners.

Teacher 12 stated that,

> Zoom has been a wonderful version of latest app that shares unique features of sharing screen and also video sharing and receiving. In this way the online assessment can be done in an effective manner.

11.3.1.2 Theme Two Analysis

The second theme presents the viewpoint that Zoom provides motivation to learn. The comprehensive use of Zoom, though, was an accidental process worldwide because of pandemic but it has become a regular feature particularly in academia. So, the teachers believed that Zoom motivates and inspires learners to indulge themselves in learning individually and collectively. Contrary to orthodox ways, Zoom app creates a comfortable learning ambiance.

In the words of teacher 24,

> I think Zoom app provided several distinctive aspects like instant connectivity, active conversation, collective learning activities. This motivates the learners to keep them intact and develop their focus in learning.

The teachers opinioned that the flexibility of Zoom app in learning venture is another hallmark that provides motivation to learners. The availability of features like audio, video, recording and texting urge learners to learn positively.

Teacher 15 stated,

> Teacher ESL learning needs a lot of motivation and passion for learning and Zoom app has indeed provided motivation and confidence.

11.3.1.3 Theme Three Analysis

Third theme views if Zoom technology has been found helpful in learning situations? After COVID-19, the preference for Zoom is still higher compared to many other apps because of the characteristics it offers to users. Teachers opinioned that in ESL settings Zoom promotes ubiquity, flexibility and also facilitates beyond classroom learning. It promotes seamless learning.

Teacher 17 has said,

> I felt lucky to use Zoom for my learning purposes. The comfort and ease provided by Zoom has given a sense of serenity to me and it also supported me to teach regularly. My course content and syllabus all were completed with continuity and regularity.

Teachers also shared that the various dimensions provided by Zoom app provided great help in sharing knowledge and providing content to the learners.

Teacher 3 has said that,

> Zoom offered independence and confidence to me and my learners. It helped me to teach after regular classes in a relaxing environment. Rather I informally taught formal contents.

11.3.1.4 Theme Four Analysis

This theme reflects the troubles confronted by ESL learners while incorporating Zoom in their classrooms. The teachers stated that in ESL setting the exploration of technology-based learning is not commonly practiced. Students are not prone to use it regularly. Moreover, there can be issues of electricity, internet and other technical problems, but surprisingly the learners utilized Zoom app on their mobile phones and found no problem in using it.

Teacher 25 stated,

> I felt that there is no issue in using Zoom app for learning purpose. The user interface, guidelines and functions are so simple that one learns it quickly. I used it in my ESL classes and my learners also learnt it quickly. I am of the view that the variety it offers is equally backed up with the simple construction of the app.

The teachers have explored that there are some other learning apps but either they have been less attractive or they have been quite complex in their use. Zoom app is found to be impactful because of its user friendliness.

Teacher 8 shared,

> Zoom experience was exciting as it was an interesting learning venture and the learners quickly adjusted themselves with it.

11.3.1.5 Theme Five Analysis

This theme shares if Zoom app is a source of comfort or trouble for ESL teachers and learners. The teachers belonging to all the three countries expressed their viewpoint and told that Zoom app has given them a lot of comfort and ease in their respective classes.

Teacher 10 expressed,

> I am of the view that Zoom app significantly helped me in the teaching process. The features of audio and video at the same time provided me with variety to teach effectively.

Having mobile phones with much larger screen it was easier for the teachers to view their students. The teachers shared that ESL learning situations are already complex and any complex mobile app can add more trouble while learning. It can distract learners and they can lose interest. This had not happened in the process of using Zoom app.

Teacher 27 said,

> I found it quite pertinent for my teaching. The user-interface is friendly and provides me with multiple options to engage my learners and monitor them effectively.

11.3.1.6 Theme Six Analysis

The next theme looks if the variety provided by Zoom is a distraction or attraction. The teachers here have opinioned that ESL learning contexts are perplexed. Already established methods are teacher-dominant and it only follow traditions. Zoom app itself has some distinctive features that are engaging for learning process. It promotes collaboration and discussion in an easy ambiance. Even the focus of the teacher is more intense as every eye is looking at him/her.

Teacher 18 has stated,

> Obviously any new app has a lot of attraction for the learners. But Zoom has created more inspiration for my learners rather than distraction. They found interest in learning. They became agile and confident in Zoom classrooms. I think Zoom worked as a positive stimulus for them.

The involvement of Zoom actually helped to learn things differently. Particularly in ESL settings the learners can also view the screen sharing by the teachers. These features were unheard by Indian, Pakistani and Sri Lankan ESL learners. Therefore, they became inspired and interested in learning.

Teacher 4 stated,

> I think Zoom is not distraction as the objective of this app is learning and teaching. Due to the feature of camera, monitoring becomes easy, moreover the virtual hand raise option allows learners to enquire without interrupting teachers. Finally, teacher can check any passive learner in the group easily.

11.3.1.7 Theme Seven Analysis

This theme is extracted to view if the teachers from India, Pakistan and Sri Lanka faced any technical problem or outrage during their Zoom sessions. All the teachers stated that there was no technical issue while using Zoom app. They shared that Zoom app is not a heavy application so it does not affect processor speed. Moreover, they also share that while changing commands such as from sharing screen to sharing camera, the app runs smoothly and thus it helps the continuity of the lecture. It actually creates a strong bonding between them and their learners.

Teacher 23 stated,

> I found Zoom app magnificent as it runs smoothly without stopping the other functions of the phone. Further, if I need to write something in chat box the app didn't stop working. The smooth running made me comfortable while teaching and I focused more on my work rather than managing the app. In this way learning and teaching went well.

Certain other apps actually share a lot of other features but if some of their functions are used simultaneously, they are slow. This creates discontinuation in the process of learning, which makes participants frustrated. Especially in ESL learning scenarios in which maintaining the interests of learners is the key for acquisition of knowledge. Zoom in this regard has been found to be smooth and continuous without any technical issue. This is the major reason the teachers all across the world loved to utilize it.

Teacher 14 has expressed,

> Zoom is user friendly with providing ease in the functioning. I could not find any technical issue. Neither my mobile stuck nor my screen was held. Overall Zoom smooth functioning is the key to its success.

11.3.1.8 Theme Eight Analysis

The teachers shared some of the advantages they explored while using Zoom app. They experienced that the video quality during the call is perfect. There is no blur image, rather it is high-definition video. Besides, during call the video never gets stuck and the meetings can be conducted smoothly. Further, it is secure and safe as without meeting ID and password no one can enter the classroom. The invitation process itself is easy. There is also a feature of sharing link, which once shared can be used regularly. Despite the availability of link and password, without the permission of the host no one can join the meeting.

Teacher 5 expressed,

> I believe that my meeting is secure and safe. Zoom does offer a lot of aspects like joining a meeting and hosting a meeting. With that there is additional option to send a link for meeting. A meeting ID and Password generated by the host can also be shared. Along with all these features, I found that it is on the host to allow someone to join meeting or not. This gives me a sense of control and power.

Another advantage is scheduling a meeting for the next day. It helps to inform the participants regarding the activities of learning you plan to utilize with your learners. Additionally, joining the meeting is quite easy and is just a click away. Another aspect is the adjustment of class size. Indian, Pakistani and Sri Lankan classrooms are usually 40–50 minimum, whereas Zoom can allow up to 1,200 persons in a single meeting. The other important feature is the screen shots which can be taken and shared instantly. The display buttons are large and easy to handle. If one wants to share a file it can be done to all participants.

Teacher 13 said,

> I can cater maximum learners with full concentration. Besides I can record my lesson for the absent participants. Learners can also take screen shots for their learning. Overall variety of interesting features are there.

11.3.1.9 Theme Nine Analysis

This theme explains the opinion about the comparison of using Zoom app on mobile phone or computers. All the teachers in India, Pakistan and Sri Lanka explicitly stated that they preferred mobile phones rather than computers for taking Zoom meetings. They believed that mobile phones offer several features like portability and convenience. Due to light weight, larger size of screen, ubiquity, and fast processor speed, mobile phones are the most favored gadget.

Teacher 11 stated,

> I believe that Computers are expensive and require regular electricity. Also, laptops are difficult to get because of their price and secondly, they also need to handle with care. My students also had these issues; most of them did not have a computer or a laptop. I prefer to use mobile phone and I used it successfully during my Zoom sessions. My students also had mobile phones (android). The variance built-in and downloadable features allow me to consume it with confidence and perfection.

All the teachers also shared their views that latest mobile phones can be easily handled and can have long-lasting battery. So even when there is no electricity students can connect through their mobile phones. The built-in feature of internet, Wi-Fi and GPRS are some distinctive features that have made the mobile phones the most demanded gadget of this century.

Teacher 26 stated,

> I cannot think a life without a mobile phone. It is a complete package of entertainment, knowledge and social connectivity. Also, this new addition of utilizing it for English language learning has given it an irreplaceable position. I cannot think of any other gadget accept mobile phones for using Zoom effectively.

11.3.1.10 Theme Ten Analysis

The final theme views the use of Zoom app as a MALL tool for anytime and anywhere learning. It is another quality of Zoom that is not designed for iOS, or computers. It equally runs on the mobile phone uninterruptedly. Therefore, as MALL tool it supports learning beyond classrooms.

Teacher 28 said,

> It has become difficult to cover syllabus in 45 minutes lecture. So, using Zoom encourages beyond classroom learning and helps me to even give instructions, assignments and tasks after my class.

The flexibility of place and time overrules the mental and psychological pressure of learning in strict formal settings. The teachers also pointed out that sometimes several factors like setting, weather and time pressure can affect comprehensive learning in classrooms yet Zoom as MALL tool helped teachers to teach anytime without having pressure of teaching. Zoom teaching through mobile phones is thus appreciated because of these features and it vividly motivates learners.

Teacher 9 said,

> My teaching has benefitted from Zoom. Now I can manage my class easily and be connected with learners even outside the class. In this way, I believe my teaching has improved a great deal.

11.3.2 Discussion

The results are analyzed through the lens of TAM and its two components; PU and PEOU. It can be extracted from the results that Zoom as an app under the umbrella of MALL is a new mode of learning for learners and teachers (Guzachchova, 2020). The utilization and efficacy of Zoom app and its compatibility with MALL has changed the overall learning patterns and methods in ESL learning. Regarding PU, all the teachers from India, Pakistan and Sri Lanka stated that Zoom app has been the most effective and impactful application for their teaching scenarios. Zoom created positive influence in their teaching affairs. It has provided a great support and facilitated them. The teachers claimed that the utility of this technology tool has made a perfect classroom ambiance and increases learners' interests (Alfadda & Mahdi, 2021). PU can be viewed from the results. Zoom app motivates learners to learn and it is helpful in the learning context. This is because of multiple features offered in this app. The aspect of audio with video facility, the feature of private texting where one learner can send a message to another with no other participants able to see them are some wonderful features. Further, all these aspects actually develop interest and motivation for the learners and create a positive impact (Kim, 2020). These findings are similar to the study by Cheung (2023) where it was viewed that Zoom does motivate the learners strongly because it is useful and effective. He shared that Zoom app is not heavy and all functions can be used comfortably on the learning device. This aspect is identical in this study and demonstrated that the learning tool must be smooth in use to maintain a successful classroom (Barbosa and Barbosa, 2019). If any app is slow the participants can get frustrated and it can create distraction in learning. Furthermore, ID, Password, user name and weblink are the pivotal checks to avoid any stranger to enter in a particular class. Through camera, the teacher can monitor every learner. This develops personalized learning ventures and students mentally are cognized that they are only attached with their peers. This is also validated by the study conducted by Souhila (2021), which expressed that Zoom environments are catchy, attractive and comfortable for the learners and instructors. Another usefulness of Zoom as a tool of MALL is that it has been preferred to use on mobile phones. Mobile phones are pocket friendly as compared to computers with features like size of screen, ubiquity, learning beyond restrictions, touch screen interface, advance processor and Wi-Fi etc. These make them more important than computers (Ali et al., 2021). Teachers used Zoom on mobile phones and it gave them ease and facility to teach, guide and help beyond classrooms.

Regarding the PEOU the results validate that the app is constructed in such a way that there is no complexity in its user interface. All its functions and buttons are easy to comprehend and use. The learners and the teachers quickly adapted to the app and it does not create hindrances and complexities while usage. Therefore, it was quite easy in use and all the teachers found it to be the most perfect app for learning and teaching ventures. Thus, this study is aptly connected with TAM and shows that all the teachers from India, Pakistan and Sri Lanka are flexible and welcoming to accept technology in their teaching methods. They appreciated the involvement of technology for education. It also shows that the PU of Zoom and PEOU have been perfect as its practical implementation and the teachers opinioned that they found it effective and motivating for the learners and for themselves. The complete findings collected here also corresponds to the study by Risma (2021) who also found the attitudes and opinions of the Algerian learners regarding the use of Zoom app and concluded that teachers and learners are welcoming this new technological change. Conclusively one can say that this study ascertains that Zoom app has a positive impact on learners' performance.

11.3.3 Conclusion

This chapter encircles upon the teachers' perspective from three important South-Asian countries, India, Pakistan and Sri Lanka, regarding the acceptability of Zoom app. The results and discussion validated that Zoom app has got identical popularity among the teachers of South Asia as around the globe. ESL learning and teaching are reciprocal, and innovative methods are always found helpful to improve them. The inclusion of Zoom is indeed a new opening and it can help learners and teachers in multiple ways and at several instances. The perspectives of teachers vividly presented the fact that unconventional methods if impactful and meaningful can be accepted by the educators and academia. The positive attitude by teachers and learners ascertained that latest technological tools are really helpful in ESL settings provided that their construction and working is smooth and efficient. It is also proved here that learners from India, Pakistan and Sri Lanka are as flexible, ready and efficient to utilize any new learning domain like the learners from other countries around the world. Thus, it can be concluded that since the outbreak of COVID-19 Zoom is really on the boom and it can be utilized by various other domains to explore desired results.

11.3.4 Future Implications

There are some implications for this study. This study was performed to gather the opinions of the teachers from India, Pakistan and Sri Lanka. These teachers had practical experience of using Zoom during COVID-19 and afterward. In this, real and practical experiences were shared by them. The results of this study have some practical implications for researchers, teachers and learners from all these three countries. Some of the implications may be:

1. Teachers must be encouraged to use Zoom even after classrooms for their teaching and discussion with learners.
2. Teachers must be equipped to utilize Zoom features perfectly.
3. The administration should provide confidence to the teachers to exploit Zoom in their teaching affairs.
4. There must be workshops among the teachers on how to use Zoom for learning.
5. Teachers must coordinate with other teachers to use Zoom for taking tests and assignments.
6. Even learners can use Zoom for group discussion, assignments and discussions.

11.3.4.1 Recommendations

The following recommendations can be made after conducting these studies:

1 Zoom can be a permanent feature for ESL classrooms.
2 Zoom can also be utilized in other disciplines regularly.
3 The institutions must promote teachers to use Zoom.

4 Under current wave of COVID-19 the curriculum designers should maintain material that supports online learning.
5 Interesting and result-oriented activities in ESL settings must be made to make Zoom classroom more effective.

REFERENCES

Abdollahpour, Z., & Asadzadeh Maleki, N. (2023). Fostering academic vocabulary learning: Opportunities for explicit learning through a mobile-assisted app in the field of applied linguistics. *International Journal of Foreign Language Teaching and Research*, *11*(44), 133–152.

Al-Adwan, A. S., Li, N., Al-Adwan, A., Abbasi, G. A., Albelbisi, N. A., & Habibi, A. (2023). Extending the technology acceptance model (TAM) to Predict University Students' intentions to use metaverse-based learning platforms. *Education and Information Technologies*, *28*(11), 15381-15413.

Alfadda, H. A., & Mahdi, H. S. (2021). Measuring students' use of zoom application in language course based on the technology acceptance model (TAM). *Journal of Psycholinguistic Research*, *50*(4), 883-900.

Ali, M. M., Malik, N., Moghal, S., Asad, Z., & Kazi, A. S. (2021). Teaching English collocations to undergraduate learners through mobile assisted language learning. Linguistica Antverpiensia, *1* (1), 326–348.

Assapari, M. M., & Hidayati, R. (2023). EFL speaking student readiness to use mobile-assisted language learning. *LLT Journal: A Journal on Language and Language Teaching*, *26*(1), 365–378.

Ayoub, J. (2019). Effective use of zoom sessions (a synchronous learning strategy) to foster success and motivation of Lebanese University Students in Bekaa: A case study. ةلجم ةيعيبطلا مولعلاو بادلآا. *Journal of Letters, Nature, and Science*. Available at https://www.awraqthaqafya.com/601.

Barbosa, T. J., & Barbosa, M. J. (2019). Zoom: An innovative solution for the live-online virtual classroom. *HETS Online Journal*, *9*(2), 144–148.

Chang, C. C., Yan, C. S., & Tseng, J. S. (2012). Perceived convenience in an extended technology acceptance model: Mobile technology and English learning for college students. Australasian Journal of Education Technology, *28*(5), 809–826.

Cheung, A. (2023). Language teaching during a pandemic: A case study of zoom use by a secondary ESL teacher in Hong Kong. RELC Journal, *54*(1), 55–70.

Davis, F. D. (1985). *A Technology Acceptance Model for Empirically Testing New End-User Information Systems: Theory and Results* (Doctoral dissertation), Massachusetts Institute of Technology.

Davis, F. D., Bagozzi, R. P., & Warshaw, P. R. (1989). User acceptance of computer technology: A comparison of two theoretical models. Management Science, *35*(8), 982–1003.

Fan, C., & Wang, J. (2023). Undergraduates' behavioral intention to use indigenous Chinese Web 2.0 tools in informal English learning: Combining language learning motivation with technology acceptance model. *Humanities and Social Sciences Communications*, *10*(1), 1–9.

Gamage, U. P., & Dilani, N. (2022). The rationale behind the influence of English on modern colloquial Sinhala language. *Advances in Social Sciences Research Journal*, *9*(9), 133–139.

Gargesh, R. (2019). South Asian Englishes. *The Handbook of World Englishes*, 105–134.

Guzachchova, N. (2020). Zoom technology as an effective tool for distance learning in teaching English to medical students. *Бюллетень науки и практики*, *6*(5), 457–460

Ilyosovna, N. A. (2020). The importance of English language. *International Journal on Orange Technologies*, *2*(1), 22–24.

Kalyanpur, M., Boruah, P. B., Molina, S. C., & Shenoy, S. (2022). *The Politics of English Language Education and Social Inequality: Global Pressures, National Priorities and Schooling in India*, Taylor & Francis.

Karakaya, K., & Bozkurt, A. (2022). Mobile-assisted language learning (MALL) research trends and patterns through bibliometric analysis: Empowering language learners through ubiquitous educational technologies. System, *2*(2), 102925.

Kessler, M., Loewen, S., & Gönülal, T. (2023). Mobile-assisted language learning with Babbel and Duolingo: Comparing L2 learning gains and user experience. *Computer Assisted Language Learning*, 1–25.

Kim, H. (2020). The efficacy of zoom technology as an educational tool for English reading comprehension achievement in EFL classroom. *International Journal of Advanced Culture Technology*, *8*(3), 198–205.

Lee, J. C., & Xiong, L. (2023). Exploring learners' continuous usage decisions regarding mobile-assisted language learning applications: A social support theory perspective. *Education and Information Technologies*, *28*(12), 1–27.

Lei, X., Fathi, J., Noorbakhsh, S., & Rahimi, M. (2022). The impact of mobile-assisted language learning on English as a foreign language learners' vocabulary learning attitudes and self-regulatory capacity. *Frontiers in Psychology*, *13*, 872922.

Metruk, R. (2022). Smartphone English language learning challenges: A systematic literature review. SAGE Open, *12*(1), 21582440221079627.

Nadhifa, K., & Kher, D. F. (2022). Students' perceptions on the use of social media for learning English based on technology acceptance model (TAM). *Journal of English Language Teaching*, *11*(3), 310–318.

Quyyum, A. (2021). English as an esteemed language in modern society. Pakistan Journal of Social Research, *3*(3), 557–564.

Ramadina, A. A., Hassan, N. S. I. C., & Nurhadi, K. (2023). Mobile-assisted English language learning for young learners during COVID-19 distance learning: Teachers' experiences and voices. *Register Journal*, *16*(1), 168–185.

Rehman, A. A., & Alharthi, K. (2016). An introduction to research paradigms. International Journal of Educational Investigations, *3*(8), 51–59.

Risma, S. N. (2021). An analysis of the utilizing zoom application to English learners' speaking skill motivation project. *Professional Journal of English Education*, *4*(3), 433–445.

Shahid, C., Gurmani, M. T., Rehman, S. U., & Saif, L. (2023). The role of technology in English language learning in online classes at tertiary level. Journal of Social Sciences Review, *3*(2), 232–247.

Sianturi, A. D., & Hung, R. T. (2022, September). The challenges of venue is china Kahoot! In Teaching and learning in higher education – A systematic review. In *Proceedings of the 6th International Conference on Digital Technology in Education* (pp. 72–77).

Silalahi, D. E., Siallagan, H., Munthe, B., & Sihombing, P. S. R. (2022). Investigating students' motivation toward the use of zoom meeting application as English learning media during COVID-19 pandemic. *Journal of Curriculum and Teaching*, *11*(5), 41–48.

Souhila, B. (2021). Zoom sessions in distant learning: Algerian EFL students' perceptions and attitudes. *Arab World English Journal* (*AWEJ*.) Special Issue on Covid, 19, 264–280.

Tan, S. Y., & Md Yunus, M. (2023). Sustaining English language education with social networking sites (SNSs): A systematic review. Sustainability, *15*(7), 5710.

Wulansari, T. P., Sudiyanto, S., & Sumaryati, S. (2022, June). Opportunities and challenges of mobile learning towards society 5.0. In *International Conference on Education* (*ICE*), Indonesia, (Vol. 1, No. 1, pp. 53–60).

Zhang, W. (2021). Non-state actors in education: The nature, dynamics and regulatory implications of private supplementary tutoring. *Background Paper for the Global Education Monitoring Report*. UNESCO.

12 Teaching–Learning through the Distance Education System

An Exploration of What, Why, and How Possibilities in the Era of Industry 4.0

Aneela Pirzado, Zafarullah Sahito, and Muhammad Kamran Abbasi

12.1 INTRODUCTION

The term education is derived from *educare* or *educere*, a Latin word, where educare means to bring up, mainly related to the mind, and educere means to bring out or lead the body physically to the upper level (Derksen, 2014; Marcus & Ralph, 2021, 17). The word education is defined by different writers, philosophers, and theorists differently, and they connected it with their fields of expertise (Curtis et al., 2013, 1; Matheson, 2014, 15), i.e., education is a purposeful activity to achieve the predetermined goals through the transmission of knowledge, skills, and traits (Chazan, 2022, 13; UNESCO, 2018). One group of theorists believes that education is a process that occurs during educational events (Peters, 2010, 7) and the other group believes that education is a product of different processes, i.e., mental states and temperaments of educated people (Peters et al., 1973). Additionally, it refers to an academic field consisting of methods, processes, and institutions facilitating the whole teaching and learning process. Generally, there are two types of definitions, i.e. (a) thick and (b) thin definitions. Thick focuses on the claim of learners' improvement while thin definitions do a value-neutral account of education (Kotzee, 2011, 549; Watson, 2016, 146) closely related to the distinction between descriptive and prescriptive conceptions to describe and express, respectively (Smith, 2020, 781). The definitions see education as an activity to achieve certain aims (Siegel, 2023), which are classified into different categories, i.e., epistemic goods and skills, respectively, which focus on knowledge and understanding; and rationality and critical thinking respectively (Watson, 2016, 146).

In the teacher-centered perspective, education is the transmission of knowledge and skills in a moral way (Beckett, 2018, 380; Peters, 2015, 23), while student-centered perspective focuses on the experience of the learning process to transform and

DOI: 10.1201/9781003400691-12

enrich the experience (Jackson, 2011, 4) to improve and enhance the problem-solving skills (Beckett, 2011, 239; Freire, 2020, 374). Peters outlined three essential features of education (Marshall, 2006, 33; Peters, 2015, 23). The transmission of (a) knowledge and its understanding; (b) worthwhile; and (c) morals and manners (Marshall, 2006, 33; Peters, 2015, 23) are forms that share some essential features, such as support, growth, and development (Biletzki & Matar, 2002; Sluga, 2006, 1); these may or may not be context-bound to understand (Curtis et al., 2013, 1) clearly in order to measure and improve (Biesta, 2015, 75; Wilson, 2003, 101) the conditions. Education is a socialization process of transmitting accumulated knowledge from one generation to the next (Haack, 1981, 289) that helps individuals to function in society as responsible citizens (Chazan, 2022, 13). Human-centered definition of education focuses on the wellbeing of the individuals to lead a good life (Haack, 1981, 289). It emphasized the critical thinking process to distinguish education from indoctrination (Curren, 2013, 231; Siegel, 2009, 3), promoting beliefs, independency, and rationality (Siegel, 2014, 120). These abilities are to be fostered (Davies & Barnett, 2015, 1) till the child's mind is sufficiently developed (Siegel et al., 2008), through the scientific and effective teaching and learning processes.

Teaching is a process of sharing knowledge or abilities with another person (Manuel et al., 2012, 115), to involve them to convey the available information and experiences through a lecture and discussion. Teaching is considered both an art and a science by different schools of thought. As an art, it emphasizes the creative and artistic skills of the teacher through creating a valuable learning environment at the classroom and school levels. As a science, it focuses on the logical, mechanical, and procedural methods needed to achieve specific and predetermined goals effectively. The art of teaching involves a close relationship between an individual with more knowledge and experience and another individual who is less knowledgeable and experienced, with the aim of enhancement of education. This idea has been conveyed by Rajagopalan (2019, 5) through the analogy that teaching is akin to selling and buying different products through organizing and controlling (Brubacher, 1970, 98) to encourage learning. Teaching is a type of interpersonal influence with the intention of modifying the potential behavior of individuals (Gauge, 1963). Teaching is considered a series of actions involving an agent, a goal, and a circumstance consisting of two groups of factors, i.e., class size, student characteristics, physical resources; and teaching methods and approaches (Smith, 2020, 781).

Teaching involves an interactive process occurring during specific activities in the classroom, mainly through verbal communication between teacher and student (Amidon & Hough, 1967). One group of scholars such as Gagne and Briggs (1974) has also made notable contributions to define teaching focusing on the teaching process that follows a scientific approach and involving three key elements, i.e. the subject matter being taught; method of conveying to students; and the feedback provided to students to employ effective teaching strategies to enhance students' learning outcomes. The system allows for modification, improvement, and development of teaching and learning activities to make it inherently flexible by providing suitable teaching–learning environments. The learner's ultimate learning outcomes can be determined through their acquired knowledge structures. As Davis and Roper (1982) have highlighted, there are four stages that make up the complete framework of teaching: (a) The process of preparing for teaching involves examining the subject matter, defining and articulating goals, and composing them in writing. (b) Teaching framework outlines the instructional techniques used to accomplish teaching goals. (c) Find the appropriate approaches for teaching and learning that will effectively convey the content and facilitate communication. (d) The process of overseeing the teaching and learning process is centered on evaluating student performance based on learning objectives and using this as feedback for both teachers and students.

Learning is a type of transformation of knowledge by a teacher to introduce modifications to his or her students using different methods to enhance particular abilities, alter certain mindsets, or comprehend specific scientific principles at workplaces in a proper setting (Sequeira, 2012, 1). To become engaged learners in higher education, students anticipate as adult learners with the ability to influence their learning environment by asking questions and seeking clarification of any uncertainties they may have (Michael & Modell, 2003, 61) to control and fulfill their responsibility for their learning experience (Munna & Kalam, 2021, 1). Students desire to interact with an instructor who is collaborative and has a good sense of humor to deliver clear lectures and incorporate relevant examples to facilitate students for better comprehension of the course materials, which is a requirement of modern classrooms (Becker et al., 1990, 221). Learning can take place anywhere through different ways, i.e.: (a) in formal education system; (b) informal education; and (c) non-formal education.

The relationship between teaching and learning can be understood through an exploration of the processes. An act of teaching occurs when someone teaches knowledge or abilities to another, such as giving can refer to imparting knowledge or sharing experiences. Both the art and the science of teaching are recognized, which places emphasis on the teacher's creative and artistic ability to create a valuable environment in the classroom to assist students in their learning. As a science, it clarifies the logic, mechanical, or operational processes to be followed to accomplish effective goals in the form of achievement. Efforts have been undertaken in the realm of education to elucidate the connections between teaching and learning, and such endeavors persist to this day. The philosophy of education focused on an analysis of success as evidence (Harpaz, 1973, 1). Scholars aimed to investigate the connections between teaching and learning by analyzing the fundamental concepts of "teaching" and "learning" alone, together, in opposition, and in advance. Teaching and learning may take place together, parallelly, and as outcomes, where teaching is input and learning is output. It consists of many items; a few are: (a) teaching and learning go together with the help of teachers and students; (b) start with the previous knowledge and learning of the students; (c) let students to observe and then perform; (d) let students to help each other; (e) teach any topic together with other topics and subjects through connectionism approach; (f) use examples from daily life to be a good example and brand for any method; and (g) make the students as a community and bring them (Harpaz and Harpaz, 2014, 57) as a part of your organization, institution, department, and classroom.

12.2 OBJECTIVES AND THE RESEARCH QUESTIONS

The objectives of this chapter are (a) to understand the concepts of Education, Teaching, and Learning; (b) to understand the concepts of distance education (e-learning), and Industry 4.0; (c) to understand the process of teaching and learning that takes place through the distance education system in the era of Industry 4.0; and (d) to explore the challenges, issues, and problems with their possibilities and solutions to implement the distance education (e-learning) in the era of Industry 4.0. The research questions are (a) What is education, teaching, and learning? (b) What is distance education (DE), and Industry 4.0? (c) How does the process of education take places through DE; and (d) What challenges, issues, and problems are faced to implement the distance education system (DES) with proper possibilities and solutions?

12.3 DE AND DES (WHAT IS DISTANCE EDUCATION?)

DE is the system of imparting and acquiring education where students may not be present within educational institutions, particularly within the classroom setting. The students are free to make use of their own learning methods to acquire education without face-to-face contact with their teachers. It is the learning system where the teaching methods are different from learning methods, where the learner works alone or in groups, and is guided by study materials to complete their assignments and pass the requirements of the examination. In the DE system, communication between the teachers and learners may take place through phones, computers, etc., with the help of new technologies, methods, and software. In DES, both preparation and teaching are done by the students themselves (Jindal & Chahal, 2018, 99). DES is the fastest growing kind of domestic and international education, consisting of structured learning in which the students and instructors are separated by time and location. Once it was thought of as a unique style of instruction using unorthodox delivery methods and is now developing into a crucial idea in conventional education.

The term distance learning (DL) describes methods that focus on trying to expand access to education and training by relieving learners of time and place constraints. Individuals and groups of learners can benefit from its flexible learning opportunities through its fast and easy track system, which is why it is known as the fastest growing area of education. According to Markova, Glazkova and Zaborova, DL is recommended for students who are focused and have self-control over their learning and available time with flexibility. As a consequence, it can prepare the student to work autonomously. Furthermore, research by supporters of DL shows that it can be more effective than traditional face-to-face learning patterns, which is supported by various pieces of evidence that quantity and quality promote better student learning awareness and contentment. Especially during the COVID-19 pandemic, Markova et al. (2017, 685) highlighted students' perceptions of DL as a solution at a time when everything was closed except education, i.e. teaching–learning through DES processes. That's why, the main goal of DL is to provide instruction to students who are not physically present in a traditional educational setting such as a classroom, often on an individual basis. The emphasis in DL is on teaching methods and technology, which has been described as a process to create and provide access to learning when the information source and the learners are separated by time, distance, or both, situations, and contexts.

The idea of DL is not new. The first significant American correspondence program with a distance between teacher and student was established at the University of Chicago in the late 1800s. In the United States, William Rainey Harper, the first president of the University of Chicago, developed the concept of extended education, in which the research university established satellite colleges of education in the surrounding community, and in 1892 he also encouraged the concept of correspondence school courses to further promote education, which idea was further implemented by the Columbia University. With the establishment of an external program in 1858, the University of London claims to be the first university to provide degrees through DE. The London School of Economics, Royal Holloway, and Goldsmiths are just a few of the universities that provide postgraduate, undergraduate, and diploma degrees through their DE programs, which are now known as the University of London International Programs. The DE programs are popular and are used for a range of reasons, i.e., flexibility, easiness, accessibility, etc. Today, many private and community, profit-oriented and non-profit-oriented institutions worldwide offer DL, from the most basic instruction to the highest level of degree, due to the rapid development of new technologies and the constant cost reduction in processing, storing, and transmitting data.

12.3.1 Teaching and Learning through DE (How Teaching and Learning Take Place through the DES)

DE is recognized as a method of giving education and training to learners who cannot attend classes in a typical classroom setting. It is fundamental to implement appropriate teaching and learning methods when instruction and comprehension occur in distance mode. The curriculum is carried out using unique resources, known as independent study resources, which are made available to students via different mediums, i.e., newspapers, magazines, radio, satellite, audio and video tapes, CD-ROMs, the internet, the World Wide Web, etc. As a result, it is possible to assert that the application of modern technology has made significant improvements to the ease of DL. Nowadays, human beings are busier than before, which is one of the reasons that traditional education is not applicable to many of us. Therefore, DE is acceptable throughout the world with open arms and high interests, in which students and teachers are dissociated, but communicate to each other through different technological devices (Bruder, 1989, 30). DE can be categorized into two types, i.e., (a) asynchronous and (b) synchronous. Asynchronous focuses on the interaction of teacher and students at different times, for example, listening to the recorded lecture after class, and learning from the material provided, or watching tutorials at their own pace of learning. On the contrary, synchronous focuses on the instantaneous interaction of teacher and student by attending live online sessions (Shahabadi & Uplane, 2015, 129). Asynchronous type of DE is older as compared to synchronous, which is possible due to the improvement in communication technology (Mayadas, 1997, 1). Both methods are popular because the selection of teaching methods depends on the learners' need to gain knowledge and skills through better understanding. Occasionally, both asynchronous and synchronous methods are used in teaching, while modern DE uses computers and the internet to deliver and gain knowledge, as 80% of the content is delivered online (Allen & Seaman, 2011, 4).

12.3.2 DES in the 21st Century (What Is Industry 4.0?)

In the 21st century, the concept of Industry 4.0 makes it possible for the manufacturing industry to go digital, with almost all production parts, goods, and equipment equipped with built-in sensing devices. Every industrial sector in the world has the potential to evolve much more quickly and with greater impact than any of the three previous industrial revolutions, namely Industry 1.0, 2.0, and 3.0, through the analysis of related data within a ubiquitous system, with the fusion of digital data and physical objects (Liu, 2021, 9). Therefore, Industry 4.0 is a current issue that is intended to revolutionize today's industrial production as a whole. Let's understand the industrial revolution through its historical evolution.

The concept of an industrial revolution is closely associated with some fundamental changes that take place in the area of industry, ranging from Industry 1.0, 2.0, 3.0, and now 4.0. Industry 1.0 was characterized by the identification of production mechanization to assist the potency and efficiency of human business. The railways and steam engine were invented in 1784, which took place around the end of the 18th century. During that time, the industry confronted mechanical production capacities with the use of steam and hydroelectric power. The functions that were powered by humans and animals were replaced by machines. Industry 2.0 emerged in the early 20th century with the identification of mass production and quality through standardization, which remained very active till 1970. Industry 3.0 was marked by the widespread adaptation and flexibility of robot-based assembling and automation, which was described as the replacement of manufacturing and physical labor (Samaranayake et al., 2017, 529). The concept of Industrial Revolution 4.0 originated from a project founded by the German government to advocate the digitalization of industry. It is described by Afrianto (2018, 1–13) that IR 4.0 is characterized by the advancement in computerization that is managed by four key aspects, i.e., (a) Raised connections, power of computing, and volume of data; (b) The advent of inspection, capacity, and business computing and intelligence; (c) The eventuality of innovative types of relatedness among machines and human beings; and (d) The digital conveying of commands to the natural or physical world, for instance, 3D printing and robotics.

12.4 METHODOLOGY

Desk review research method was used to discover the existing literature (Sileyew, 2019) available offline (Bolander & Miriam, 2014) and online (Hewson et al., 2015; Kara, 2015). The discovered literature is known as the secondary data or information (Clark et al., 2021), which make available the solid arguments about the topic to explain well (Kraus et al., 2020) with better understanding and interpretation (Hennink & Bonnie, 2022). It focuses on the modern trends of knowledge, understanding, skill development, and finally the research (Sovacool et al., 2018). The latest information about societies and communities provides valuable insights into emerging new and old markets, enabling stakeholders to address the situations and the behaviors of different stakeholders to bring proper solutions of the problems (Kozinets et al., 2014); to influence the processes, products, and their success (Grover & Arpan, 2017); and to grow and develop with the scientific knowledge and development through authentic conclusion (Merriam & Elizabeth, 2015).

12.5 DISCUSSION

12.5.1 Pros and Cons of the DES in the Era of Industry 4.0 (What Are the Advantages and Disadvantages of DE?)

The pros and cons of the DE are as follows: (a) Flexibility is a major benefit of DE as it caters to individuals who may face challenges attending a conventional institution. DE includes individuals who are full-time workers, stay-at-home mothers, or members of the military, and are full-time busy in different personal and professional works. (b) Numerous internet-based educational programs enable them to continue their study at a location that is convenient for them. (c) Thus DE permits them to incorporate and increase their level of education into their scheduled timetable. (d) Having flexibility in a Distance Learning Program (DLP) provides them an opportunity to study without disrupting their personal life routines. (c) Job holders and businessmen can easily schedule their learning activities around and parallel to their other commitments without causing any disturbance to their personal or professional life. (f) DL courses enable students to fulfill their coursework from almost any location, as long as they have access to a computer and the internet. (g) One of the main benefits of DL is that individuals have the freedom to select the education program of their preference and can do so at any point in their lives. Additionally, there are a multitude of school options available to choose from. It is possible to live and study from any location by opting for DLP without any restrictions regarding the city, college, country, etc. (h) One can save money by not having to commute because this program doesn't mandate frequent in-person sessions, which reduces the travel time and cost. (i) As a remote education, there is no requirement to physically attend classes at a specific time or location, which allows students to complete their coursework at their own convenience and schedule. (j) Acquiring additional knowledge is another benefit, that is, the computer and internet skills acquired through DL can be applied to various other aspects of life to complete their assignment. (k) Numerous individuals encounter a challenge with physical accessibility when attending traditional classes due to limited mobility. However, this problem can be resolved through online classes, as individuals can participate from the comfort of their own homes using their own furniture, etc. (l) Pursuing further education can thus be achieved while also enjoying the freedom of movement at home; (m) Both slow and quick learners have options available to them, which can reduce stress and provide a greater sense of satisfaction for learners while completing their degree.

The drawbacks of DE are as follows: (a) The absence of opportunities to engage and interact with fellow students in the classroom. (b) Participating in group studies can be challenging, as students are typically required to work independently without the benefit of communication with classmates. (c) This can result in a solitary learning experience with limited opportunities for social interaction. (d) There is no opportunity to verbally communicate with peers and instructors; however, the use of messaging, email, video conferencing, and online message boards has been reducing this issue gradually. (e) These forms of interaction aid in the development of crucial abilities such as critical thinking and problem-solving. Several DL programs have created virtual discussion forums or chat rooms where students can exchange thoughts and communicate. However, such forums and chat rooms cannot fully replace the kind of interaction that takes place in a physical classroom with a teacher and classmates. If your favorite aspect of learning is the classroom setting, you might want to think again about pursuing DE. (f) Technology that is challenging and not easily reachable for everyone, while it is accurate that DL offers excellent opportunities to acquire knowledge and skills, can be complex, particularly for those who are afraid of technology. (g) To engage in DE, one must have a computer with constant internet connectivity. Furthermore, setting up necessary tools and facilities, such as live video communication, which is essential in this type of education, requires careful planning and significant expenses. (h) Unseen costs like there are certain hidden fees are unavoidable. For example, if a student resides or works in a remote location where there is an irregular supply of goods, they may need to have their study materials mailed in advance. This will undoubtedly result in additional fees and expenses, which can add to the discomfort of many individuals. For effective outcomes in remote learning, both the educator and student should prepare a well-thought-out plan beforehand. Utilizing virtual courses eliminates the need for building physical classrooms, dormitories, and related expenses, thus reducing overhead costs. Beginning a DL program with advanced technology can still incur significant expenses. Even after the program is up and running, there will be ongoing costs such as maintaining up-to-date technology, creating new educational materials, revising courses, and promoting the program, which must be taken into account in the yearly budget. Consequently, the provider must have both the financial resources and technical assistance readily accessible to fulfill current requirements and additionally possess the capacity to enhance their technical competencies in tandem with the internet's expansion of services (Porter, 1997). (i) No immediate feedback is available in DE that differs from traditional classroom learning. Instead, students must wait for their teacher to review their work and provide comments.

The format of DE may not be suitable for every type of learner. Online learning may not be suitable for everyone. If an individual struggles with motivation, tends to procrastinate, and requires significant teacher guidance, they should carefully consider whether enrolling in an online program is the right decision for them. Because there are certain employers who do not recognize or approve the degrees earned through online education programs. Although most employers accept DL, there are still a few who view it negatively. It's important to acknowledge that your online degree may not be suitable for certain professions or further education. DE necessitates the ability to adjust to unfamiliar technologies. If someone does not enjoy using technology, then they might not be benefited as much as others who are tech-savvy from an online course. Before enrolling in a class, ensure a level of comfort while working on computers and online applications.

12.5.2 Challenges, Issues, and Problems with Their Possibilities and Solutions to Implement the DE (E-Learning) in the Era of Industry 4.0 (Why and How DES Can Be Implemented in the 4.0 Industry)

The terms "Industry 4.0" or "digital transformation" are frequently used to describe the era in which we live, and that continue to have an impact on practically all facets of contemporary society. Digital change is viewed as a process of change that typically can be realized through the use of technical aspects, particularly in the wake of Industry 4.0. Industry 4.0 is defined as an institution's digitalization of all its work-related activities and procedures. It is also known as the digitalization of industry or the digital age (Horák, 2016, 574). Many research investigations are conducted to meet the educational needs of people in educational institutions' ever-evolving technological environments. In order to suit people's educational needs, many higher education institutions today also feature a DL center. With directives and rules published for specific goals and purposes to build a national standard in DE, the Council of Higher Education in Turkey is the organization that decides the processes and guiding principles of DL. Universities set up their distant learning centers in compliance with the policies and guidelines established by the Council of Higher Education (Oz & Ozdamar, 2020, 58). Multiple research studies have been conducted on the effectiveness of Industry 4.0 in DL, which is becoming more common in educational settings, and the extent to which DL can adapt to these changes and transformations should be evaluated in this context. According to Klaus Schwab, the founder and current head of the World Economic Forum, humanity is at the start of a period that will fundamentally alter how we live, work, and relate to the Fourth Industrial Revolution, and the speed to measure the changes that have already occurred should not be ignored (Schwab, 2017). Machines began to self-manage at this time, eliminating the necessity for human labor in the production process (Schwab, 2017). By merging computer, interaction, and internet technologies, machines have developed this structure (Oz & Ozdamar, 2020, 58).

DL was entitled as a worldwide strategy to support the education system and that gives universal access to people of the world to get education as per their wish, will, and available resources, i.e., time, energy, money, etc. The main benefit of the DE is to cover and control the dropout rates and ratios of any program through facilitating the students (Bilgiç & Tuzun, 2020, 143; Sinecen, 2018, 1) to avoid any obstacles. Students may face situational, attitudinal, psychological, and pedagogical problems (Montgomery, 2020, 85; Prinsloo et al., 2022, 34) in their regular programs. They may face sociocultural and instructional problems (Bristol, 2010; Yilmaz, 2017, 39) like poor learning environment, may not be able to handle their coursework well (Beaudoin, 2016, 9), may not have time management skills to manage the demands of their family, job, studies (Kember, 1989, 196), study materials, social and economic duties, and insufficient learning assistance (Badat, 2005, 183; Williams, 2003, 45).

The following issues and problems have been faced by DE:

1. Unavailability of educational leaders to follow and implement the philosophy of the DE system properly.
2. Unavailability of educational managers to implement the philosophy of the DES properly.
3. Unavailability of organizations to make available all needed items to follow and implement the philosophy of the DE system properly.
4. Non-serious attitude of organizations, leaders, and managers toward the implementation of the system of DE.
5. Untrained curriculum experts to design the scientific curriculum keeping the needs and the requirements of the learners, system, and the situations to implement the DES properly.

6. Unavailability of the writers to meet the needs and the requirements of the appropriate contents of the books or materials to publish.
7. Untrained faculty to follow the instructions properly to use the published materials efficiently and effectively to fulfill the needs and requirements of the learners and the systems.
8. Untrained faculty to follow the instructions of the technological systems to implement the required ICT instruments properly, efficiently, and effectively to fulfill the needs and requirements of the learners and the systems.
9. Inappropriate required systems, i.e., computers, online platforms, and other ICT instruments.
10. Unavailability of internet with good speed to deal with the needs and requirements of the system.
11. Inappropriate availability of pedagogies and their utilization.
12. Unavailability of technical experts of content, pedagogies, and technologies.

Unavailability of researchers of the fields to follow the philosophy of the DES properly.

The solutions and the recommendations of all the above issues and challenges are to take the system of DE seriously through understanding of its philosophy and the needs and requirements of the learners, teachers, organizations, and all other concerned stakeholders. Financial support must be available to purchase every type of tool, methods, and materials to upgrade and update the systems. The required experts must be appointed and professional growth and development facilities and opportunities must be available to upgrade the line of experts and professionals. The research team must be developed to focus the interventions on their results and the incorporation of the new materials and methods and instruments for future endeavors. Through research new trends and techniques can be identified and implanted for further value addition and the smooth running of the systems. Best practices of the best institutions of the world must be focused and used by taking help and support through having interaction in the conferences, symposiums, discussions, online seminars, faculty and student exchange system, signing of MoUs, etc.

Changes in the learning environment and atmosphere have a positive impact on the roles and the responsibilities of all the concerned stakeholders, i.e., instructors, students, etc., in the execution of the system, especially in the teaching and learning processes. Learning must be student-centered in the era of the 21st century, which is known as the digital era, directly connected with Society 5.0 and the Industry 4.0. However, the role of a teacher cannot be ignored but it will be given more weightage due to leading the students who facilitate and support them at every level. Teachers engage their students through different methods and techniques for learning and facilitating, supporting and encouraging them to be active and creative as the knowledge is widely reachable and the teacher is no longer the prime source of information. Explicit learning models and approaches are available to use, which are required for the student-centered learning systems and approaches, i.e., project-based learning and collaborative learning methods are the prominent methods of this digital era of learning.

The primary variables of DL or DE are distance, communication, and infrastructure, which can be minimized only through the ICT tools and techniques to find proper and scientific solutions of the problems in Society 5.0 and Industry 4.0. The innovations and experiments are the key indicators of the solutions of the problems in Society 5.0 and Industry 4.0, which can solve the problems of education of the adult learners of huge populations and distances. Therefore, the above challenges and issues have been categorized based on the expenditures of the programs, motivators of all stakeholders, students facilitation, teacher contact, feedback system, separation and isolation of all stakeholders, lack of experiences, lack of training, inexperience to course development, non-technicality to use technology, lack of help and support for remote learning, and insufficient facilities to select faculty for DL courses, because the regular and distance coursework has some limitations to teach. As blended or hybrid learning (Kumar et al., 2021, 85151) and e-learning are the methods in line with the digital education and the world, learning does not take place through traditional ways such as face-to-face sessions in a classroom, etc. But it is through technology-based distant education mode to provide excellent education that can be accessible to everyone, at any time, and from any distance and location. Therefore, we can say that e-learning and blended learning methods and techniques are the best learning strategies to pursue the education of the majority of the individuals of the huge populations of the world. Blended learning encourages greater learning as it has a great capacity to bridge the gap between face-to-face sessions and distant education to adjust the unique needs of the concerned individuals.

12.5.3 The Future of DE in the Era of Industry 4.0 and Society 5.0

The future of DE is directly connected with the needs, requirements, and the demands of Industry 4.0, as it has defined a new age of digitization, information transparency, connectivity, and automatism. As the digitalization, information transparency, connectivity, and automatism are the key components of Society 5.0 known as the smart society of smart people, the future of modern society poses unique difficulties to educational institutions. These unique difficulties of Society 5.0 and Industry 4.0 will definitely be resolved through any system and systematic approaches (Kumar & Nayyar, 2020, 1), which are the availability of a strong education system consisting of the merger of the synchronous and asynchronous systems of interventions. If it will be done with the limited passage of time and resources, it will be a significant shift that impacts positively on the education, work, and the business of the inhabitants of Society 5.0 and Industry 4.0. This transformation of systems, methods, techniques, and approaches involves the integration of manufacturing facilities, material supply networks, and service systems that definitely add some value to the goods or services delivered to consumers all over the globe (Murphy, 2018, 433; Fresen, 2018, 224). Society 5.0 may be considered as a smart community in which the physical and cyber worlds are highly interconnected (Alpaydın & Demirli, 2022, 3; Salgues, 2018, 1). The advancement of scientific and technical inventions and innovation to achieve the predetermined goal of economic growth and development through usage of information in the technological era on a priority basis is by personalization, interactivity, and user control (Roberts, 2018, 119).

The learning environment and atmosphere in the digital age are rapidly changing every day where the internet is considered as one of the most significant items to use in daily life. Students use the internet to access information and learning tools to fulfill the needs and the requirements of their day-to-day assignments and studies.

For that they may use search engines, i.e., Google, Bing, Google Scholar, and learning management systems, i.e. Google Classroom, One Note, Moodle, Zoom, Facebook, WhatsApp, etc., to promote, improve, and enhance the fast-paced interaction to meet the deadlines of their studies and programs. Students try their level best to get information, exchange it quickly through various social media and communication platforms, and work efficiently and effectively without meeting their teachers. They also purchase any books, materials, and equipment through paying online quickly and easily through their virtual accounts, especially paying their tuition and examination fees to speed up the interaction and educational transactions to save their time and energy, which became possible through internet known as the digital revolution, the prime need of Society 5.0 and Industry 4.0. Distance or open universities must maintain a balance between environmental, economic, and social issues through navigating the digital revolution to deal with political and geopolitical reservations (Ergul et al., 2021, 124). Universities strive to maintain their DESs or institutes to: (a) deliver the knowledge as a part of environmental cooperation; (b) help and support the production of innovation through sharing of tangible and intangible assets; (c) foster the economic growth and development through digital and social innovation with the help and support of social innovators and interested experts and parties. The digital and green transitions necessitate future-oriented strategies and organizational reforms to support the digital culture knowledge and its adoption (Benneworth & Cunha, 2015, 508), which focuses the shift of technology with human behavior and organizational structures (Ahmada & Fakhr, 2022; Cascio & Montealegre, 2016, 349; Erdurmazlı, 2021, 125) in order to meet the challenges and issues of Society 5.0 and Industry 4.0.

12.6 CONCLUSION

DE is the instruction that occurs while students and teachers are located apart from one another but connected through technical and technological sources. The main purpose of technology is to facilitate interaction among people. Through emails or texts, the instructors provide reading and study materials, assignments, and other program-related information. On the other hand, by reading the course materials provided by the institutions and instructors, the students also pick up knowledge and information about various topics and concepts. Students frequently use internet sources to gain understanding of a topic normally and when they run into issues or hurdles while learning processes. However, they typically need and want to speak with their professors frequently to clear up any questions and advance their learning, but they cannot. Alternatively, students interact with the provided sources to get solutions to their problems.

Cognitive abilities, personal involvement, diverse modes of learning, good relationships with instructors, mode of training, indirect instruction, multidisciplinary fields, revision of teaching–learning methods, enhancing collaboration, and improving problem-solving are the factors that highlight the significance of DE. The characteristics of remote learning include the detachment of teacher and student, the function of the educational institution, the importance of technology, two-way communication, the detachment of students from their peers, and industrialization. The use of monitoring and assessment techniques is crucial in DL. These techniques' primary goals are to recognize their shortcomings and make enhancements. DL has mostly benefited those who belong to the underprivileged, outcast, and economically disadvantaged segments of society. They can develop their academic talents later if they aren't able to take advantage of opportunities sooner. Lastly, it can be reported that learners who maintain positive relationships with their teachers and develop their technical skills get maximum benefits from remote learning.

REFERENCES

Afrianto, Afrianto. (2018). Being a professional teacher in the era of industrial revolution 4.0: Opportunities, challenges and strategies for innovative classroom practices. *English Language Teaching and Research, 2*(1): 1–13.

Ahmada, Halima Ali, and Rubina Fakhr. (2022). Relationship between organizational structure and employee behavior. *Journal of Positive School Psychology, 6*(8): 1959–1963.

Allen, I. Elaine, and Jeff Seaman. (2011). *Going the Distance: Online Education in the United States, 2011*, Sloan Consortium, PO Box 1238, Newburyport, MA 01950.

Alpaydın, Yusuf, and Cihad Demirli. (2022). *Educational Theory in the 21st Century: Science, Technology, Society and* Education. Springer Nature.

Amidon, Edmund J., and John J. Hough. (1967). Interaction Analysis: Theory, Research and Application. ERIC.

Badat, Saleem. (2005). South Africa: Distance higher education policies for access, social equity, quality, and social and economic responsiveness in a context of the diversity of provision. *Distance Education, 26*(2): 183–204.

Beaudoin, Michael. (2016). Issues in distance education: A primer for higher education decision makers. *New Directions for Higher Education,* 2016(173): 9–19.

Becker, William, William Greene, and Sherwin Rosen. (1990). Research on high school economic education. *The Journal of Economic Education, 21*(3): 231–245.

Beckett, Kelvin Stewart. (2011). RS Peters and the concept of education. *Educational Theory, 61*(3): 239–255.

Beckett, Kelvin. (2018). John Dewey's conception of education: Finding common ground with RS Peters and Paulo Freire. *Educational Philosophy and Theory, 50*(4): 380–389.

Benneworth, Paul, and Jorge Cunha. (2015). Universities' contributions to social innovation: Reflections in theory & practice. *European Journal of Innovation Management, 18*(4): 508–527.

Biesta, Gert. (2015). What is education for? On good education, teacher judgement, and educational professionalism. *European Journal of Education, 50*(1): 75–87.

Biletzki, Anat, and Anat Matar. (2002). *Ludwig Wittgenstein: 3.4 Language-Games and Family Resemblance. The Stanford Encyclopedia of Philosophy*. Metaphysics Research Lab, Stanford University.

Bilgiç, Hatice Gökçe, and Hakan Tuzun. (2020). Issues and challenges with web-based distance education programs in Turkish higher education institutes. *Turkish Online Journal of Distance Education, 21*(1): 143–164.

Bolander, Brook, and Miriam A. Locher. (2014). Doing sociolinguistic research on computer-mediated data: A review of four methodological issues. *Discourse, Context & Media, 3*: 14–26.

Bristol, Tim. (2010). Issues in implementing online education in a developing country. In: Mehdi Khosrow-Pour (Ed), *Web-Based Education: Concepts, Methodologies, Tools and Applications*, pp. 1342–1346. *IGI Global.*

Brubacher, John S. (1970). The theory of higher education. *The Journal of Higher Education, 41*(2): 98–115.

Bruder, Isabelle. (1989). Distance learning: What's holding back this boundless delivery system? *Electronic Learning, 8*(6): 30–35.
Cascio, Wayne F., and Ramiro Montealegre. (2016). How technology is changing work and organizations. *Annual Review of Organizational Psychology and Organizational Behavior, 3*: 349–375.
Chazan, Barry. (2022). *Principles and Pedagogies in Jewish Education*, pp. 13–21. Springer Nature.
Clark, Tom, Liam Foster, Alan Bryman, and Luke Sloan. (2021). *Bryman's Social Research Methods*, Oxford University Press.
Curren, Randall. (2013). A neo-Aristotelian account of education, justice, and the human good. *Theory and Research in Education, 11*(3): 231–249.
Curtis, Will, Les Hankin, John Sharp, and Stephen Ward. (2013). Education studies: An issue based approach. *Education Studies*, 1–264.
Davies, Martin, and Ronald Barnett, (Eds.). (2015). *The Palgrave Handbook of Critical Thinking in Higher Education*, pp. 1–25. Springer.
Derksen, Rick. (2014). Etymological dictionary of the Baltic inherited lexicon. In *Etymological Dictionary of the Baltic Inherited Lexicon*. Brill.
Erdurmazlı, Eser. (2021). Effects of information technologies on organizational culture: A discussion based on the key role of organizational structure. In S.D. Göker (Ed.), *A Closer Look at Organizational Culture in Action*, pp. 125–139. London: IntechOpen.
Ergul Aydin, Zeliha, Zehra Kamisli Ozturk, and Zeynep Idil Erzurum Cicek. (2021). Turkish sentiment analysis for open and distance education systems. *Turkish Online Journal of Distance Education, 22*(3): 124–138.
Freire, Paulo. (2020). Pedagogy of the oppressed. *In Toward a Sociology of Education*, pp. 374–386. Routledge.
Fresen, Jill W. (2018). Embracing distance education in a blended learning model: Challenges and prospects. *Distance Education, 39*(2): 224–240.
Gage, Nathaniel Lees. A method for "improving" teacher behavior. *Journal of Teacher Education, 14*(3): 261–266.
Gagne, Robert M., and Leslie J. Briggs. (1974). *Principles of Instructional Design*. Holt, Rinehart & Winston.
Grover, Purva, and Arpan Kumar Kar. (2017). Big data analytics: A review on theoretical contributions and tools used in literature. *Global Journal of Flexible Systems Management, 18*: 203–229.
Haack, Robin. (1981). Education and the good life. *Philosophy, 56*(217): 289–302.
Harpaz, Yoram, and Yoram Harpaz (2014). Teaching and learning in a community of thinking: The theory. *Teaching and Learning in a Community of Thinking: The Third Model*, 57–87.
Hennink, Monique, and Bonnie N. Kaiser. (2022). Sample sizes for saturation in qualitative research: A systematic review of empirical tests. *Social Science & Medicine, 292*: 114523.
Hewson, Claire, Carl Vogel, and Dianna Laurent. (2015). *Internet Research Methods*. Sage.
Horák, Josef. (2016). Does industry 4.0 influence efficiency of financial management of a company. In *The 10th International Days of Statistics and Economics*, Prague, September 8–10, pp. 574–582.
Jackson, Philip W. (2011). *What is Education?* University of Chicago Press.
Jindal, Aman, and B. P. S. Chahal. (2018). Challenges and opportunities for online education in India. *Pramana Research Journal, 8*(4): 99–106.
Kara, Helen. (2015). *Creative Research Methods in the Social Sciences: A Practical Guide*. Policy Press.
Kember, David. (1989). An illustration, with case studies, of a linear-process model of drop-out from distance education. *Distance Education, 10*(2): 196–211.

Kotzee, Ben. (2011). Education and "thick" epistemology. *Educational Theory, 61*(5): 549–564.

Kozinets, Robert V., Pierre-Yann Dolbec, and Amanda Earley. (2014). Netnographic analysis: Understanding culture through social media data. In *Sage Handbook of Qualitative Data Analysis*, pp. 262–275.

Kraus, Sascha, Matthias Breier, and Sonia Dasí-Rodríguez. (2020). The art of crafting a systematic literature review in entrepreneurship research. *International Entrepreneurship and Management Journal, 16*: 1023–1042.

Kumar, Adarsh, Rajalakshmi Krishnamurthi, Surbhi Bhatia, Keshav Kaushik, Neelu Jyothi Ahuja, Anand Nayyar, and Mehedi Masud. (2021). Blended learning tools and practices: A comprehensive analysis. *IEEE Access, 9*: 85151–85197.

Kumar, Akshi, and Anand Nayyar. (2020). si 3-Industry: A sustainable, intelligent, innovative, internet-of-things industry. In *A Roadmap to Industry 4.0*: *Smart Production, Sharp Business and Sustainable Development*, pp. 1–21.

Liu, Tao Zheng. (2021). *China-US Competition: The Chinese Perspective*. PhD diss., The University of Waikato.

Marcus, Benjamin P., and Allison K. Ralph. (2021). Origins and developments of religious literacy education. *Religion & Education, 48*(1): 17–36.

Markova, Tatiana, Irina Glazkova, and Elena Zaborova. (2017). Quality issues of online distance learning. *Procedia-Social and Behavioral Sciences, 237*: 685–691.

Marshall, James D. (2006). The meaning of the concept of education: Searching for the lost arc. *Journal of Thought, 41*(3): 33–37.

Matheson, David. (2014). "Now that I have read about education, how do I write about it?" *An Introduction to the Study of Education*, 4th ed., p. 334. Routledge.

Manuel Musial; Fabienne Pradere; André Tricot. (2012). *How to Design a Teaching Course*? Brussels: De Boeck.

Mayadas, Frank. (1997). Asynchronous learning networks: A Sloan Foundation perspective. *Journal of Asynchronous Learning Networks, 1*(1): 1–16.

Merriam, Sharan B., and Elizabeth J. Tisdell. (2015). *Qualitative Research: A Guide to Design and Implementation*. John Wiley & Sons.

Michael, Joel, and Harold I. Modell. (2003). *Active Learning in Secondary and College Science Classrooms: A Working Model for Helping the Learner to Learn*. Routledge.

Montgomery, Diane. (2020). *Teaching for Learning Gain in Higher Education*: *Developing Self-Regulated Learners*, pp. 85–103. Routledge.

Munna, Afzal Sayed, and Md Abul Kalam. (2021). Teaching and learning process to enhance teaching effectiveness: A literature review. *International Journal of Humanities and Innovation* (*IJHI*), *4*(1): 1–4.

Murphy, David. (2018). Open and distance non-formal education in developing countries. *Distance Education, 39*(3): 433–434.

Oz, Omer, and Nilgun Ozdamar. (2020). Academic's views on industry 4.0 within the scope of open and distance education. *Asian Journal of Distance Education, 15*(2): 58–85.

Peters, Richard S., John Woods, and William H. Dray. (1973). Aims of education: A conceptual inquiry. In *The Philosophy of Education*. Oxford University Press.

Peters, R. S. (2010). What is an educational process? In Peters, R. S. (Ed.). *The Concept of Education*. Routledge.

Peters, Richard Stanley. (2015). *Criteria of Education: Ethics and Education* (Routledge revivals). Routledge.

Porter, Lynnette R. (1997). Creating the virtual classroom: Distance learning with the Internet, pp. 216–229. John Wiley & Sons.

Prinsloo, Paul, Sharon Slade, and Mohammad Khalil, (Eds.) (2022). *Learning Analytics in Open and Distributed Learning*: *Potential and Challenges*. Springer Nature.

Rajagopalan, Isola. (2019). Concept of teaching. *Shanlax International Journal of Education, 7*(2): 5–8.

Roberts, Jennifer. (2018). Personalised learning in developing countries – is higher education ready? In *European Distance and E-Learning Network (EDEN) Conference Proceedings*, no. 2, pp. 119–132. European Distance and E-Learning Network.
Salgues, Bruno. (2018). *Society 5.0: Industry of the Future, Technologies, Methods and Tools.* John Wiley & Sons.
Samaranayake, Premaratne, Krishnamurthy Ramanathan, and Tritos Laosirihongthong. (2017). Implementing industry 4.0-A technological readiness perspective. In *2017 IEEE International Conference on Industrial Engineering and Engineering Management (IEEM)*, pp. 529–533. IEEE.
Sequeira, Aloysius Henry. (2012). Introduction to concepts of teaching and learning. *Social Sciences Education E-Journal*, 1–6.
Schwab, Klaus. *The fourth industrial revolution*. Crown Currency, 2017.
Shahabadi, Mehdi Mehri, and Megha Uplane. (2015). Synchronous and asynchronous e- learning styles and academic performance of e-learners. *Procedia-Social and Behavioral Sciences, 176*: 129–138.
Siegel, Harvey. (2009). Introduction: Philosophy of education and philosophy. In *The Oxford Handbook of Philosophy of Education*, pp. 3–9.
Siegel, Harvey. (2014). John White on philosophy of education and philosophy. *Theory and Research in Education, 12*(1): 120–127.
Siegel, Harvey. (2023). *History and Society: Philosophy of Education.* Archived at 1.08 pm, on 04 August 2023 from https://www.britannica.com/topic/philosophy-of-education
Sileyew, Kassu Jilcha. (2019). Research design and methodology. *Cyberspace*, *7*, 1–12.
Sinecen, Mahmut, (Ed.) (2018). Trends in E-learning. In *BoD-Books on Demand.*
Sluga, Hans. (2006). Family resemblance. *Grazer Philosophische Studien, 71*(1): 1–21.
Smith, Sharon. (2020). Forms of education: Rethinking educational experience against and outside the humanist legacy: By Emile Bojesen. Pp 141. Abingdon: Routledge. (2020). £ 120.00 (hbk). ISBN 9781138481251 (hbk). *British Journal of Educational Studies, 68*(6): 781–783. https://doi.org/10.1080/00071005.2020.1785788.
Sovacool, Benjamin K., Jonn Axsen, and Steve Sorrell. (2018). Promoting novelty, rigor, and style in energy social science: Towards codes of practice for appropriate methods and research design. *Energy Research & Social Science, 45*: 12–42.
UNESCO. (2018). *SDG Resources for Educators – Quality Education*. UNESCO.
Watson, Lani. (2016). The epistemology of education. *Philosophy Compass, 11*(3): 146–159.
Williams, Peter E. (2003). Roles and competencies for distance education programs in higher education institutions. *The American Journal of Distance Education, 17*(1): 45–57.
Wilson, John. (2003). The concept of education revisited. *Journal of Philosophy of Education, 37*(1): 101–108. https://doi.org/10.1111/1467-9752.3701007.
Yilmaz, Ramazan. (2017). Problems experienced in evaluating success and performance in distance education: A case study. *Turkish Online Journal of Distance Education, 18*(1): 39–51.

AUTHORS' BIOGRAPHY

1. **Zafarullah Sahito** is working as an Associate professor at the Department of Education, Sukkur IBA University, Sukkur, Sindh, Pakistan. Dr. Sahito has also honorary affiliation with the Faculty of Economic and Financial Sciences, North-West University, South Africa. He has more than 23 years of experience in the field of School and Teacher Education. There are more than 35 research papers and book chapters to his credit.

2. **Aneela Pirzado** is a B.Ed. student at the Department of Education, Sukkur IBA University, Sukkur, Sindh, Pakistan.
3. **Muhammad Kamran Abbasi** is working as an Associate professor at the Department of Distance, Continuing and Computer Education, Faculty of Education, University of Sindh, Elsa Kazi Campus, Hyderabad, Sindh, Pakistan. Dr Abbasi has more than 20 years of experience in the field of Computer and Teacher Education. There are more than 20 papers and book chapters to his credit.

13 The Concept of Higher Education Teaching Design toward Future TVET Flexible Learning

An Analysis of Heuta-Peera-Cyber Approaches

M.M. Mohamad, N. Mafarja, H. Mohd Affendi, R.H. Raja Mohd. Ali, and A. Ahmad

13.1 INTRODUCTION

Higher education teaching design is a crucial component of educating learners in preparation for their future jobs and enabling them to acquire significant proficiencies and knowledge to be successful in their chosen fields. With the advent of recent technologies and changing workplace demands, there is a growing need for more flexible and adaptable Technical and Vocational Education and Training (TVET) programs that can be delivered through online and distance learning methods. This concept of higher education teaching design toward future TVET flexible learning involves creating learning environments that are adaptable to the changing needs of students and the workforce. It involves incorporating innovative instructional design principles, such as heutagogy, peer learning, and cyberlearning, to create engaging and effective learning experiences for learners. TVET is an educational field that focuses on providing students with relevant industry knowledge and efficient skills. In recent years, there has been an increasing demand for more flexible and adaptable TVET programs that can be delivered through online and distance learning methods. By leveraging technology and fostering a collaborative learning community, higher education institutions can offer TVET students flexible learning options that allow them to balance their academic pursuits with other commitments. This strategy has many benefits that are not only limited to enhancing students' learning experiences but also contribute to meeting the growing need for a workforce that is more skilled and flexible. Consequently, designing and delivering TVET programs through flexible and adaptable learning

DOI: 10.1201/9781003400691-13

environments is becoming increasingly essential, and this analysis aims to explore the key principles, benefits, and challenges of this approach. It will also examine the role of technology in enabling flexible educating options, the importance of creating a supportive and collaborative learning community, and the potential impact of higher education teaching design toward future TVET flexible learning using Heuta-Peera-Cyber approaches on education and the workforce.

Utilizing andragogy, peeragogy, heutagogy, and cybergogy techniques through the promotion of self-regulated learning can successfully convey the focus from being confined to the teachers alone to also involving students in the learning process, which highly improves the learning outcomes (Chan et al., 2019; Hase, 2016; Muresan, 2014). Self-regulated learning, which employs technology and provides ample opportunities for active and meaningful learning, has been found to be more effective than peer learning (Azevedo et al., 2012; Lysenko et al., 2021). Despite the advantages of online learning, the unexpected pandemic situation has led to some challenges. To address these challenges, several studies have investigated the potential interaction between andragogy, peeragogy, heutagogy, and cybergogy (Amanina et al., 2022; Blaschke, 2021, 2019; Blaschke & Hase, 2019).

However, the main objective for this chapter is to:

1. To provide an overview of the concept of higher education teaching design in the context of TVET and its significance in addressing future learning needs.
2. To analyze the Heuta-Peera-Cyber approaches framework and its application in higher education teaching design for TVET flexible learning.
3. To identify best practices and case studies that exemplify the successful implementation of Heuta-Peera-Cyber approaches in TVET flexible learning, highlighting the positive outcomes for learners, educators, and institutions.
4. To propose recommendations for educators, administrators, and policymakers in designing and implementing effective higher education teaching strategies for TVET flexible learning based on the Heuta-Peera-Cyber approaches.

Meanwhile, the research questions for this chapter in book are:

1. What is the current understanding of higher education teaching design in the context of TVET, and how does it align with the future learning needs?
2. How does the Heuta-Peera-Cyber approaches framework contribute to the design of effective teaching strategies for TVET flexible learning?
3. How do successful case studies and best practices exemplify the effective implementation of Heuta-Peera-Cyber approaches in TVET flexible learning, and what positive outcomes have been observed for learners, educators, and institutions?
4. What recommendations can be proposed for educators, administrators, and policymakers in designing and implementing effective higher education teaching strategies for TVET flexible learning based on the Heuta-Peera-Cyber approaches?

13.2 CONCEPT OF HIGHER EDUCATION

Higher education is a topic of great importance as it is essential for the development of any society. The caliber of a country's human resources is closely tied to the quality of its higher education system. Higher education encompasses the teaching and learning activities that take place at colleges and universities, which students must complete to attain higher educational qualifications. Through higher education, students gain in-depth knowledge and understanding, which allows them to explore new frontiers of knowledge. Additionally, higher education fosters lifelong learning, authorizing learners to frequently modernize the knowledge they have and skills in response to develop social needs (De Wit et al., 2015).

Higher education is typically referred to as the postgraduate stage of pedagogy, which is considered the highest level of the educating system. It comprises universal education, non-technical education, and training in its technical professional form. Higher level of educating system is structured into three different degrees: the first level which is Bachelor, and then the second level which is Master, followed by the third level which is Master of Philosophy or Doctor of Philosophy (De Wit et al., 2015).

Malaysia boasts a well-established and diverse higher education system that offers extensive arrangements of educational opportunities to education seekers globally. Furthermore, regulating and accrediting higher education institutions throughout the country is the responsibility of the Ministry of Higher Education. Presently, there are approximately 20 government universities and over 500 private educational institutions in Malaysia, including foreign universities which also have abroad campuses, such as Nottingham, Monash, and Southampton universities (Nair & Munusami, 2020). The Malaysian higher education system places a significant emphasis on TVET (T for technical, V for vocational, E for education, and T for training) to prepare students for specialized industries. Both public and private institutions offer TVET programs, which are highly valued by the government and industry in Malaysia (Nair & Munusami, 2020).

13.3 TEACHING DESIGN TOWARD FLEXIBLE LEARNING

Teaching design toward flexible learning involves creating learning environments that are adaptable to the changing needs of students and the workforce. It requires incorporating innovative instructional design principles that can create effective and engaging learning experiences for students, including those pursuing TVET programs. In recent years, there has been an increasing demand for more flexible and adaptable learning options in Malaysia, such as online and distance learning programs. This has led to the development of various teaching design approaches that are geared toward meeting students' needs in these environments. One such approach is blended learning, which involves online and traditional learning to create a flexible and interactive learning experience for students (Goodyear, 2015). Another approach is the flipped classroom, which involves students learning content online before attending in-person classes where they can apply their learning through hands-on activities and discussions. The Malaysian government has also placed significant emphasis on developing TVET programs that are flexible and adaptable to the needs of students (Halik Bassah, 2022). This has led to the implementation of various initiatives, such as the National Dual Training System, which combines on-the-job training with classroom learning, and the Malaysia Technical University Network, which offers TVET programs in various fields of study. In conclusion, teaching design toward flexible learning in Malaysia is becoming increasingly important in creating effective and engaging learning environments for students. By embracing innovative instructional design principles and adapting to changing demands, higher education institutions can create effective and engaging TVET programs that achieve the students' needs in the 21st century (Ahmad et al., 2019).

13.4 FUTURE TVET FLEXIBLE LEARNING

Future TVET flexible learning involves creating learning environments and experiences that are adaptable to the changing needs of students and the workforce. It requires incorporating innovative instructional design principles that can create effective and engaging learning experiences for students, including those pursuing TVET programs (Joan, 2013). One of the key principles of future TVET is its flexibility, which learning with the use of technology could enhance the learning experience for students. This involves utilizing various types of technology, such as online learning platforms, virtual reality, and artificial intelligence, to deliver content and engage learners and grab their attention during the educating processes. The principle of providing flexible and adaptable learning options, such as blended learning and flipped classrooms, is crucial to enable students to study at their own pace and on their own schedule. This approach facilitates a balance between academic pursuits and other commitments, such as work or family responsibilities. Blended learning can also be addressed as a hybrid learning method that could mix traditional physical, face-to-face instruction with online instruction. At the same time, flipped classrooms involve learners watching lectures or accessing course materials online before attending in-person classes for discussion and collaboration (Betaubun, 2021). By incorporating these methods, higher education institutions can provide a more personalized and effective learning experience for students. Future TVET flexible learning also emphasizes the importance of creating a supportive and collaborative learning community for students. This involves fostering a culture of collaboration and feedback, where students can learn from and with each other and receive support from their instructors and peers. In conclusion, future TVET flexible learning is essential to develop and formulate students for the changing stipulation of the workforce and enable them to acquire fundamental expertise and the awareness to succeed in their chosen fields. By embracing innovative instructional design principles, utilizing technology, and creating a supportive learning community, higher education institutions can create effective and engaging TVET programs that satisfy students in the 21st century.

The model of heutagogy learning aims to enhance the innovative capacities of vocational learners by promoting the ability to think about tasks or problems in innovative ways. As technology is incorporating in the educating process, it can handle a pivotal position in developing students' creative abilities (Valiyev & Ibrahimova, 2021; Nussbaum et al., 2020). During the realm of vocational educational system, the increasing use of digital technology significantly impacts the creativity of vocational students. In the digital age, vocational students have the opportunity to use their expertise to create unique and innovative solutions that contribute to technological advancements. The heutagogy model prizes independent learning and aligns with the objective of fostering student creativity. By encouraging students to approach tasks with a creative mindset, this model facilitates the enhancement and improvement of innovative thinking capacities.

13.5 HEUTAGOGY APPROACH

In today's world, educators must overcome many obstacles to prepare students for a lifetime of learning and success in the world's knowledge economy. This requires equipping educators with skills and competencies that they can apply proficiently and innovatively to new circumstances in a constantly evolving and complicated world (Kuit & Fell, 2010). Traditional pedagogical and andragogical approaches are not effectively adequate for this purpose. As opposed to this, a self-directed and self-determined approach is required, in which students reflect on what and how they learn it, and teachers show students how to educate themselves (Kamenetz, 2010). The heutagogy approach empowers learners to be more self-sufficient, independent, and capable of navigating the complex demands of the modern world. With andragogy falling out of favor with some educators who view it as "outmoded in light of recent rapid development in new teaching methods, learning resources, and digital media", new technologies have also prompted the need for thinking about alternative pedagogical approaches (Bansal et al., 2020).

Heutagogy is a concept that has emerged in response to advances in higher education, proposing new practices and concepts to promote capable learners. The environment that uses a heutagogical learning prioritizes the enhancement of both learners' competencies. Additionally, it enhances their potentiality and ability to receive new knowledge (Bhoryrub et al., 2010; Tunstall & Neergaard, 2022). The emergence of Web 2.0 has returned the dividends in the concept of heutagogy due to advantages and benefits of social network and media, which complement and promote this learning paradigm. Heutagogy is often referred to as a "net-centric" theory that leverages many key features of the internet. Further, it can be considered a pedagogical approach because it is able to be utilized to emerging technologies in distance education, serving as a structure for instruction and learning in the era of digital (Blaschke, 2012).

Heutagogic design elements refer to the principles and strategies that could be utilized to build effective heutagogic learning experiences. Heutagogy is a learning approach that shifts the learning focus from teachers to students and places a strong emphasis on the acquisition of abilities for self-learning. Heutagogic design elements are essential to promoting effective and self-directed learning. These elements include self-directed learning, active learning, personalized learning, reflective learning, and technology-enabled learning. Self-directed learning emphasizes the importance of learners taking ownership of their learning process and setting goals for themselves. Active learning encourages learners to engage in hands-on activities, collaboration, and problem-solving. Personalized learning enables students to customize their educational experiences to meet their individual needs and interests. Reflective learning involves learners reflecting on their learning process and evaluating outcomes using self-assessment, peer feedback, and journaling. Technology-enabled learning involves leveraging technology to enhance learning, with access to instructional videos, interactive simulations, and online communities. Incorporating these elements into the learning environment can help promote and support self-directed and effective learning, enabling learners to improve the skills and competencies required to be successful in a constantly evolving world. Overall, heutagogic design elements are focused on developing learners' self-directed learning skills, promoting active and personalized learning, and leveraging technology to enhance the learning experience. By incorporating these elements into the design of learning experiences, educators can create more effective, engaging, and learner-centered learning environments (Figure 13.1).

The heutagogical learning approach provides a framework for promoting effective learning, creativity, self-assuredness, the capacity of utilizing knowledge in real-life situations, and working well with others. Moreover, Mohammad et al. (2019) have identified nine key heutagogical approaches, including an adaptable curriculum that acknowledges the fluid identity of learning, students as drivers of their learning, collaborative learning, guidance and support for students, and questions asked by students to drive their learning. In addition, students are involved in designing assessments and creating contextually relevant content, reflective practice is encouraged through the use of learning journals, and questions asked by students facilitate genuine collaboration between educators and learners. By incorporating these approaches into the learning environment, educators can promote self-directed and effective learning, enabling learners to improve their essential skills in order to thrive in this fast-changing and diverse world. In addition to its pedagogical principles,

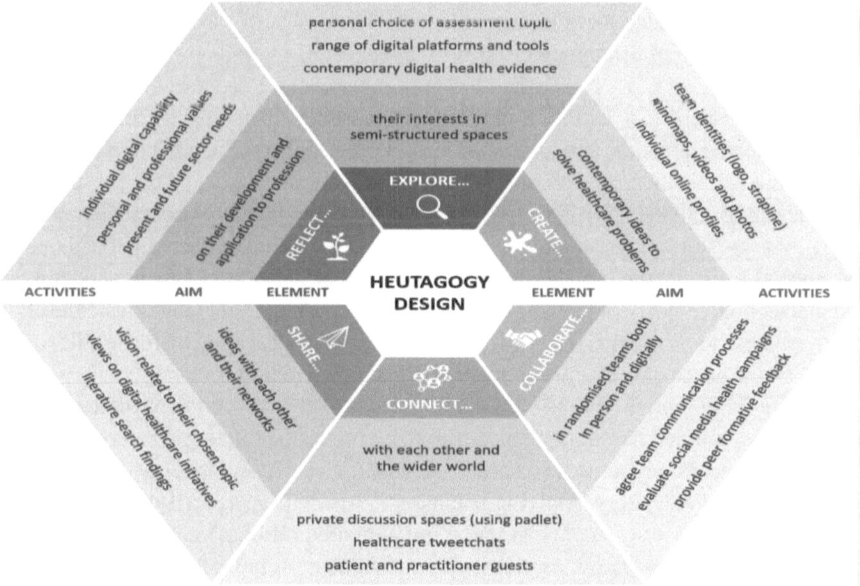

FIGURE 13.1 Elements of heutagogy design.

heutagogy can be enhanced by the affordances of digital technologies, as noted by Blaschke and Hase (2015) and Jaakkola (2015), particularly social media, as highlighted by Blaschke (2012, 2014b), Cochrane et al. (2014), and Cochrane and Narayan (2014). These digital tools can provide learners with access to a wealth of information, facilitate interaction and collaboration, and offer opportunities for independent education. Indeed, heutagogy is considered a suitable learning methodology for the learners of 21st century who desire to participate fully and actively in their education and leverage the benefits of digital technologies to create effective and engaging learning experiences. By prioritizing learners' autonomy, creativity, and capacity to learn, heutagogy can serve as an effective pedagogical approach to address the challenges of the digital age. Additionally, by leveraging the affordances of digital technologies, heutagogy is able to give learners personalized and tailored education practical knowledge, enabling them to acquire the capabilities and abilities necessary to thrive in this rapidly changing and complex world.

Distance education has a particular interest in heutagogy due to its critical features of learner autonomy and self-directedness, which have their roots in adult teaching and learning. Self-directed education is a prevalent feature of remote education as well as online learning forms like contract education and prior knowledge evaluation. The mature adult students' demographic is targeted by both distance education and heutagogy. The theory of distance education has the chance to be reinforced by heutagogy because of the variety of ways that heutagogy can expand the andragogical approach and the opportunities it offers to develop distance learning approaches such as Web 2.0. Therefore, heutagogy has significant implications for the development of effective distance education programs that cater to the unique necessities and characteristics of adult learners.

Social media and Web 2.0 have taken a crucial part in promoting confabulations about heutagogy in the level of higher education. The design of Web 2.0 platforms promotes the approach of heutagogy by permitting educators to guide their own educating paths, also taking an active role in their learning experiences. Recent studies have shown that social network can facilitate independent learning, including learning through mobile, reflective practice, and the use of virtual philosophers. The virtual philosopher is an online application that promotes problem-solving and critical thinking through active learning activities. Twitter users have been found to be more occupied with their educational pursuits and have more qualified GPAs. The use of learner-generated material, such as active media use, has the potential to enhance the development of cognitive and metacognitive skills. Whereas further researches are required to investigate the efficiency of these tools in fostering effective and self-directed learning, there are still some studies on the usage of social network and its function in aiding heutagogy.

Despite its potential benefits, academic educational institutions have been hesitant to fully embrace heutagogy since it may be impractical to set up a comprehensive heutagogical pedagogical structure. According to McAuliffe et al. (2008), due to the absence of educators, applying heutagogy is difficult in a certifying institution, and learner-guided assessment cannot be implemented rationally. However, instructors in the fields of nursing, engineering, and education have found heutagogy to be a reliable response to the pressing problems that their students face in the workplace and have centered their learning environments around it (Bhoyrub et al., 2010;

Ashton & Newman, 2006; Gardner et al., 2008). For instance, according to Bhoyrub et al. (2010), heutagogy offers a learning structure that caters to the requirements of nursing students, who should be educated in a dynamic, complicated, and unpredictably changing environment. A heutagogical approach to learning helps nursing students be a long-term learner and "makes sense of the necessary uncertainties that define nursing". Despite the challenges, heutagogy has the potential to consider the unique necessities of students in various professions and may serve as a crucial framework for lifelong learning.

According to Sugiyono and Setiawati (2020), heutagogy is an educational approach that prioritizes the human character of learners, their self-worth, and capacities, and recognizes ecological systems, environmental interactions, and learning exercises over teaching. Additionally, heutagogy is concerned with issues of people's adjustment to the new century. The heutagogy model challenges instructors to understand the world better than just the teacher's world, requires teachers to immerse themselves in the world of learners, and enables teachers to see beyond their discipline and preferred theories. Heutagogical approach has a greater focus on the requirement of the learners and their abilities, while empowering them to handle their learning responsibilities. By focusing on learner's individual strengths and interests, heutagogy can help in inventing more engaging and effective educating experiences that is tailored to the needs of each student.

13.6 PEERAGOGY APPROACH

Easy access to open education resources, as well as the availability of free and low-cost communication platforms that operate around the clock, has allowed people to learn together outside of formal institutions without any time constraints. The term used to describe this mode of teaching and learning is "peeragogy" (Rheingold, 2014), which combines the words "peer" and "agogy". The Latin term for this concept is "pearagogy", which has a similar meaning in English. In peeragogy, individuals learn and teach each other in a collaborative and self-directed manner, which allows having a great personalized and versatile learning experience. This educational method can be particularly effective for learners who may not have access to traditional educational resources, or who prefer to learn in a more informal and collaborative setting. Peeragogy has the potential to democratize education and provide opportunities for lifelong learning and skills development outside of formal educational institutions.

Once a group is formed, whether in a formal classroom setting or informally through online platforms, they become peers or friends with people of different ages or the same age. The internet allows these peers to learn from one another, leading to collaborative and self-directed learning experiences. While studies show that regular classroom meetings result in more advanced learning, the advantages of peeragogy are that it allows for more personalized and flexible learning experiences. If a teacher is familiar with the system, they can instruct students both at school and at home using the internet's online digital system. This can also be done in the teacher's spare time, depending on the students' readiness. Peeragogy allows students to pace themselves in learning, whether individually or in a group, which can be particularly effective for learners who may have different learning styles or needs. Therefore, peeragogy has the potential to enhance the learning experience for students by providing opportunities for collaborative and self-directed learning.

Indeed, with others and a strong emphasis on peer learning, peeragogy is an active kind of education. As described by Corneli et al. (2015) and Okewole and Knokh (2016), peeragogy emphasizes the importance of students working together to produce and use common knowledge, in a collaborative and participatory manner. This approach to learning has the potential to promote deeper understanding, critical thinking, and aptitude for solving problems as learners are involved in meaningful interactions with their peers and co-create knowledge. Peeragogy can also encourage a sense of belonging and community among students, as they work together toward shared goals and support each other in their learning journeys.

When COVID-19 was widespread, distance learning and online education became increasingly important, and peeragogy has emerged as a relevant approach in this context. Peeragogy involves active learning with a strong emphasis on learning from one another, where peers work together to produce and utilize shared knowledge. The use of digital teaching resources and group collaboration techniques has become more prevalent, providing learners with access to learning equipment, which is not limited by time schedule and a certain venue. By working together and utilizing digital materials, learners can engage in a more meaningful and interactive learning experience, which can enhance their critical thinking, problem-solving, and communication skills. Peeragogy can also foster a sense of responsibility and accountability among learners, as they work collaboratively toward a common goal. Therefore, peeragogy, combined with digital teaching materials and cooperative learning, has the potential to support educators with functional and engaging experiences in the context of online learning (Eliastuti et al., 2020; Etik et al., 2020).

13.7 CYBERGOGY APPROACH

Cybergogy is a teaching approach that focuses on promoting and encouraging learner-centered, independent, and teamwork learning in a virtual setting. It involves the use of technology and digital tools to create a dynamic and interactive learning environment that promotes self-directed learning and encourages collaboration among learners. Cybergogy recognizes the unique challenges and opportunities of online learning and seeks to leverage technology to enhance the learning experience. It emphasizes the importance of learner autonomy and encourages learners to actively participate in their own education path by setting their own goals, managing their own progress, and seeking feedback from their peers and instructors. The ultimate goal of cybergogy is to create a virtual learning environment that is engaging, effective, and meaningful for learners, while also preparing them for the demands of the digital age. According to Wang (2007), cybergogy is a descriptive label for strategies that create engaging online learning activities. Daud et al. (2019) suggested that cybergogy is aimed at helping both adults and young students to learn in a virtual environment, with a focus on learner-centered and collaborative learning. The root of cybergogy is a recognition of the approaches used, which may differ from those used in face-to-face learning environments (Matsuyama et al., 2019).

Cybergogy is a structured learning model that includes the concepts of pedagogy and andragogy. The education process is centered around the learner, who is encouraged to be independent and collaborate with others using online learning resources. Cybergogy is particularly relevant in the form of e-learning, where thousands of programs and materials are available on the internet for teachers to direct their students to learn from. The aim of cybergogy is to create engaging and beneficial learning activities in a virtual environment, where the strategies used may differ from those utilized in traditional learning environments (Wang & Kang, 2006a).

Cybergogy is a virtual learning environment that aims to build students' affective, cognitive, and psychomotor skills. With the onset of the COVID-19 pandemic, cybergogy has become increasingly prevalent in Indonesia and other countries as a means of continuing education and learning while adhering to social distancing guidelines. This approach to learning is facilitated by technological advances, which have simplified the process of connecting and collaborating in various contexts. As Brown et al. (2020) suggested, technology-mediated learning opportunities are becoming increasingly common, with an increasing percentage of learning occurring outside of formal learning environments. The use of cybergogy in online learning environments has the chance to improve the learning skills and promote the evolution of significant abilities, such as critical thinking, problem-solving, collaboration, and digital literacy. In essence, cybergogy is a teaching approach that leverages technology to create a dynamic and interactive virtual learning environment that promotes self-directed learning and encourages collaboration among learners.

The symbiotic relationship between technology and heutagogy is increasingly apparent in cybergogy, where learners are encouraged to take more control over their learning experiences. Technological developments have initiated new chances for learners to specify how and where their learning takes place, making learning a continuous process (Blaschke, 2021). As a result, cybergogy has become an important tool for learners, enabling them to engage in self-directed and collaborative learning experiences that are facilitated by technology. By combining heutagogy and technology, cybergogy has the chance to enhance the educational process and create more engaging and effective learning experiences for learners in virtual environments.

The content resulting from heutagogical principles can be applied to teaching, particularly in e-learning environments. Gregory and Bannister-Tyrrell (2017) and Praherdhiono et al. (2018) suggest that integrating heutagogical principles into online teaching can enhance collaboration, interaction, and engagement among learners. The e-learning environment is particularly suited to heutagogical approaches, as it provides opportunities for learners to collaborate and interact with one another. Agonács and Matos (2019) and Keserwani et al. (2019) highlight the importance of cooperation, conversation, and involvement in heutagogy, which are key features of e-learning environments.

Heutagogical principles promote an interactive and collaborative approach to learning, with a focus on learner independence and self-directed education. In the context of e-learning, heutagogy encourages learners to be in charge of their own education experiences by setting their own learning target, identifying their own learning needs, and engaging in self-directed and collaborative learning activities. When giving learners an engaging and personalized education adventure, heutagogical principles are able to enhance the influence of online instruction and promote lifelong education. This approach to learning fosters the growth of crucial abilities including interaction, problem-solving, and analytical thinking, which are essential for success in the digital age. By emphasizing collaboration and giving different chances for educators to assess and provide a realistic review on their progress, heutagogical principles can help to invent an interactive and evolving educational adventure, which promotes deeper learning and more solid feeling of independence over the educational progress (Hase & Kenyon, 2013; Wangid, 2014).

The cybergogy model encompasses learning aspects such as cognitive, emotional, and social, aiming to improve students' abilities and skills in these areas. Blaschke (2012) suggests that this model is built on an online platform application, which includes a learning simulation designed to enhance students' cognitive, social, and emotional intelligence as shown in Figure 13.2. The cognitive dimension of the cybergogy model is closely related to students' social life and greatly affects their personal abilities and emotional engagement, both consciously and unconsciously (Wang & Kang, 2006b).

To achieve the goals of cybergogy learning, learners must have a broad-minded character, high levels of motivation, and establish good relationships with the elements and the education progress (Blaschke, 2012). Also, it is important to consider the educating environment during the cybergogy modeling process, as learners must feel comfortable and supported in a high-social characteristic environment and have a strong desire to obtain results from the learning process they are following. Overall, the cybergogy model provides a comprehensive approach to online learning that considers cognitive, emotional, and social aspects of the learning process. By promoting collaboration, interaction, and engagement, cybergogy has the potential to enhance the influence of online education and improve pupils' abilities and skills in various areas.

The cybergogy model emphasizes the importance of engaging learners in the e-learning process to increase the success of the learning program. This is particularly

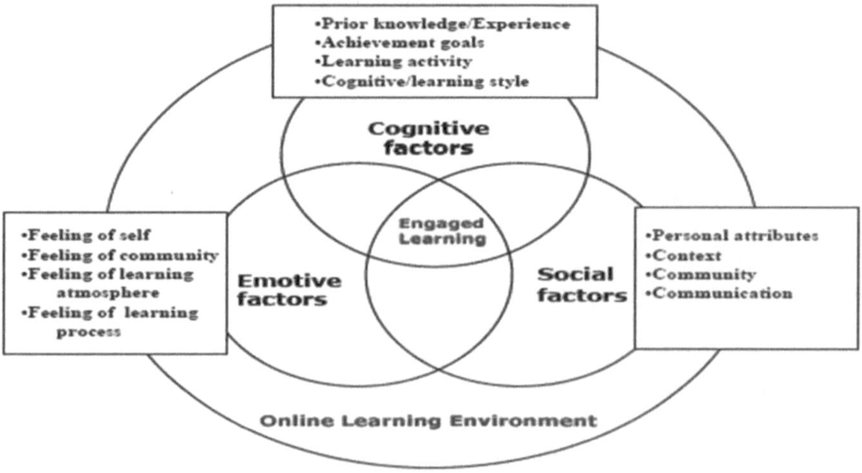

FIGURE 13.2 Cybergogy model of engaged learning (Wang & Kang, 2006a).

relevant when considering the development of technological self-efficacy. Chen (2017) suggests that engagement is an essential element in increasing success and performance in developing technological self-efficacy. Engagement fosters a favorable and receptive mindset toward learning to utilize computer technology, which can help learners feel more motivated, capitalize on resources, and meet challenges. Cybergogy satisfies the need for learning procedures that permit total learner involvement for more efficient learning (Muresan, 2015). By promoting collaboration, interaction, and engagement, cybergogy can help learners develop their technological self-efficacy by providing opportunities for learners to have charge of their educational experiences, establish their own educational objectives, determine their own educational needs, and engage in self-directed and collaborative learning activities. This approach can contribute to a higher likelihood of increased technological self-efficacy, which can be beneficial for learners' future academic and professional success.

The cybergogy approach to teaching and learning emphasizes that the educational process may happen at any place and at any moment, as long as learners have access to internet and the computers. The accessibility of a comprehensive and diverse subject depending on the network makes it easy for learners to access the information they need. With the advent of Technology 4.0, communication between lecturers and students has been facilitated through e-learning platforms, enabling collaborative learning strategies through video and audio sharing, blogging, and web-based tools like GitHub, which can support teamwork and project collaboration. Cybertechnology also supports community-based learning by encouraging active participation in conversations, idea-sharing, negotiation, and problem-solving within a community of peers (Bilfaqih & Qomarudin, 2015). This approach to teaching and learning leverages digital tools and technology to create a dynamic and interactive learning experience that allows educators take control of their education and collaborate with others to achieve their learning goals. However, Education 4.0 is a term that narrates the current phase of instruction that integrates technology and digital learning into traditional educational practices. Heutagogy, peeragogy, and cybergogy are all pedagogical principles that can be applied in Education 4.0 (see Table 13.1).

However, we can provide an example of how the heutagogy, peeragogy, and cybergogy approaches can be used in TVET. Heutagogy is a learner-centered approach to education where learners are in command of their own education. Peeragogy is a collaborative approach where learners work together to create knowledge and solve problems. Cybergogy is an approach that integrates technology into teaching and learning.

Heutagogy helps learners to: Encourage pupils to plan and draw their own educational aims and objectives, and to plan and organize their learning activities, provide resources and guidance to help students learn independently, such as online tutorials and self-assessment tools, and allow students to choose their own learning methods and strategies that work best for them, such as hands-on practice or video tutorials.

Peeragogy help learners to: Create opportunities for students to work collaboratively on welding projects, such as building a metal sculpture or a small structure like a table or chair, encourage students to share their knowledge and skills with each other, and to give and receive feedback on their work and use group discussions and problem-solving activities to help students develop critical thinking, communication, and teamwork skills.

TABLE 13.1
Description of Heutagogy, Peeragogy, and Cybergogy Principles

Education 4.0 Pedagogy	Principle	Description
Heutagogy (Blaschke, 2012; Hase, 2016)	Human agency	The learner's desire to deliberately acquire new knowledge through their favorite way of learning.
	Capability	Incorporating knowledge and abilities acquired in novel contexts or circumstances into real-world scenarios.
	Self-reflection and double-loop learning	In the educational process, learners are reflective to evolve the skills of metacognition and self-regulation and modify their attitudes and values.
	Nonlinear learning	The educational process is fluid and adaptive.
Peeragogy (Corneli et al., 2015)	Context as decentered center	The learner makes a personal contribution that modifies the environment or context.
	Meta-learning as a font of knowledge	Practice of awareness by students during their educational progress
	Peers provided feedback that wouldn't be there otherwise	A society (paragogues) that connects and shares the same interests online could change each other's perceptions of knowledge
	Learning is distributed and nonlinear	The capacity of peer educators to collaborate on developing content in a virtual community on an open learning platform
	Realize the dream and then wake up!	Each knowledge community member should be able to finish their educational goal and then transition to the following knowledge community.
Cybergogy (Wang & Kang, 2006a)	Cognitive factor	Learners' capacity to understand and use knowledge can be impacted by a variety of circumstances. Prior knowledge, accomplishment objectives, learning activities, and cognitive or learning style of the learner are some of these variables.
	Emotive factor	The psychological state of learners and the relationships within their learning community can also have a big impact on how well they can learn and comprehend new information. Learners' capacity to engage with and apply new information can be impacted by a variety of factors, including their self-efficacy, self-confidence, and self-competence; their ties to their community; and the design of the learning environment.
	Social factor	The socioeconomic background, sense of community, and community identity of the learners, all of which could be strengthened through cooperative activities.

Cybergogy help learners: Use online resources and tools to enhance learning, such as virtual welding simulations or instructional videos, integrate technology into hands-on welding activities, such as using digital welding machines or robotic welding systems, and use online platforms, such as discussion forums or social media groups, to facilitate communication and collaboration among students, and to provide access to additional resources and support. Overall, using the heutagogy, peeragogy, and cybergogy approaches can help TVET students become more self-directed, collaborative, and technology-savvy learners, which is essential in today's rapidly changing job market.

13.8 DISCUSSION

The concept of higher education teaching design plays a crucial role in preparing learners for the demands of the future workforce in TVET fields. In this chapter, we have explored the application of Heuta-Peera-Cyber approaches in designing effective teaching strategies for TVET flexible learning environments. The integration of the Cyber approach, which leverages digital technologies and online resources, further enhances the design of higher education teaching for TVET flexible learning. The Cyber approach enables learners to access a wide range of learning materials, collaborate with peers, and engage in interactive simulations and virtual environments. These digital tools not only facilitate personalized learning experiences but also provide opportunities for real-world application and skill development.

Implementing Heuta-Peera-Cyber approaches in TVET flexible learning brings numerous benefits. Learners are more actively engaged in the learning process, fostering a sense of ownership and motivation. The flexible nature of these approaches allows learners to access educational resources at their own pace and convenience, accommodating their personal and professional commitments. Additionally, the integration of digital technologies promotes digital literacy and prepares learners for the technology-driven demands of the future workforce.

However, challenges must also be considered when implementing these approaches. Adequate infrastructure, including access to reliable internet connections and appropriate digital devices, is essential for effective implementation. Additionally, the design and delivery of assessments in flexible learning environments require careful consideration to ensure fairness, validity, and reliability. Successful case studies and best practices demonstrate the positive outcomes of implementing Heuta-Peera-Cyber approaches in TVET flexible learning. These examples highlight increased learner engagement, improved learning outcomes, and enhanced employability of TVET graduates. Institutions and educators can draw inspiration from these examples to inform their own teaching design strategies and adapt them to their specific contexts. To optimize the benefits of Heuta-Peera-Cyber approaches, educators, administrators, and policymakers should collaborate to develop comprehensive guidelines and policies. These guidelines can address issues such as curriculum alignment, faculty development, technical support, and quality assurance (Blaschke & Hase, 2016). Furthermore, ongoing research and exploration are crucial to staying abreast of emerging technologies, pedagogical approaches, and instructional design principles that can further enhance TVET flexible learning.

In conclusion, the concept of higher education teaching design toward future TVET flexible learning, analyzed through the lens of Heuta-Peera-Cyber approaches, offers a holistic framework for designing effective teaching strategies. By leveraging the HEUTA framework, incorporating the PEERA model, and integrating digital technologies, educators can create engaging, learner-centered, and future-oriented learning experiences for TVET learners. The successful implementation of these approaches requires collaboration, appropriate infrastructure, and ongoing research to continually improve and adapt to the evolving needs of learners and the workforce.

13.9 CONCLUSION

The concept of higher education teaching design toward future TVET (T for Technical, V for Vocational, E for Education, and T for Training) flexible learning refers to the enhancement of innovative and effective teaching approaches that take into account the adjustments requirements and expectations of learners, besides advances in technology and education research.

Heuta-Peera-Cyber are three approaches that can be used to design such teaching strategies. HEUTA (Heuristic, Experiential, and Active Teaching) emphasizes the importance of active learning, where students engage in hands-on experiences and problem-solving activities to develop their skills and knowledge. PEERA (Problem-based, Engaging, Exploratory, Reflective, and Authentic) focuses on creating a learning environment that is relevant, engaging, and reflective of real-world scenarios. CYBER (Collaborative, Yielding, Blended, Engaging, and Reflective) emphasizes the utilization of technology to encourage collaborative education, blended learning, and reflective thinking.

Together, these approaches can be used to design teaching strategies that are flexible, adaptable, and responsive to the needs of individual learners. By incorporating elements of active learning, problem-based learning, and technology-mediated learning, educators can create learning experiences that are engaging, relevant, and effective. Additionally, these approaches can help foster a culture of lifelong learning, where students are encouraged to continue learning beyond the classroom and into their professional lives.

However, implementing Heuta-Peera-Cyber approaches in TVET institutions requires a deliberate and systematic effort on the part of educators and administrators. Here are some steps that can be taken to implement these approaches:

1. Carry out a needs assessment: Before designing any teaching strategies, it is important to grasp the needs and expectations of students and employers. This can involve conducting surveys, focus groups, or other forms of needs assessment to determine which knowledge and skills are most in demand in the sector.
2. Design learning outcomes: Based on the needs analysis, educators can design learning outcomes that are aligned with knowledge and different capacities, which are related to the industry. These outcomes should be specific, measurable, and achievable, and should guide the development of teaching strategies.

3. Develop Heuta-Peera-Cyber teaching strategies: Using the learning outcomes as a guide, educators can design teaching strategies that incorporate elements of HEUTA, PEERA, and CYBER approaches. This can involve designing hands-on activities, problem-based learning scenarios, and technology-mediated learning experiences.
4. Incorporate industry partnerships: To ensure that the teaching strategies are relevant to the needs of the industry, it is important to incorporate industry partnerships into the design process. Industry representatives can give a realistic review on relevance of education outcomes and effectiveness of the teaching strategies.
5. Evaluate and revise: After implementing the teaching strategies, it is important to evaluate their effectiveness and revise them as needed. This can involve collecting feedback from students, monitoring student performance, and identifying areas for improvement.

While Heuta-Peera-Cyber approaches can be effective in improving the quality of TVET education, educators may face several challenges when implementing these approaches. Some of these challenges include:

1. Resistance to change: Educators may be resistant to change and may prefer traditional teaching methods. This can make it difficult to implement Heuta-Peera-Cyber approaches, which require a more active and student-centered approach to learning.
2. Lack of resources: Implementing Heuta-Peera-Cyber approaches may require the use of technology, hands-on activities, and other resources that may not be readily available in some TVET institutions. This can limit the effectiveness of the teaching strategies and make it more difficult to engage students.
3. Limited training and professional development: Educators may require additional training and professional development to effectively implement Heuta-Peera-Cyber approaches. This may be difficult to obtain in some TVET institutions, which may have limited resources for professional development.
4. Difficulty in assessing student learning: Heuta-Peera-Cyber approaches emphasize active learning and problem-based learning, which can make it difficult to assess student learning using traditional assessment methods. Educators may need to develop new assessment methods that are more aligned with these teaching strategies.
5. Limited industry partnerships: Heuta-Peera-Cyber approaches require close collaboration with industry partners to ensure that the teaching strategies are relevant to the needs of the industry. However, some TVET institutions may have limited industry partnerships, which can limit the effectiveness of these approaches.

In addition, based on the concept of higher education teaching design toward future TVET flexible learning using the heutagogy, peeragogy, and cybergogy approaches, we recommended some recommendations for future research:

1. Explore how these approaches impact student learning outcomes, engagement, and satisfaction in TVET flexible learning environments. This could involve conducting quantitative or qualitative studies, including surveys, focus groups, or interviews with students and instructors.
2. Explore how different technologies, such as virtual reality, simulations, and online platforms, can be used to enhance TVET flexible learning. This research could also explore the challenges and opportunities associated with using technology in TVET flexible learning environments.
3. Explore the role of peer collaboration in TVET flexible learning: Peeragogy emphasizes the importance of collaboration and self-directed learning, and future research could explore how peer collaboration can be facilitated in TVET flexible learning environments. This could involve examining how group work, peer feedback, and other collaborative learning activities impact student learning outcomes and engagement.
4. Explore how heutagogy impacts instructor roles, including how instructors can best support student learning in a more self-directed environment.
5. Explore the challenges associated with implementing flexible learning in TVET, including institutional barriers, instructor resistance, and student readiness.

REFERENCES

Agonács, N., & Matos, J. F. (2019). Heutagogy and self-determined learning: A review of the published literature on the application and implementation of the theory. *Open Learning, 34*(3), 223–240. https://doi.org/10.1080/02680513.2018.1562329

Ahmad, S. A., Yoke, S. K., Yunos, R. M., & Amin, J. M. (2019). Exploring lecturers' readiness for 21st century education in Malaysian higher learning institutions. *European Journal of Teaching and Education, 1*(1), 15–29. https://doi.org/10.33422/EJTE.2019.10.27

Amanina, N., Zaidatun, B., Kew, T., & Na, S. (2022). Innovative pedagogical principles and technological tools capabilities for immersive blended learning: A systematic literature review. *Education and Information Technologies* (Issue 0123456789). Springer US. https://doi.org/10.1007/s10639-022-11243-w

Ashton, J., & Newman, L. (2006). An unfinished symphony: 21st century teacher education using knowledge creating heutagogies. *British Journal of Educational Technology, 37*(6) 825–840. https://doi.org/10.1111/j.1467-8535.2006.00662.x.

Azevedo, R., Behnagh, R. F., Duffy, M., Harley, J. M., & Trevors, G. (2012). Metacognition and self-regulated learning in student-centered learning environments. In S. Land, & D. Jonassen (Eds), *Theoretical Foundations of Learning Environments* (2nd Edition, pp. 171–197). Routledge. https://doi.org/10.4324/9780203813799

Bansal, A., Jain, S., Sharma, L., Sharma, N., Jain, C., & Madaan, M. (2020). Students' perception regarding pedagogy, andragogy, and heutagogy as teaching-learning methods in undergraduate medical education. *Journal of Education and Health Promotion, 9*, 301. https://doi.org/10.4103/jehp.jehp_221_20

Betaubun, M. (2021). The students' attitude and learning experience toward flipped classroom implementation during COVID-19 outbreak: A survey study. *Jurnal Pendidikan Progresif, 11*(1), 54–62.

Bhoyrub, J., Hurley, J., Neilson, G. R., Ramsay, M., & Smith, M. (2010). Heutagogy: An alternative practice-based learning approach. *Nurse Education in Practice, 10*(6), 322–326. https://doi.org/10.1016/j.nepr.2010.05.001

Bilfaqih, Y., & Qomarudin, M. N. (2015). *Esensi penyusunan materi pembelajaran daring.* Deepublish.

Blaschke, L.M. (2012). Heutagogy and lifelong learning: A review of heutagogical practice and selfdetermined learning. *The International Review of Research in Open and Distributed Learning, 13*(1), 56–71. https://doi.org/10.19173/irrodl.v13i1.1076

Blaschke, L. M. (2014). Using social media to engage and develop online learners in self-determined learning. *Research in Learning Technology, 22*. https://doi.org/10.3402/rlt.v22.21635

Blaschke, L. M. (2019). The pedagogy-andragogy-heutagogy continuum and technology-supported personal learning environments. *In SpringerBriefs in Open and Distance Education* (pp. 75–84). Springer Singapore. https://doi.org/10.1007/978-981-13-7740-2_9

Blaschke, L. M. (2021). The dynamic mix of heutagogy and technology: Preparing learners for lifelong learning. *British Journal of Educational Technology, 52*(4),1629–1645. https://doi.org/10.1111/bjet.13105

Blaschke, L.M., & Hase, S. (2015). Heutagogy: A holistic framework for creating 21st century selfdetermined learners. In M. M. Kinshuk & B. Gros (Eds.), *The Future of Ubiquitous Learning: Learning Designs for Emerging Pedagogies* (pp. 25–40). Heidelberg, Germany: Springer Verlag.

Blaschke, L. M., & Hase, S. (2016). Heutagogy: A holistic framework for creating twenty-first-century self-determined learners. In: Gros, B., Kinshuk, & Maina, M. (Eds), *The Future of Ubiquitous Learning* (pp. 25–40). Springer. https://doi.org/10.1007/978-3-662-47724-3_2

Blaschke, L. M., & Hase, S. (2019). Heutagogy and digital media networks. *Pacific Journal of Technology Enhanced Learning, 1*(1), 1–14

Brown, M., McCormack, M., Reeves, J., Brook, D. C., Grajek, S., Alexander, B., Bali, M., Bulger, S., Dark, S., & Engelbert, N. (2020). *2020 Educause Horizon Report Teaching and Learning Edition.* EDUCAUSE.

Chan, C. G., Embi, M. A. Bin, & Hashim, H. (2019). Primary school teachers' readiness towards heutagogy and peeragogy. *Asian Education Studies, 4*(1), 11. https://doi.org/10.20849/aes.v4i1.602

Chen, I.-S. (2017). Computer self-efficacy, learning performance, and the mediating role of learning engagement. *Computers in Human Behavior, 72*, 362–370. https://doi.org/10.1016/j.chb.2017.02.059

Cochrane, T., & Narayan, V. (2014). Cultivating creative approaches to learning. In L. M. Blaschke, C. Kenyon, & S. Hase (Eds.), *Experiences in Self-Determined Learning* (pp. 149–169). Center for Open Education Research (COER), University of Oldenburg: Oldenburg, Germany. https://uol.de/fileadmin/user_upload/coer/Experiences-in-self-determined-learning.pdf.

Cochrane, T., Antonczak, L., Guinibert, M., & Mulrennan, D. (2014). Developing a mobile social media framework for creative pedagogies. In *10th International Conference on Mobile Learning*, Madrid, Spain. https://files.eric.ed.gov/fulltext/ED557241.pdf.

Corneli, J., Danoff, C. J., Pierce, C., Ricaurte, P., & Macdonald, L. S. (2015). Patterns of peeragogy citation. *Pattern Languages of Programs Conference, 17*(1), 27–43. https://doi.org/10.1080/13645579.2014.854012

Daud, W. A. A. W., Teck, W. K., Ghani, M. T. A., & Ramli, S. 2019. The needs analysis of developing mobile learning application for cybergogical teaching and learning of Arabic language proficiency. *International Journal of Academic Research in Business and Social Sciences, 9*(8), 33–46.

De Wit, H., Hunter, F., Howard, L., & Egron-Polak, E. (2015). *Internationalisation of Higher Education*: A study for the European parliament. http://www.europarl.europa.eu/RegData/etudes/STUD/2015/540370/IPOL_ST U(2015)540370_EN.pdf

Eliastuti, M., Rasyid, Y., & Lustyantie, N. (2020). Digital teaching materials for Indonesian scientific writing teaching in college. *International Journal of Advanced Science and Technology, 29*(5 Special Issue), 1498-1502.

Etik, E., Lustyantie, N., & Emzir, E. (2020). The effect of cooperative learning and thinking ability toward writing skill of exposition paragraph. *SALTeL Journal (Southeast Asia Language Teaching and Learning), 3*(2), 22–31. https://doi.org/10.35307/saltel.v3i2.50

Gardner, A., Hase, S., Gardner, G., Dunn, S.V., & Carryer, J. (2008). From competence to capability: A study of nurse practitioners in clinical practice. *Journal of Clinical Nursing, 17*(2), 250–258. https://doi.org/10.1111/j.1365-2702.206.0188.x

Goodyear, P. (2015). Teaching as design. *Herdsa Review of Higher Education, 2*(2), 27–50.

Gregory, S., & Bannister-Tyrrell, M. (2017). Digital learner presence and online teaching tools: Higher cognitive requirements of online learners for effective learning. *Research and Practice in Technology Enhanced Learning, 12*(1), 1–17.

Halik Bassah, N. A. S. (2022). The issues and challenges of TVET in Malaysia: From the perspective of industry experts. In: *TVET@Asia*, issue 18, 1–15. Online: https://tvet-online.asia/issue/18/the-issues-and-challenges-of-tvet-in-malaysia-perspective-of-industry-experts/ (retrieved 31.05.2023).

Hase, S. (2016). Self-determined learning (heutagogy): Where have we come since 2000? *Southern Institute of Technology Journal of Applied Research*, Special Ed(May), 1-21. https://sit.ac.nz/Portals/0/upload/documents/sitjar/Heutagogy-One.pdf%0Ahttps://www.sit.ac.nz/Portals/0/upload/documents/sitjar/Heutagogy-One.pdf

Hase, S., & Kenyon, C. (2013). *Self-Determined Learning: Heutagogy in Action*. A&C Black.

Jaakkola, M. (2015). Teacher heutagogy in the network society: A framework for critical reflection. In P. Jandric, & D. Boras (Eds.), *Critical Learning in Digital Networks* (pp. 163–178). Switzerland: Springer International Publishing.

Joan, R. (2013). Flexible learning as new learning design in classroom process to promote quality education. *i-Manager's Journal on School Educational Technology, 9* (1), 37–42.

Kamenetz, A. (2010). A is for app: How Smartphones, handheld computers sparked an educational revolution. *Fast Company, 144*, 60

Keserwani, P., De, K., Roy, P. P., & Pal, U. (2019). Zero shot learning based script identification in the wild. In *2019 International Conference on (Document Analysis and Recognition ICDAR)*, University of Technology Sydney (UTS), Australia and International Convention Centre (ICC) Sydney, 987–992.

Kuit, J.A., & Fell, A. (2010). Web 2.0 to pedagogy 2.0: A social-constructivist approach to learning enhanced by technology. *In Critical Design and Effective Tools for E-Learning in Higher Education: Theory into Practice*, 310–325. United States: IGI Global.

Lysenko, L., Kiforo, E., Wade, C. A., Abrami, P. C., Iminza, R., & Kiforo, E. (2021). Self-regulated learning in Kenyan classrooms: A test of a process e-portfolio. *International Journal of Instruction, 15*(3), 63–82. https://doi.org/10.29333/iji.2022.1534a

Matsuyama, Y., Nakaya, M., Okazaki, H., Lebowitz, A. J., Leppink, J., & Van Der Vleuten, C. (2019). Does changing from a teacher-centered to a learner-centered context promote self-regulated learning: A qualitative study in a Japanese undergraduate setting. *BMC Medical Education, 19*(1), 1–12.

McAuliffe, M., Hargreaves, D., Winter, A., & Chadwick, G. (2008). *Does pedagogy still rule? In Proceedings of the 2008 AAEE Conference*, December 7–10, 2008. Yeppoon, Queensland. Retrieved from: https://www.engineersmedia.com.au/journals/aaee/pdf/AJEE_15_1_McAuliffe%20F2.pdf

Mohammad, S., Siang, T. C., Osman, S., Jamaluddin, N. Y., Alfu, N. A. M., & Huei, L. Y. (2019). A proposed heutagogy framework for structural steel design in civil engineering curriculum. *International Journal of Emerging Technologies in Learning, 14*(24), 96–105. https://doi.org/10.3991/ijet.v14i24.12091

Muresan, M. (2014). Using cybergogy and andragogy paradigms in lifelong learning. *Procedia – Social and Behavioral Sciences, 116,* 4722–4726. https://doi.org/10.1016/j.sbspro.2014.01.1015

Muresan, M. (2015). Collaborative learning and cybergogy paradigms for the development of transversal competences in higher education. *Euromentor Journal, 6*(2), 21–29.

Nair, B. V., & Munusami, C. (2020). Knowledge management practices: An exploratory study at the Malaysian higher education institutions. *Journal of Research in Innovative Teaching & Learning, 13*(2), 174–190.

Nussbaum, M., Barahona, C., Rodriguez, F., Guentulle, V., Lopez, F., Vazquez-Uscanga, E., & Cabezas, V. (2020). Taking critical thinking, creativity and grit online. *Educational Technology Research and Development.* https://doi.org/10.1007/s11423-020-09867-1

Okewole, J. T., & Knokh, I. (2016). Curating to creating: An open pedagogy for quality learning people product. In *Conference Proceedings & Working Papers*, Commonwealth of Learning (COL) and Open University Malaysia (OUM), Malaysia, 1–6.

Praherdhiono, H., Adi, E. P., & Devita, R. N. (2018). Understanding of digital learning sources with the heutagogy approach using the k-means and naive bayes methods. In 4th *International Conference on Education and Technology, ICET 2018*, 23–27. https://doi.org/10.1109/ICEAT.2018. 8693929

Rheingold, H. (Ed.) (2014). *The Peeragogy Handbook* (2nd Edition). Chicago: PubDomEd.

Sugiyono, B., & Setiawati, S. (2020). *Becoming an Agile Facilitator: menjadi Faislitator Daring yang Lincah dan Tangkas.* Malang: Media Nusa Creative.

Tunstall, R., & Neergaard, H. (2022). Flashmob: A heutagogical tool for social learning in entrepreneurship education. *Entrepreneurship Education and Pedagogy, 5*(3), 472–492. https://doi.org/10.1177/25151274211017547

Valiyev, A. N., & Ibrahimova, D. (2021). Opportunities for the development of creativity skills of students in the process of teaching drawing science. *Academica: An International Multidisciplinary Research Journal, 11* (3). https://doi.org/10.5958/2249-7137.2021.00843.0

Wang, M. J. (2007). Designing online courses that effectively engage learners from diverse cultural backgrounds. *British Journal of Educational Technology, 38*(2), 294–311.

Wang, M. J. & Kang, J. (2006aa) *Cybergogy of Engaged Learning through Information and Communication Technology: A Framework for Creating Learner Engagement.* New York: Springer Publishing.

Wang, M., & Kang, M. (2006bb). Cybergogy for engaged learning: A framework for creating learner engagement through information and communication technology. *Engaged Learning with Emerging Technologies*, 225–253. https://doi.org/10.1007/1-4020-3669-8_11

Wangid, M. N. (2014). Student-centered learning: Self-regulated learning. In *International Conference on Fundamentals and Implementation of Education (ICFIE)*, Graduate School Yogyakarta State University Indonesia In Collaboration With Faculty Of Education University Of Auckland, Indonesia, 161–165.

14 Innovative Pedagogical Approaches for Sustainable Development

Implications for Business Students' Engagement and Learning Performance in Higher Education

Song Bee Lian, Tee Poh Kiong, Wong Ling Chai, Devinder Kaur, and Liew Chee Yoong

14.1 INTRODUCTION

Education for sustainable development has emerged as a focus for higher education institutions to support the mission of sustainable development goals established by the United Nations. The aims of education related to sustainable development are to develop learners' knowledge, abilities, values, competencies and skills to address the economic and socio-ecological complexities (Kioupi & Voulvoulis, 2019). There are various initiatives and strategies undertaken by the higher education institutions in aligning curricula, learning outcomes and assessments towards sustainability (Žalėnienė & Pereira, 2021; Stough et al., 2018). To achieve effectiveness in embedding sustainable development in education, appropriate pedagogical approaches are crucial to engage learners to achieve desired learning performance (Burns et al., 2019). The pedagogical approaches are evolving to more innovative and increasing learners' engagement, such as collaborative learning, transformative learning, research-based learning and active learning. Hence, higher institutions and instructors are constantly finding the best pedagogical approaches to be applied in sustainable development education, particularly in the business education context (Dziubaniuk et al., 2023). The selection of the suitable pedagogical approach in business education for sustainable development is critical as it has significant impact on achieving the desired learning outcomes (Chiang & Chen, 2022).

Past studies have highlighted the importance of active learning (Hui et al., 2021; MacVaugh & Norton, 2012), formative analytics (Chiang & Chen, 2022), research-based learning (Gamage et al., 2021), transformative learning (Corazza

 DOI: 10.1201/9781003400691-14

et al., 2022) and collaborative learning (Hariharan & Williams, 2011) in business education related to sustainable development, but there are limited studies that link these pedagogies to the effects of student engagement and learning performance. Understanding student engagement and learning performance is important for instructors to create and facilitate more appealing learning experiences and outcomes for students (Sukor et al., 2021; Wong et al., 2023b). Therefore, there is a gap in knowledge related to the appropriate pedagogical approaches to be used that can contribute to more effective student engagement (Cottafava et al., 2019) and learning performance (Hyytinen et al., 2023) in sustainable development related to business education.

This chapter aims to examine the innovative pedagogical approaches of active learning, formative analytics, research-based learning, transformative learning and collaborative learning for sustainable development, and its implications on business students' engagement and learning performance in the higher education context. The present study will develop a new framework integrating innovative pedagogical approaches that are linked to student engagement and learning performance in business education related to sustainable development. The proposed innovative pedagogical approaches for sustainable development can be applied by the higher education institutions through embedding in the teaching and learning activities, curriculum design, educators' training and assessments.

14.2 LITERATURE REVIEW

14.2.1 Student Engagement

Student engagement relates to the students' effort, willingness, desire and commitment to involve or participate in learning activities that contribute to the desired learning outcomes (Gray & Diloreto, 2016; Song et al., 2023). The learning techniques of teaching, mentoring and coaching are useful to improve student engagement (Gamage et al., 2021; Arjomandia et al., 2021).

Innovative pedagogical styles used in the higher education focusing on sustainable development can improve business students' engagement (Perkiss et al., 2020). For example, transformative learning experiences have empowered students' engagement to effectively co-creating and co-managing projects with the focus on sustainable development goals (Cottafava et al., 2019). In a study conducted on the undergraduate business programme with sustainable development in its curriculum by Collier et al. (2022), it was reported that assignments focused on sustainable development goals within the business programme have improved students' understanding, interest and engagement with the subject matter.

14.2.2 Active Learning

Active learning refers to the active learning participation by learners focusing on cognitive development through discussions, evaluation, role-play, projects and other active learning activities to develop higher-order thinking skills (Gogus, 2012). Problem-based learning is an active learning pedagogy which has proven to improve the learners' sustainability competences (Kioupi et al., 2022). The emphasis on achieving sustainability competences extends to the ability to perform capabilities embedded in knowledge and skills, further relating to their application in real business environments. The active learning will equip learners with higher-order thinking skills and sustainability change agent skills (Demsie et al., 2023).

In a study conducted by Howell (2021) it was shown that flipped classroom approach contributed to better learners' engagement. Learners can easily relate information learnt in real practice in the contemporary environment. Supporting the findings from Howell's study, Claro and Esteves (2021) asserted that active learning pedagogy is beneficial in developing learners' sustainability-oriented capabilities in acquiring knowledge, practices, interactions and improvements as it enhanced student engagement and learning performance. Hence, hypothesis 1 is developed:

H1: Active learning has a significant positive impact on student engagement.

14.2.3 Formative Analytics

Formative analytics involves learners' reflection on the learning acquired and developing suggestions on what can be improved to achieve further learning success (Sharples, 2019). Learners apply learning achievement information based on previously set learning goals to form suggestions for improvement. In business education directed towards sustainable development, the learning processes encompass acquiring knowledge and evaluation of critical issues related to the environment, economy and society; subsequently, students use self-reflective approach to make recommendations for improvement (Pauw et al., 2015).

Chiang and Chen (2022) reported that self-reflection capability is the learner's ability to reflect on their role in community or society, leading to further action. In a study conducted on Chinese students in a business course, it was found that embedded learners' responsible learning in the sustainable development education resulted in better sustainability learning outcomes and performances. With that, the next hypothesis is postulated:

H2: Formative analytics has a significant positive impact on student engagement.

14.2.4 Research-Based Learning

Research-based learning focuses on the advancement of existing knowledge, with the combination of theoretical knowledge base, and usage of data collection and analysis procedures to study a phenomenon (Arora et al., 2017). Research-based learning involves research and inquiry skills, as well as methods to enable improvement in assimilation and application of knowledge (Strachan et al., 2019). Research-based learning helps the development of critical, reflexive and independent learners (Brew & Saunders, 2020).

Research-based learning leads to better student engagement as learners have to engage with instructors, as well as various stakeholders such as the public or customers to obtain data. This learning pedagogy encourages the learners to explore real-world issues, obtain relevant evidence to identify the issues and interpret the findings, which leads to better engagement (Amos & Levinson, 2019). For example, sustainable development related to business education involves the impact of sustainable development strategies towards the real business environment or the effectiveness in implementation of sustainable development practices and achieving the goals. The below hypothesis is developed:

H3: Research-based learning has a significant positive impact on student engagement.

14.2.5 Transformative Learning

Transformative learning is an important pedagogical approach in business education related to sustainable development to develop new generation of young leaders with contemporary sustainable behaviours (Cottafava et al., 2019). Transformative learning emphasises the learners' ability to make decisions and create actions in their environment (Corazza et al., 2022). Based on transformational learning theory, Mezirow has outlined that the transformative learning process involves reflection and reasoning, and decision-making for change (Mezirow, 2018). As capacity building, transformative learning focuses on building students' relevant skills needed for sustainability development.

Transformative learning, as a pedagogical method for sustainable development, has positively influenced student engagement in business education (Corazza et al., 2022). Transformative learning contributed to the transfer of knowledge and interaction between instructors and students that influenced students' attitudes and behaviours on sustainable development matters (Nousheen & Kalsoom, 2022; Dziubaniuk et al., 2023). The subsequent hypothesis is formed:

H4: Transformative learning has a significant positive impact on student engagement.

14.2.6 Collaborative Learning

Collaborative learning focuses on facilitating learners working together in a group to obtain and share knowledge, and solve problems (Laal & Ghodsi, 2012). Collaborative learning facilitates learners' communication skills that demonstrate effective engagement between them (Baanqud et al., 2020; Wong et al., 2023a). Hariharan and Williams (2011) asserted that collaborative learning in the business education related to sustainable development has contributed to students being able to identify the relevance of sustainability and deeper understanding on the sustainability context in business, which also enhanced their engagement.

Sustainable development in education embedding social value collaborations between students, instructors and communities has contributed to better engagement (Kalemaki et al., 2021). Collaborative learning can enhance learners' social skills through engagement performed and helps learners in developing their capabilities (Chiang & Chen, 2022). The subsequent hypothesis is postulated:

H5: Collaborative learning has a significant positive impact on student engagement.

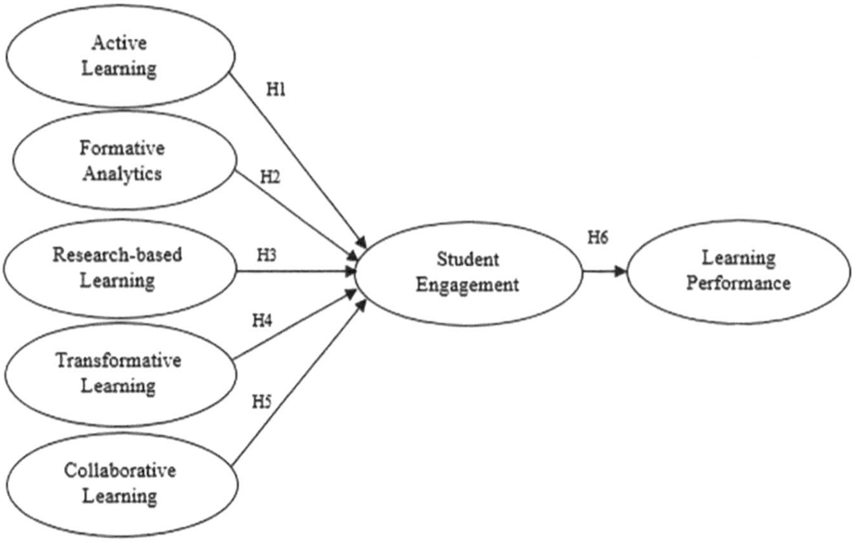

FIGURE 14.1 Conceptual Framework.

14.2.7 Learning Performance

Learning performance refers to the measurement of students' performance on academics (Oyinloye & Imenda, 2019). Learning performance involves the desired achievement of a learning activity. Effective student engagement contributed to better learning performance (Delfino, 2019). In sustainable development for business education, the pedagogical approaches can motivate learners to perform better in academics as they engage with their peers, instructors and external parties (Dziubaniuk et al., 2023). Finally, the below hypothesis is postulated:

H6: Student engagement has a significant positive impact on learning performance.

The research model is presented in Figure 14.1. Innovative pedagogical approaches for sustainable development comprising active learning, formative analytics, research-based learning, transformative learning and collaborative learning are correlated to student engagement. Subsequently, student engagement is hypothesised towards learning performance.

14.3 RESEARCH METHODOLOGY

14.3.1 Sampling Procedures, Data Collection and Measurement Items

The sample consists of 427 undergraduate and postgraduate business students who have experienced education for sustainable development for the past six months from 10 higher education institutions in Malaysia. As this study has seven constructs in the structural model, the chosen sample size of 427 has fulfilled the recommended minimum sample size of 300 (Hair et al., 2010). Using quantitative approach, we have gathered the responses from convenience sampling, whereby the samples were targeted personally by the researchers during their presence at the selected 10 higher education institutions in Malaysia. The convenience sampling method provides more flexibility and efficiency (Jager et al., 2017). We have used the closed-ended survey questionnaires with 29 measurement items and Likert scales from 1 (constitutes strongly disagree) to 6 (constitutes strongly agree). The 6-point Likert scales were used as it provides more reliable and accurate responses as neutral point was eliminated (Komorita and Graham, 1965).

The survey questionnaires were divided into two sections. The first section includes the demographic questions related to gender, age, and local and international students. The second section includes the survey questionnaires for the seven constructs of active learning, formative analytics, research-based learning, transformative learning, collaborative learning, students' engagement and learning performance.

14.3.2 Validity and Reliability Assessment

Assessment of the measurement variables' consistency and accuracy is required through validity and reliability assessment. Based on the pilot study conducted on 35 respondents, the Exploratory Factor Analysis (EFA) findings showed a Kaiser-Meyer-Olkin (KMO) value of 0.917, which exceeded the base value of 0.5 and was categorised as excellent as it is above 0.9 (Field, 2013). The factor loadings for all 29 items have achieved construct validity with values between 0.686 and 0.879, above the required minimum value of 0.3 (Hair et al., 2010). Bartlett's test of Sphericity recorded significant $p=0.000$ (Barlett, 1951). Referring to Cronbach's alpha (CA), the internal reliability results achieved exceeded the minimum threshold value of 0.7 (Saunders et al., 2009). The Composite Reliability (CR) and Average Variance Extracted (AVE) values have also met the requirements of 0.7 (Hair et al., 2019) and 0.6 (Bagozzi & Yi, 1988) respectively (Table 14.1).

TABLE 14.1
Validity and Reliability Assessment

Construct (Source)	Item	Statement	Factor Loading	AVE, CR, CA
Active learning (Carr et al., 2015)	AL1	Participated in group discussions	0.724	0.630(AVE)
	AL2	Participated in community projects related to the study	0.797	0.872(CR) 0.868(CA)
	AL3	Involved in class presentation	0.860	
	AL4	Discuss and evaluate ideas obtained from reading with the instructors	0.789	

(Continued)

TABLE 14.1 (*Continued*)
Validity and Reliability Assessment

Construct (Source)	Item	Statement	Factor Loading	AVE, CR, CA
Formative analytics (Pauw et al., 2015; Sharples, 2019)	FA1	Able to apply self-reflective approach to recommend for improvement	0.735	0.585(AVE) 0.875(CR)
	FA2	Able to obtain reflection on the learning acquired	0.807	0.868(CA)
	FA3	Developed suggestions on what can be improved to achieve further learning success	0.778	
	FA4	Embedded responsible learning in sustainable development education	0.811	
	FA5	Able to reflect on personal role in society	0.686	
Research-based learning (Wesselsa et al., 2021; Amos & Levinson, 2019)	RL1	Acquired the learning on research process	0.792	0.706(AVE)
	RL2	Acquired inquiry skills to improvement in assimilation of knowledge	0.859	0.906(CR) 0.903(CA)
	RL3	Acquired inquiry skills to improvement in application of knowledge	0.879	
	RL4	Learning exposure to the real business environment issues	0.828	
Transformative learning (Walker, 2018)	TL1	Adopted new ways of thinking	0.752	
	TL2	Made different decisions to change	0.818	0.627(AVE)
	TL3	situations or solve problems	0.812	0.870(CR)
	TL4	Able to perform critical self-reflection for improvement	0.784	0.870(CA)
		Instructor creates opportunities to demonstrate change in behaviour		
Collaborative learning (Laal & Ghodsi, 2012; Kalemaki et al., 2021)	CL1	Worked in a group to obtain knowledge	0.757	
	CL2	Worked in a group to share knowledge	0.848	0.688(AVE)
	CL3	Worked in a group to solve problem	0.865	0.898(CR)
	CL4	Embedded social value collaborations with peer, instructors or communities	0.844	0.896(CA)
Student engagement (Song et al., 2022)	SE1	Enhanced engagement with peers	0.787	0.637(AVE)
	SE2	Enhanced engagement with instructors	0.802	0.875(CR)
Learning performance (Song et al., 2022)	SE3	Enhanced engagement with communities	0.788	0.876(CA)
	SE4	Effective internal and external	0.814	0.641(AVE)
	LP1	engagement	0.819	0.877(CR)
	LP2	Performed well in assessment results	0.821	0.876(CA)
	LP3	Increased understanding/knowledge on	0.795	
	LP4	the subject	0.767	
		Obtained the skills required for learning performance		
		Achieved the desired learning outcomes for the subject		

14.4 RESULTS

14.4.1 Profile of Samples

This research encompassed N=427 students, comprising 51% of males and 49% of females. As for the age group, most of the samples were between 21 and 24 years (43%), subsequently between 18 and 20 years (38%), 25 and 30 years (19%). The respondents of international students represented 40% and local students represented 60%.

14.4.2 Model Compatibility and Hypotheses Testing

The data analysis was implemented using structural equation modelling (SEM) method from SPSS AMOS version 27. SEM enables hypotheses testing based on multiple constructs (Awang, 2014). Figure 14.2 illustrates the structural model. Overall, the fit indices exhibited good fit. The model is significant at p value of 0.00. The incremental fit values (incremental fit index (IFI)=0.945, comparative fit index (CFI)=0.945 and Tucker-Lewis index (TLI)=0.938) exceeded the 0.9 minimum requirement (Awang, 2014). RMSEA value was achieved at 0.053, which met the base value of 0.08 (Kenny et al., 2014). Finally, the ratio of x^2/df was 2.191 and chi-square value was 790.993. The findings concluded that all hypotheses (H1 to H6) were supported as stipulated in Table 14.2.

H1 indicated that active learning had a significant positive effect on student engagement with $p < 0.001$ and estimated value of 0.236. The finding had supported

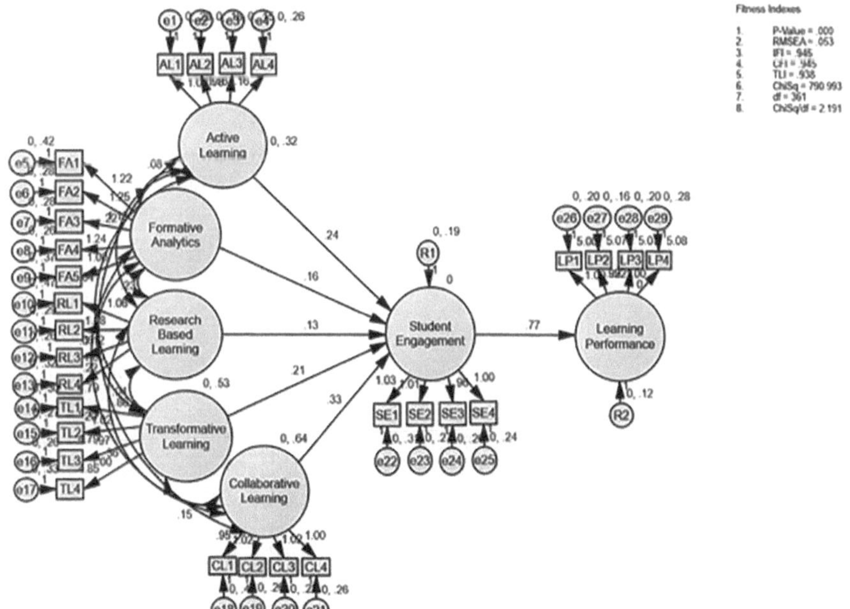

FIGURE 14.2 Structural Model.

TABLE 14.2
Hypotheses Testing Results

Hypothesised Relationships	Estimate	p Values	Result
H1 Active learning → student engagement	0.236	***	Supported
H2 Formative analytics → student engagement	0.165	0.004	Supported
H3 Research-based learning → student engagement	0.129	0.003	Supported
H4 Transformative learning → student engagement	0.206	***	Supported
H5 Collaborative learning → student engagement	0.327	***	Supported
H6 Student engagement → learning performance	0.774	***	Supported

Note: ***$p < 0.001$.

past studies of Howell (2021) and Claro and Esteves (2021), which concluded that business students were actively involved in discussions, projects, presentations and evaluation of ideas when experiencing education for sustainable development. Formative learning also influenced student engagement and H2 was supported, with $p = 0.004$ and estimated value of 0.165. Subsequently, H3 was also supported which proved research-based learning had influenced student engagement with $p = 0.003$ and estimated value of 0.129, consistent with the past studies of Amos and Levinson (2019). This result showed that research-based learning was carried out effectively among the business students in the form of research process, exposure to real business environment issues and knowledge application that enhanced engagement. Transformative learning significantly influenced student engagement with $p < 0.001$ and estimated value of 0.206, confirming H4 was supported.

Collaborative learning had the strongest correlation value of 0.327 among other independent constructs, with H5 being supported. Therefore, we can conclude that education for sustainable development has contributed to better social values between the relevant stakeholders. Finally, student engagement had a significant impact on learning performance with H6 being supported, $p < 0.001$ and correlation value of 0.774. Hence, studies of Delfino (2019) and Dziubaniuk et al. (2023) were consistent with the results from the present research.

14.5 DISCUSSION

The innovative pedagogical approaches suggested for sustainable development in this study are critically important to be applied in enhancing the business students' engagement and learning performance. To increase the effectiveness of active learning, instructors can leverage on the combination of group discussions, class presentation, case study analysis and community project involvement to assist learners develop sustainability competencies. Formative analytics involves deep learning process by learners which require further efforts on self-reflection on learning content, critical thinking and evaluation on the strategies to be undertaken for improvement, and ability to reflect on the role in community. In this context, instructors should relate the learning process with the real business environmental problems on sustainability so that learners can improve on self-reflection aspect.

Research-based learning can be embedded in sustainable development for business education through exposure to the learners on the research process, which is important when undertaking real research to find out the actual existence of the business environmental problems related to sustainability and propose for solutions. For example, students should be exposed to research-based learning through investigation of the real business environment issues such as waste management and inefficiency in resources management, and subsequently relate to solve the issues through the application of sustainable development goals (SDGs) 12 on sustainable consumption and production.

Transformative learning involves embedding leadership, managerial and soft skills to enhance the learners' autonomy in making changes through effective decision-making. The need is to establish an appropriate model of collaborative value learning that relates to acquiring and sharing of knowledge and collective problem-solving. Business students should collaborate with external parties such as business organisations, government and non-government organisations in sustainability projects. It is important to expose the learners to more effective engagement involving various stakeholders (i.e. peers, instructors, business organisations, government and communities) when applying these innovative pedagogical approaches in education.

To maximise learners' learning performance, the assessments, learning outcomes and skills attainment of the subjects can be linked to the goals for business education related to sustainable development. Furthermore, the learning outcomes should be well defined and focused on the objectives to achieve in education for sustainable development. Hence, the proposed innovative pedagogical approaches could be aligned appropriately to the learning outcomes to drive better student engagement and learning performance.

14.6 CONCLUSION

This chapter has covered the innovative pedagogical approaches for sustainable development education that can be utilised to influence business students' engagement and learning performance. As for the limitations in this study, the samples were confined to business students in higher education context. Future studies should expand the samples to other courses such as information technology and applied sciences. The pedagogical approaches were active learning, formative analytics, research-based learning, transformative learning and collaborative learning. Therefore, we propose future studies to consider other pedagogical approaches as comparative study to compare the effectiveness of different learning approaches.

REFERENCES

Amos, R., & Levinson, R. (2019). Socio-scientific inquiry-based learning: An approach for engaging with the 2030 sustainable development goals through school science. *International Journal of Development Education and Global Learning*, *11*(1), 29–49. https://doi.org/10.18546/IJDEGL.11.1.03.

Arjomandia, A., Seufertb, J. H., O'Brienc, M., & Rossettod, L. C. (2021). An empirical study of business student engagement with active teaching strategies: A comparison of first year and senior students. *American Business Review*, *24*(1), 67–89. https://doi.org/10.37625/abr.24.1.67-89.

Arora, V., Saxena, P., & Gangwar, N. (2017). Higher education faculty career orientation and advancement. In M.S. Manna (Ed.). *Project-based learning (PBL) and research-based learning* (pp. 76–91). Delhi: CEGER

Awang, Z. (2014). *Research Methodology and Data Analysis* (2nd ed.). Universiti Teknologi Mara: UiTM Press.

Baanqud, N. S., Al-Samarraie, H., Alzahrani, A. I., & Alfarraj, O. (2020). Engagement in cloud-supported collaborative learning and student knowledge construction: A modeling study. *International Journal of Education Technology in Higher Education*, *17*(56), 1–23. https://doi.org/10.1186/s41239-020-00232-z.

Bagozzi, R., & Yi, Y. (1988). On the evaluation of structural equation models. *Journal of the Academy of Marketing Sciences*, *16*(1), 74–94.

Bartlett, M. S. (1951). The effect of standardization on a Chi-square approximation in factor analysis. *Biometrika*, *38*(3/4), 337–344.

Brew, A., & Saunders, C. (2020). Making sense of research-based learning in teacher education. *Teaching and Teacher Education*, *87*, 102935. https://doi.org/10.1016/j.tate.2019.102935

Burns, H., Kelly, S. S., & Spalding, H. E. (2019). Teaching sustainability: Recommendations for best pedagogical practices. *The Journal of Sustainability Education*, *19*, 1–16.

Carr, R., Palmer, S., & Hagel, P. (2015). Active learning: The importance of developing a comprehensive measure. *Active Learning in Higher Education*, *16*(3), 1–14. https://doi.org/10.1177/1469787415589529.

Chiang, M., & Chen, P. (2022). Education for sustainable development in the business programme to develop international Chinese college students' sustainability in Thailand. *Journal of Cleaner Production*, *374*, 134045. https://doi.org/10.1016/j.jclepro.2022.134045.

Claro, P. B., & Esteves, N. R. (2021). Teaching sustainability-oriented capabilities using active learning approach. *International Journal of Sustainability in Higher Education*, *22*(6), 1246–1265. https://doi.org/10.1108/IJSHE-07-2020-0263.

Collier, E., Odell, K. E., & Rosenbloom, A. (2022). Teaching sustainable development: An approach to rapidly introducing the UN sustainable development goals into an undergraduate business curriculum. *Journal of Global Responsibility*, *13*(4), 361–379. https://doi.org/10.1108/JGR-11-2021-0100.

Corazza, L., Cottafava, D., & Torchia, D. (2022). Education for sustainable development: A critical reflexive discourse on a transformative learning activity for business students. *Environment, Development and Sustainability*. https://doi.org/10.1007/s10668-022-02335-1.

Cottafava, D., Cavaglià, G., & Corazza, L. (2019). Education of sustainable development goals through students' active engagement: A transformative learning experience. *Sustainability Accounting, Management and Policy Journal*, *10*(3), 521–544. https://doi.org/10.1108/SAMPJ-05-2018-0152.

Delfino, A. P. (2019). Student engagement and academic performance of students of partido State University. *Asian Journal of University Education*, *15*(3), 42–55. https://doi.org/10.24191/ajue.v15i3.05.

Demssie, Y. N., Biemans, H. J. A, Wesselink, R., & Mulder, M. (2023). Fostering students' systems thinking competence for sustainability by using multiple real-world learning approaches. *Environmental Education Research*, *29*(2), 261–286. https://doi.org/10.1080/13504622.2022.2141692.

Dziubaniuk, O., Ivanova-Gongne, M., & Nyholm, M. (2023). Learning and teaching sustainable business in the digital era: A connectivism theory approach. *International Journal of Education Technology in High Education*, 20(20), 1–23. https://doi.org/10.1186/s41239-023-00390-w.

Field, A. (2013). *Discovering Statistics Using IBM SPSS Statistics: And Sex and Drugs and Rock "N" Roll* (4th ed.). Sage: Los Angeles, London, New Delhi.

Gamage, K. A. A., Perera, D. A. S., & Wijewardena, M. A. D. N. (2021). Mentoring and coaching as a learning technique in higher education: The impact of learning context on student engagement in online learning. *Education Sciences*, *11*(10), 574. https://doi.org/10.3390/educsci11100574.

Gogus, A. (2012). Active learning. In: Seel, N. M. (Eds.) *Encyclopedia of the Sciences of Learning*. Springer: Boston, MA. https://doi.org/10.1007/978-1-4419-1428-6_489.

Gray, J. A., & Diloreto, M. (2016). The effects of student engagement, student satisfaction, and perceived learning in online learning environments. *International Journal of Educational Leadership Preparation*, *11*(1), 1–20.

Hair, J. F., Black, W. C., Babin, B. J., & Anderson, R. E. (2010). *Multivariate Data Analysis: A Global Perspective*. Prentice Hall: New Jersey.

Hair, J. F., Risher, J. J., Sarstedt, M., & Ringle, C. M. (2019). When to use and how to report the results of PLS-SEM. *European Business Review*, *31*(1), 2–24. https://doi.org/10.1108/EBR-11-2018-0203.

Hariharan, K., & Williams, J. B. (2011). Greening the business curriculum through collaborative learning spaces: Theoretical and practical perspectives. *Malaysian Journal of Distance Education*, *13*(1), 35–45.

Howell, R. A. (2021). Engaging students in education for sustainable development: The benefits of active learning, reflective practices and flipped classroom pedagogies. *Journal of Cleaner Production*, *325*, 129318. https://doi.org/10.1016/j.jclepro.2021.129318.

Hui, T., Lau, S. S. S., & Yuen M. (2021). Active learning as a beyond-the-classroom strategy to improve university students' career adaptability. *Sustainability*, *13*(11), 6246. https://doi.org/10.3390/su13116246

Hyytinen, H., Laakso, S., Pietikäinen, J., Ratvio, R., Ruippo, L., Tuononen, T., & Vainio, A. (2023). Perceived interest in learning sustainability competencies among higher education students. *International Journal of Sustainability in Higher Education, 24*(9), 118–137. https://doi.org/10.1108/IJSHE-06-2022-0198.

Jager, J., Putnick, D. L., & Bornstein, M. H. II. (2017). More than just convenient: The SCIENTIFIC merits of homogeneous convenience samples. *Monographs of the Society for Research in Child Development*, *82*(2), 13–30. https://doi.org/10.1111/mono.12296.

Kalemaki, I, Garefi, I., & Protopsaltis, A. (2021). Assessing the impact of social innovation education on student's engagement. *European Journal of Sustainable Development*, *10*(1), 389–400. https://doi.org/10.14207/ejsd.2021.v10n1p389.

Kenny, D., Kaniskan, B., & McCoach, D. (2014). The performance of RMSEA in models with small degrees of freedom. *Sociological Methods & Research*, *44*(3), 486–507. https://doi.org/10.1177/0049124114543236.

Kioupi V., & Voulvoulis N. (2019). Education for sustainable development: A systemic framework for connecting the SDGs to educational outcomes. *Sustainability*. *11*(21), 6104. https://doi.org/10.3390/su11216104

Kioupi, V., Vakhitova, T. V., & Whalen, K. A. (2022). Active learning as enabler of sustainability learning outcomes: Capturing the perceptions of learners during a materials education workshop. MRS Energy & Sustainability, *9*, 64–78. https://doi.org/10.1557/s43581-021-00019-3

Komorita, S. S., & Graham, W. K. (1965). Number of scale points and the reliability of scales. *Educational and Psychological Measurement*, *25*, 987–995. https://doi.org/10.1177/0013164465.02500404.

Laal, M., & Ghodsi, S. M. (2012). Benefits of collaborative learning. *Procedia – Social and Behavioral Sciences*, *31*, 486–490, https://doi.org/10.1016/j.sbspro.2011.12.091.

MacVaugh, J., & Norton, M. (2012). Introducing sustainability into business education contexts using active learning. *International Journal of Sustainability in Higher Education, 13*(1), 72–87. https://doi.org/10.1108/14676371211190326.

Nousheen, A., & Kalsoom, Q. (2022). Education for sustainable development amidst COVID-19 pandemic: Role of sustainability pedagogies in developing students' sustainability consciousness. *International Journal of Sustainability in Higher Education, 23*(6), 1386–1403.

Oyinloye, O. M., & Imenda, S. N. (2019). The impact of assessment for learning on learner performance in life science. *EURASIA Journal of Mathematics, Science and Technology Education, 15*(11), 1–8. https://doi.org/10.29333/ejmste/108689.

Pauw, J. B. d., Gericke, N., Olsson, D., & Berglund, T. (2015). The effectiveness of education for sustainable development. *Sustainability, 7*(11), 15693–15717. https://doi.org/10.3390/su71115693.

Perkiss, S., Anastasiadis, S., Bayerlein, L., Dean, B., Jun, H., Acosta, P., Gonzalez-Perez, M.A., Wersun, A., & Gibbons, B. (2020). Advancing sustainability education in business studies through digital service learning. *American Business Review, 23*(2), 283–299.

Saunders, M., Lewis, P., & Thornhill, A. (2009). *s*. Pearson, New York.

Sharples, M. (2019). *Practical Pedagogy – 40 New Ways to Teach and Learn* (1st ed.). Routledge: London.

Song, B. L., Lee, K. L, Liew, C. Y., Ho, R. C., & Lin, W. L. (2022). Business students' perspectives on case method coaching for problem-based learning: Impacts on student engagement and learning performance in higher education. *Education and Training, 64*(3), 416–432. https://doi.org/10.1108/ET-03-2021-0106.

Song, B. L., Lee, K. L., Liew, C. Y., & Subramaniam, M. (2023). The role of social media engagement in building relationship quality and brand performance in higher education marketing. *International Journal of Educational Management, 37*(2), 417–430. https://doi.org/10.1108/IJEM-08-2022-0315.

Stough, T., Ceulemans, K., Lambrechts, W., & Cappuyns, V. (2018). Assessing sustainability in higher education curricula: A critical reflection on validity issues. *Journal of Cleaner Production, 172*, 4456–4466. https://doi.org/10.1016/j.jclepro.2017.02.017

Strachan, S. M., Marshall, S., Murray, P., Coyle, E. J., & Sonnenberg-Klein, J. (2019). Using vertically integrated projects to embed research-based education for sustainable development in undergraduate curricula. *International Journal of Sustainability in Higher Education, 20*(8), 1313–1328. https://doi.org/10.1108/IJSHE-10-2018-0198

Sukor, R., Ayub, A. F. M., Rashid, N. K. M. A., & Halim, F. A (2021). Relationship between students' engagement with academic performance among non-food science students enrolled in food science course. *Journal of Turkish Science Education, 18*(4), 638–648. https://doi.org/10.36681/tused.2021.95.

Walker, S. L. (2018). Development and validation of an instrument for assessing transformative learning: The transformative learning environments survey (TLES). *Journal of Transformative Learning, 5*(1), 23–46.

Wessels, I., Rueß, J., Gess, C., Deicke, W., & Ziegler, M. (2021). Is research-based learning effective? Evidence from a pre-post analysis in the social sciences. *Studies in Higher Education, 46*(12), 2595–2609. https://doi.org/10.1080/03075079.2020.1739014.

Wong, L. C., Tee, P. K., Cham, T. H., & Lim, M. F. (2023a). Online learning during Covid-19 pandemic: A view of undergraduate student perspective in Malaysia. In: Al-Emran, M., Al-Sharafi, M. A., Shaalan, K. (Eds.), *International Conference on Information Systems and Intelligent Applications. ICISIA 2022. Lecture Notes in Networks and Systems, 550.* Springer, Cham. https://doi.org/10.1007/978-3-031-16865-9_32.

Wong, L. C., Tiong, Y. Y., Tee, P. K. & Chan, B. Y. F. (2023b). Exploring a tricomponent attitude model for the application of "forced" online learning in the post-covid era. *Malaysian Journal of Learning and Instruction*, *20*(2), 233–266. https://doi.org/10.32890/mjli2023.20.2.

Žalėnienė, I., & Pereira, P. (2021). Higher education for sustainability: A global perspective. *Geography and Sustainability*, *2*(2), 99–106. https://doi.org/10.1016/j.geosus.2021.05.001.

15 Exploring the Role of AI Platforms in Improving English-Speaking Skills in Malaysian Higher Education Institutions

Sheiladevi Sukumaran and Nurul Salwana Khair

15.1 INTRODUCTION

The ability to communicate effectively in English is becoming increasingly important in today's globalized world. However, many students in Malaysian higher education institutions struggle with English proficiency, especially speaking skills. This highlights the need for innovative solutions to enhance English language education. The rise of globalization and the prevalent use of digital platforms underscore the importance of linguistic proficiency. Being multilingual doesn't only open more career avenues but also fosters cultural understanding, diminishes divides, and encourages mutual respect among diverse populations. Language education empowers learners to engage in international business, access a broader scope of educational materials, and comfortably navigate different cultural landscapes (Novita & Purwati, 2021). Effective language skills bridge communication gaps, facilitating seamless interactions even between those from varied linguistic backgrounds (Menggo et al., 2019). These competencies are especially vital for international partnerships or transnational dealings, where clear understanding and collaboration are paramount (Munoz Luna & Taillefer, 2014).

Artificial intelligence (AI) has notably transformed classroom dynamics and the broader educational landscape. The inclusion of AI tools in language education offers custom-tailored lessons, making language learning more efficient and personalized. This innovation has refined the process of learning languages (Wang et al., 2023). With the aid of immersive tools like virtual reality (VR) and augmented reality (AR), AI-driven platforms present authentic linguistic scenarios. This enables students to hone their conversational abilities, experience cultural immersion, and conquer language-related challenges using interactive simulations (Tedre et al., 2021). The infusion of technology in the classroom has redefined the educator's role. They've evolved from mere instructors to active facilitators, leveraging AI to amplify instructional

 DOI: 10.1201/9781003400691-15

outcomes. AI-driven analytics offer educators valuable insights into students' progress, allowing for more targeted support and individualized teaching approaches.

Technology plays a pivotal role in fostering students' engagement by offering diverse avenues for collaboration on both global and local scales. The rise of online forums, collaborative documents, and video conferencing tools has transformed the way students engage with each other. These digital platforms offer a favorable setting that actively fosters peer-to-peer learning, facilitates effortless exchange of knowledge, and enables cultural interaction among individuals from diverse backgrounds (Al-Maroof et al., 2021). This integration of technology into educational settings significantly amplifies the overall learning experience by transcending traditional classroom boundaries to foster a dynamic level of interactivity among learners worldwide (Gyasi et al., 2021).

The importance of language acquisition in the technological age is immense, as it fosters global communication, cultural understanding, and personal development. The incorporation of AI has significantly altered classroom activities and the learning process. In addition to providing personalized learning experiences, fostering cultural immersion, and providing immediate feedback, AI-powered tools enable educators to adapt their teaching strategies and more effectively engage students. While technology disrupts traditional teaching practices, it creates new opportunities for collaborative and adaptive learning, empowering students to flourish in a globally interconnected world. The dynamic synergy between language learning and AI in education promises enhanced linguistic proficiency and enhanced educational outcomes in the future (Rui & Badarch, 2022).

15.2 RESEARCH OBJECTIVES AND RESEARCH QUESTIONS

This study aims to assess perspectives on using AI to improve English proficiency among university students. The research questions are: (1) How receptive are students to adopting AI tools for writing, tutoring, assessment, and other applications? (2) Do they perceive benefits of AI for accessibility, engagement, and personalized learning? (3) What reservations exist regarding over-reliance on automation versus human teachers? The objectives are to quantify openness to AI integration, determine perceived advantages and risks, identify effective AI applications for skill development, and evaluate concerns about replacement of instructors. Research methods will include statistical analysis of survey ratings and open-ended questions to measure attitudes toward AI-enabled English learning. This will provide insights into how AI platforms could be designed and incorporated to enhance proficiency based on student needs and perspectives.

15.3 LITERATURE REVIEW

15.3.1 The Role of AI in Language Learning

AI in Language Acquisition and Instruction: AI is rapidly transforming the field of language acquisition and instruction. Now, AI is primarily implemented in the form of online tools or applications that are designed to assess the performance of students and offer feedback (Zhao, 2023). However, the potential of AI in education extends far beyond the realm of straightforward web applications.

While there is no denying that AI has made significant advancements in language learning, there are opposing arguments regarding its role in developing English-speaking skills in Malaysian higher education institutions. Some critics argue that relying too heavily on AI for language learning may hinder the development of essential skills such as oral communication and interpersonal interaction. They believe that face-to-face interactions with instructors and peers are crucial in acquiring fluency and cultural understanding (Taillefer & Munoz-Luna, 2014). Moreover, some argue that AI chatbots, although designed to simulate human conversation, cannot replicate the nuances and complexities of real-life communication (Yun & Park, 2022). The lack of emotional intelligence and contextual understanding can result in limited language proficiency.

AI has the potential to revolutionize the way we learn languages. By providing personalized, adaptive, and engaging learning experiences, AI can help students to learn languages more effectively and efficiently (Moreno-Guerrero et al., 2020). AI tools such as VR and AR can immerse students into virtual worlds, creating a realistic language learning environment where they can practice their speaking skills through interactions with AI chatbots (Han et al., 2022). In addition to these benefits, AI can also be used to provide access to language learning resources. AI can be used to create virtual libraries of language learning resources, such as textbooks, audio recordings, and video lessons. This can help to make language learning more accessible to people who live in remote areas or who cannot afford traditional language learning programs (Rebolledo Font de la Vall & González Araya, 2023). AI also has the ability to facilitate translation of both written and spoken text from one language to another, which proves beneficial in aiding students with their language acquisition. By providing access to resources in their native tongue, AI assists learners in expanding their knowledge of new languages (Fan, 2016). The integration of AI-powered language learning tools has paved the way for personalized and adaptable learning experiences, specifically tailored to meet individual student requirements. This has significantly enhanced the educational landscape and shifted it toward a student-centered approach (Awang et al., 2018).

15.3.2 English Language Learning in Malaysian Higher Education

In Malaysia, the significance of English language proficiency is widely recognized owing to its status as a global language and its crucial role in diverse fields including business, science, and technology. Malaysia strives to have one of the top higher education systems in the world, thus both local Malaysians and overseas students are encouraged to speak English (Shah et al., 2017). Although numerous higher education institutions in Malaysia provide English language courses and programs, students encounter various difficulties in enhancing their proficiency in spoken English. Limited exposure to English-speaking environments and opportunities for practice can also contribute to the struggle with spoken English (Selan, 2021). While the Malaysian school system may focus on teaching English, speaking the language is often a different issue altogether.

A large number of Malaysian students are afraid of making mistakes when they speak English. This can prevent them from speaking up in class or participating in conversations. English is considered a global lingua franca and a crucial skill for Malaysian graduates to thrive in an increasingly interconnected world. However, studies suggest that many students struggle with spoken English due to limited exposure and opportunities to practice outside the classroom (Wang & Rajprasit, 2015). Grammar and pronunciation can be two of the most challenging aspects of learning English. Teachers worldwide tend to focus more on grammar and vocabulary that are tested in examinations, neglecting the development of speaking skills (Nadesan & Shah, 2020). As a result, students may have a good understanding of rules and vocabulary but face challenges when it comes to conversing in English beyond the classroom.

The education system in Malaysia places a significant emphasis on the teaching of English, aiming to equip students with the necessary language skills. However, when it comes to spoken English, many Malaysian students encounter difficulties and tend to be hesitant due to fear of making mistakes. This apprehension can hinder their active participation in classroom discussions and other social interactions involving speaking in English. It is crucial for Malaysian graduates to effectively communicate in this global lingua franca as they navigate an increasingly interconnected world (Norhasim & Mohamad, 2020).

To address these challenges, studies suggest that more exposure and opportunities outside the traditional classroom setting are needed for Malaysian learners to practice their spoken English skills confidently. Engaging in AI-driven education represents a distinctive learning journey for students. As they advance in their academic pursuits, learners often experience a learning curve while becoming more adept and proficient in utilizing diverse technological tools (Rebolledo & Araya, 2023).

A great majority of employers in multiple nations and territories where English is not the official language see English language abilities as critical (Murray, 2010). Because of its widespread use as a medium of communication, companies place a high value on candidates' competence in the language. Aside from that, English competence is strongly linked to increased educational chances (Swee Heng, 2012). Expanding beyond mere proficiency in written communication or listening comprehension, developing fluency and confidence with oral expression is equally important. Enhancing students' ability not only benefits academic engagement but also empowers them professionally upon entering the workforce (Tuzlukova & Usha-Prabhukanth, 2018). By offering spaces where learners can freely engage in conversations without fear of being judged or criticized for their mistakes, educators can help foster a supportive learning environment conducive to overcoming anxieties associated with speaking English.

15.3.3 Research Methodology

This quantitative, cross-sectional study utilizes a survey to explore perspectives on adopting AI platforms to improve English-speaking skills. As Creswell (2008) notes, proper methodology provides valid research. The questionnaire consists of closed-ended Likert scale questions to measure attitudes toward AI for learning English, along with open-ended exploratory items. This standardized data facilitates statistical analysis for comparison, as Glaser (1978) highlights. The survey assesses key variables like perceived usefulness, accessibility, personalization, and automation concerns based on technology acceptance theories. By quantitatively measuring student opinions, this methodology aims to produce objective insights. The cross-sectional design efficiently collects data from a representative sample at one time point. This can yield initial findings to guide future longitudinal approaches. The study views proficiency as learned through experience, not innate. The survey is administered to university students in Malaysia to quantify perspectives on integrating AI linguistics platforms into English curriculums. Descriptive analysis will be applied to identify patterns in closed- and open-ended responses. In summary, this quantitative, exploratory survey aims to reliably measure student attitudes toward AI adoption for improving English communication skills. The standardized methodology and statistical analyses seek to produce generalizable findings on the role and potential of AI in language learning.

15.3.4 Research Design

This study utilizes a survey methodology to collect quantitative data on perspectives toward using AI to improve English-speaking skills. As supported by research, surveys can strengthen data credibility and integrate responses from a variety of participants. The survey aims to assess students' explicit or implicit views on the role of AI in enhancing English proficiency. Closed-ended questions will quantitatively measure opinions on how AI platforms can support learning complexity, align with key objectives, and enable competitive strategies for success. While limitations exist regarding survey depth, this standardized approach aims to reliably capture students' overall perspectives on the value and insights AI may provide in developing critical English communication skills. The survey methodology will facilitate collecting consistent data across a sample to yield generalizable conclusions.

In this research, a quantitative method was employed using a self-administered questionnaire, aiming to connect findings with a conceptual framework. The study targeted private universities in Malaysia, employing purposive sampling. Using the model of Krejcie and Morgan (1970), a sample size of 200 was determined. The questionnaire consisted of both closed- and open-ended questions, with the latter inviting feedback for future improvements. The research's duration is expected to be around seven months.

15.4 DATA ANALYSIS

In this study, SPSS 26.0 was utilized for quantitative data analysis, specifically focusing on descriptive statistics. Respondent sentiments were captured using a Likert rating scale, with the means and standard deviations calculated for each statement. The survey instrument demonstrated good reliability based on the high Cronbach's alpha of 0.872 for the sample of 200 university student respondents. This indicates that the questionnaire can produce stable, repeatable results over time with this population. The scale can be considered internally consistent and appropriate for drawing valid inferences about perspectives toward AI-enabled English learning tools (Table 15.1).

TABLE 15.1
Reliability Statistics

Cronbach's Alpha	N of Items
0.872	20

TABLE 15.2
Descriptive Statistics

Items	Mean	Std. Deviation
A1	4.15	0.837
A2	4.08	0.725
A3	3.52	1.121
A4	3.53	0.826
A5	3.39	0.781
A6	3.48	0.946
A7	3.93	0.780
A8	3.68	0.901
A9	4.19	0.880
A10	3.64	1.173
A11	4.20	0.837
A12	4.17	1.088
A13	3.95	0.822
A14	3.90	0.964
A15	3.91	0.852
A16	3.95	0.941
A17	3.28	1.214
A18	3.75	0.913
A19	3.08	1.035
A20	4.16	0.944

Based on Table 15.2, the descriptive statistics for the 20 survey items indicate moderately positive perspectives toward using AI to improve English proficiency and skills among university students. The sample of 200 respondents shows average agreement levels ranging from 3.08 to 4.20 on a 5-point scale. Looking at specific items, the highest agreement is for the need to improve English proficiency among students (A11) (mean 4.20) and AI's potential benefits for accessibility and affordability of learning (A9) (mean 4.19). This shows strong agreement on improving English skills and openness to AI's advantages. Moderate agreement exists for AI's helpfulness for writing assistance (A1) (mean 4.15), AI tutors providing personalized guidance (A2) (mean 4.08), and AI identifying knowledge gaps (A7) (mean 3.93). This indicates receptiveness to AI for writing help, adaptive tutoring, and diagnostic assessment.

However, more uncertainty is evident for AI exam scoring (A5) (mean 3.39), automated pronunciation feedback (A8) (mean 3.68), and AI assessment of speaking skills (A6) (mean 3.48). This suggests hesitancy about relying fully on AI for evaluation versus human judgment. Opinions are divided on the risks of over-relying on AI versus human teachers (A10) (mean 3.64). Still, openness to adopting AI tools is shown, with disagreement that costs are prohibitive (mean 3.28) or that students would resist AI platforms (A19) (mean 3.08).

In summary, the willingness to integrate AI to boost English proficiency and engagement is apparent, but some reservations remain around the limitations of automation for assessment versus human teaching. AI applications like writing assistance and personalized tutoring appear most favorably viewed.

15.5 DISCUSSION

This study provides timely insights into student perspectives on leveraging AI linguistics technologies to enhance English proficiency development. The survey results reveal cautious but growing openness to integrating respectful, human-centered AI tools to augment traditional teaching. Students' recognition of the need to improve English communication skills reflects the increasingly globalized academic and professional landscape requiring advanced language mastery. Their positivity toward AI's potential benefits for accessibility, personalization, and affordability aligns with evidence that adaptive systems can make learning more inclusive (Holmes et al., 2019).

However, the uncertainty toward fully automated assessment underscores limitations of current AI in evaluating nuanced linguistic and conversational aptitude. Students still place greater trust in human judgment for holistic, qualitative feedback – a perspective echoed by scholars critiquing automated scoring. Nonetheless, embrace of AI assistance for writing and tutoring indicates readiness to utilize technologies to enrich competencies, not replace human instructors. This presents opportunities to develop ethical, transparent AI tools that provide individualized guidance at scale, freeing up educators for higher-order instruction (Klimova et al., 2023). While minimizing risks of overdependence will remain crucial, findings imply integrating respectful AI use cases could meaningfully complement teaching and empower learners. The results will help institutions weigh appropriate integration strategies that balance pedagogical enhancement with ethical considerations around learner autonomy and data privacy . Further research on instructors' readiness is warranted.

15.6 RECOMMENDATION AND FUTURE DIRECTION

The rise of AI-powered language learning platforms presents a timely opportunity to transform English-speaking outcomes in Malaysian higher education. There is a great deal of potential overall for enhancement of English-speaking skills with AI. More effective and engaging AI-powered language learning platforms that can help students learn English more effectively and efficiently can be created through further pursuit of research and development in this area.

There is potential for developing more adaptive language learning platforms that dynamically adjust according to the student's progress levels. This adaptability could involve monitoring and evaluating the student's performance across different tasks or assignments. Structured on these observations, difficulty levels within subsequent tasks would then automatically respond to match those specific needs accordingly – ensuring ongoing challenge without overwhelming them (Bertolini et al., 2023). Moreover, the interaction between students and an AI-powered platform can reveal individual weaknesses and areas for improvement.

Current AI-powered language learning platforms can be effective, but they can also be quite dry and boring. There is a need for more engaging and interactive AI-powered language learning platforms that can keep students motivated and engaged in the learning process. This could be done by using games, simulations, and other interactive activities to make learning English more fun and engaging. For example, a platform could be developed that uses a game-based approach to teach grammar. The game could involve students completing challenges that require them to use the grammar rules in context. This would help students to learn the rules in a fun and engaging way while also reinforcing their understanding and application of the grammar rules.

Overall, there is a great deal of potential for AI to be used to enhance English-speaking skills. By pursuing further research and development in this area, we can create more effective and engaging AI-powered language learning platforms that can help students to learn English more effectively and efficiently.

In recent years, the proliferation of AI has significantly transformed language learning methods. Various AI-powered language learning platforms offer interactive and engaging experiences for millions of users worldwide. This comparative table aims to assess the strengths and weaknesses of six popular platforms to provide learners with insights into their features and capabilities.

A student who relies too heavily on AI platforms may not develop the critical thinking and problem-solving skills that are essential for language learning. The excessive reliance on AI platforms can have a substantial negative impact on a student's capacity to develop and nurture essential critical thinking and problem-solving abilities, which are fundamental components of the language acquisition process (Chai et al., 2020). AI platforms undoubtedly offer efficient and convenient resources for language acquisition; however, it is imperative to exercise caution and maintain a balanced approach by incorporating other learning methods. While AI platforms can provide guidance and practice in language skills, they may not specifically target critical thinking and problem-solving skills. These skills involve analyzing and evaluating information, making connections, and finding creative solutions, which may not be the primary focus of AI language learning tools (Sumakul et al., 2022). Effective collaboration among researchers, educators, and policymakers is essential to foster the integration of AI into language education. Joint research projects are a key aspect of this collaboration, aiming to explore the various possibilities of AI applications in language learning. These projects encompass diverse areas within the field, including the creation of innovative AI-driven platforms for language instruction, rigorous evaluation of their effectiveness, and analysis of their impact on student learning outcomes. The collective efforts of researchers, educators, and policymakers hold the key to revolutionizing language education with AI. By collaborating on professional development programs, they empower educators to leverage AI's potential in language learning, understanding its benefits and drawbacks (Abdullah & Abdulatif, 2022). Collaboratively, they can craft impactful professional development programs, empowering educators to harness the potential of AI for language learning. Such programs will illuminate the benefits and challenges of AI-powered language learning platforms and equip educators with the knowledge to wield these tools effectively within their classrooms.

15.7 CONCLUSION

Incorporating student perspectives is crucial in facilitating the successful integration of AI platforms for language acquisition. As ultimate beneficiaries, students' perceptions of these platforms significantly affect their willingness to engage with them and determine their efficacy in promoting English proficiency. Furthermore, students' viewpoints provide valuable insights into areas requiring enhancement and guarantee that the platform aligns with cultural norms. By considering student perceptions, educators can better ensure seamless incorporation of AI platforms within language instruction and cater to students' individual requirements effectively. Apart from students' perceptions, there are several other considerations that must be taken into account when integrating AI platforms for language learning. These considerations include the students' proficiency level, their specific learning objectives, and the availability of resources. By carefully considering these factors, educators can ensure that AI platforms are seamlessly integrated into the language learning process and effectively address the needs of individual students. The integration of AI platforms has immense potential to significantly contribute to English teaching and learning in higher education institutions in Malaysia. Excitement is growing as we anticipate exploring how these platforms will be utilized in the future to facilitate the development of students' English-speaking skills.

REFERENCES

Abdullah Sharadgah, T., & Abdulatif Sa'di, R. (2022). A systematic review of research on the use of artificial intelligence in English language teaching and learning (2015–2021): What are the current effects? *Journal of Information Technology Education: Research*, *21*, 337–377. https://doi.org/10.28945/4999

Al-Maroof, R. S., Alnazzawi, N., Akour, I. A., Ayoubi, K., Alhumaid, K., AlAhbabi, N. M., Alnnaimi, M., Thabit, S., Alfaisal, R., Aburayya, A., & Salloum, S. (2021). The effectiveness of online platforms after the pandemic: Will face-to-face classes affect students' perception of their behavioural intention (BIU) to use online platforms? *Informatics*, *8*(4), 83. https://doi.org/10.3390/informatics8040083

Awang, H., Aji, Z. M., Mohd Yakoob, M. F., Osman, W. R. S., Mukminin, A., & Habibi, A. (2018). Teachers' intention to continue using virtual learning environment (VLE): Malaysian context. *Journal of Technology and Science Education*, *8*(4), 439. https://doi.org/10.3926/jotse.463

Bertolini, R., Finch, S. J., & Nehm, R. H. (2023). An application of Bayesian inference to examine student retention and attrition in the STEM classroom. *Frontiers in Education*, *8*. https://doi.org/10.3389/feduc.2023.1073829

Chai, C. S., Wang, X., & Xu, C. (2020). An extended theory of planned behavior for the modelling of Chinese secondary school students' intention to learn artificial intelligence. *Mathematics*, *8*(11), 2089. https://doi.org/10.3390/math8112089

Creswell, J. W. (2008). *Educational Research Planning, Conducting, and Evaluating Quantitative and Qualitative Research* (3rd ed.). Upper Saddle River, NJ Pearson Education, Inc. https://www.scirp.org/(S(oyulxb452alnt1aej1nfow45))/reference/ReferencesPapers.aspx?ReferenceID=713513

Glaser, P. E. (1978). Editorial. *Solar Energy*, *20*(3), iv. https://doi.org/10.1016/0038-092x(78)90097-x

Gyasi, J. F., Zheng, L., & Long, M. (2021). Reflecting on the past to shape the future: A systematic review on cross-cultural collaborative learning from 2011 to 2020. *Sustainability*, *13*(24), 13890. https://doi.org/10.3390/su132413890

Han, J.-W., Park, J., & Lee, H. (2022). Analysis of the effect of an artificial intelligence chatbot educational program on non-face-to-face classes: A quasi-experimental study. *BMC Medical Education*, *22*(1). https://doi.org/10.1186/s12909-022-03898-3

Holmes, W., Bialik, M., & Fadel, C. (2019). *Artificial Intelligence in Education: Promises and Implications for Teaching and Learning*. The Center for Curriculum Redesign.

Klimova, B., Pikhart, M., & Kacetl, J. (2023). Ethical issues of the use of AI-driven mobile apps for education. *Frontiers in Public Health*, *10*. https://doi.org/10.3389/fpubh.2022.1118116

Krejcie, R. V., & Morgan, D. W. (1970). Determining sample size for research activities. *Educational and Psychological Measurement*, *30*(3), 607–610. https://doi.org/10.1177/001316447003000308

Menggo, S., Suastra, I. M., Budiarsa, M., & Padmadewi, N. N. (2019). Needs analysis of academic-English speaking material in promoting 21st century skills. *International Journal of Instruction*, *12*(2), 739–754. https://doi.org/10.29333/iji.2019.12247a

Moreno-Guerrero, A.-J., López-Belmonte, J., Marín-Marín, J.-A., & Soler-Costa, R. (2020). Scientific development of educational artificial intelligence in Web of science. *Future Internet*, *12*(8), 124. https://doi.org/10.3390/fi12080124

Munoz Luna, R., & Taillefer, L. (2014). Academic writing in English as a foreign language. *International Journal of Applied Linguistics*, *24*(2), 291–292. https://doi.org/10.1111/ijal.12066_5

Murray, N. (2010). Conceptualising the English language needs of first year university students. *The International Journal of the First Year in Higher Education*, *1*(1). https://doi.org/10.5204/intjfyhe.v1i1.19

Mustapha, S. M., Abdullah, N., Devarajoo, K., Ibrahim, F., Suid, S. H., & Amirrudin, S. (2021).

Nadesan, N. K., & Md. Shah, P. (2020). Non-linguistic challenges faced by Malaysian students in enhancing speaking skills. *Creative Education*, *11*(10), 1988–2001. https://doi.org/10.4236/ce.2020.1110145

Norhasim, N. A., & Mohamad, M. (2020). Teachers' challenges on the implementation of classroom-based assessment for oral proficiency: A literature review. *Creative Education*, *11*(10), 1922–1929. https://doi.org/10.4236/ce.2020.1110140

Novita, D., & Purwati, O. (2021). Incorporating culture and textbook adaptation practice for promoting intercultural competence in ELT: Teachers' perspectives. *JEES (Journal of English Educators Society)*, *6*(1), 96–104. https://doi.org/10.21070/jees.v6i1.1059

Rebolledo Font de la Vall, R., & González Araya, F. (2023). Exploring the benefits and challenges of AI-language learning tools. *International Journal of Social Sciences and Humanities Invention*, *10*(01), 7569–7576. https://doi.org/10.18535/ijsshi/v10i01.02

Rui, Z., & Badarch, T. (2022). Research on applications of artificial intelligence in education. *American Journal of Computer Science and Technology*, *5*(2), 72. https://doi.org/10.11648/j.ajcst.20220502.17

Selan, S. (2021, February 16). *Sorry, What? Speaking English Still a Struggle for Many Despite Mandatory Classes*. MalaysiaNow. https://www.malaysianow.com/news/2021/02/16/sorry-what-speaking-english-still-a-struggle-for-many-despite-mandatory-classes

Shah, S. S. A., Othman, J., & Senom, F. (2017). The pronunciation component in ESL lessons: Teachers' beliefs and practices. *Indonesian Journal of Applied Linguistics*, *6*(2), 193. https://doi.org/10.17509/ijal.v6i2.4844

Sumakul, D. T. Y. G., Hamied, F. A., & Sukyadi, D. (2022, February 10). *Students' Perceptions of the Use of AI in a Writing Class*. www.atlantis-Press.com; Atlantis Press. https://doi.org/10.2991/assehr.k.220201.009

Swee Heng, C. (2012). *Defining English Language Proficiency for Malaysian Tertiary Education: Past, Present and Future Efforts*. https://doi.org/10.7575/aiac.alls.v.3n.2p.150

Taillefer, L., & Munoz-Luna, R. (2014). Developing oral skills through Skype: A language project analysis. *Procedia – Social and Behavioral Sciences*, *141*, 260–264. https://doi.org/10.1016/j.sbspro.2014.05.045

Tedre, M., Toivonen, T., Kahila, J., Vartiainen, H., Valtonen, T., Jormanainen, I., & Pears, A. (2021). Teaching machine learning in K-12 classroom: Pedagogical and technological trajectories for artificial intelligence education. *IEEE Access*, *9*, 110558–110572. https://doi.org/10.1109/access.2021.3097962

Tuzlukova, V., & Usha-Prabhukanth, K. (2018). Critical thinking and problem solving skills: English for science foundation program students' perspectives. *Зборник радова Филозофског факултета у Приштини*, *48*(3), 37–60. https://doi.org/10.5937/zrffp48-18664

Wang, T., & Rajprasit, K. (2015). Identifying affirmative beliefs about English language learning: Self-perceptions of Thai learners with different language proficiency. *English Language Teaching*, *8*(4). https://doi.org/10.5539/elt.v8n4p1

Wang, T., Lund, B. D., Marengo, A., Pagano, A., Mannuru, N. R., Teel, Z. A., & Pange, J. (2023). *Exploring the Potential Impact of Artificial Intelligence (AI) on International Students in Higher Education: Generative AI, Chatbots, Analytics, and International Student Success*. www.preprints.org. https://doi.org/10.20944/preprints202305.0808.v1

Yun, J., & Park, J. (2022). The effects of Chatbot service recovery with emotion words on customer satisfaction, repurchase intention, and positive word-of-mouth. *Frontiers in Psychology*, *13*. frontiersin. https://doi.org/10.3389/fpsyg.2022.922503

Zhao, Y., Yin, D., Wang, L., & Yu, Y. (2023). The rise of artificial intelligence, the fall of human wellbeing? *International Journal of Social Welfare*. https://doi.org/10.1111/ijsw.12586

16 Revolution in the Indian Education System from Traditional to Digital Transformation in Higher Education

Mubeen Zehra and Sukanya Singh

16.1 INTRODUCTION

Education is the cornerstone of every nation for the holistic development of an individual. It plays a pivotal role in the development of human beings to reach their full potential. Education helps a person to compete on a global scale. Education's ultimate purpose is to develop students' intellect, curiosity, and logical thinking, which in turn develops their creativity and research mindset. Every country may improve its citizenship standards through education (Hammond & Snyder, 2015). Education fosters or exacerbates precariousness among people, which contributes to the formation of nations (Ross, 1979). Quality education is a legitimate demand of not exclusive, common masses and significant stakeholders. So, to acquire quality education, learners must go beyond their boundaries to migrate. Learners move or migrate from one region to another to attain education for an integrated approach toward the different education systems and address the gaps therein. Contributions from students and scholars of different educational backgrounds and subjects enrich the available literature, devise new ones, and fill the hiatus found in various research, schemes, and programs (Darra, 2006).

Now information and communication technology (ICT) is considered a revolutionary apparatus successfully used to shape nation building (Jorgenson & Vu, 2016). ICT consists of computer-based technology, multimedia, network, and internet. As a result of its development, ICT is now available to people not only in developed and developing countries but also in third-world countries and is deeply ingrained in society. ICT reconstructs organizations, enhances collaboration, increases citizens' active participation in exercising democratic rights, makes government responsible and accountable, enhances individual social interaction, and advances education and healthcare to make it easily accessible (Kozma, 2005).

Numerous concepts have been created and revised, as evidenced by the literature (Levin, 1993; Darra, 2006), committees and commissions (Shankdher, 1967;

DOI: 10.1201/9781003400691-16

Mathew, 2016; Biswas, 2021), policies (Bajwa, 2003; Bassi, 2011; Rao & Siva, 2016; Sarkar et al., 2017; Mahashevta, 2017), interventions (UNESCO, 2005, 2009, 2011), and frameworks related to education. Higher education research and development is a significant source of innovation, and the average education level of the workforce is connected with gains in student productivity. Enhancing educational production requires evaluating efficacy and efficiency.

The scope of ICT is tremendously increasing and is considered the crucial constituent of the education system. ICT usage is in the era of neoliberalism where learning and preparing for the examination are fashioned through computer and computer-based learning (Volman, 2005). Over time, ICT has played a vital role in higher education and has become important and significant. Various other models that evolved were responsible for providing opportunities through advanced technologies (Barak & Rafaeli, 2004). ICT intervention in education and policies creates opportunities and alternative ways to acquire a degree (Light et al., 2006). Anderson and Plomp (2009) thought that the contestations regarding the integration of ICT in the learning environment become complex and contentious. However, the expansion of ICT in all significant spheres laid the ongoing demand for the national ICT policy and its implementation. In the meddling of ICT waves, many countries have developed their national policy and ICT became the crux of committees and commissions, policies, and other international interventions (Bajwa, 2003; Rao & Siva, 2016; Sarkar & Mohapatra, 2017; Mahashevta, 2017; Mathew, 2016, ICT@school 2010, SMART 2011, National Policy on ICT in school 2012, Digital India 2015 and NPE, 2020). Also, there were more inventions related to ICT in education at a global level, i.e., UNESCO, 2005, 2009. ICT application in education can be traced from global to cross-national committees and policies to higher education level. Kozma (2005) offers a classification model of ICT's uses in the pre-service teacher education program. The creation of a conducive teaching–learning environment cannot be solely attributed to ICT. Teachers must possess the knowledge and skills to construct lesson plans, choose appropriate resources and materials, facilitate activities, and provide guidance throughout the learning process. However, most of the teachers who have undergone traditional training may not be equipped to undertake these responsibilities. Bransford, Brown and Cocking (1999) emphasize the need for a shift in the pedagogical paradigm toward a more student-centered approach to effectively utilize technology. This shift is not a simple task, especially in a nation or state with a teacher-centered approach in the educational system. The review of literature found that the integration of ICTs to adopt a constructivist paradigm in teaching–learning should be accompanied by a 21st-century teaching–learning model of style in the country.

16.1.1 ICT Landscape in Education through Various Policies

When India unveiled its Fourth Five-Year Plan in 1972, it recognized the value of technology (1969–74). India implemented the Educational Technology Scheme under this plan, which provided the states with 100% help in purchasing resources like color television, radios, cassette players, etc. The conclusions were reached after analyzing policy papers and committee/commission reports about the integration of

ICT into teaching–learning processes. The Kothari Commission's recommendations, which set the stage for science and knowledge's crucial role, may be traced back to the beginning of the period.

The Kothari Commission established the guidelines to enhance the role of educators. The objectives of the first national policy on education were incorporated into the second policy, which was created in 1986. Electronic tools and audiovisual aids were added to the teaching–learning process, and ICT integration has become more prevalent. By offering training programs with a technical focus to teachers (pre-service and in-service), the succeeding policy of 1992 improved upon the earlier policies. The ICT orientation meant that these teachers were no longer just teachers; rather, their position was redefined as a result of training during that time, and there was only a slight improvement. ICT utilization in the teaching–learning process was wisely boosted throughout 2004–2005 as the journey toward ICT integration grew and improved. International interventions are important for implementation of ICT into the teaching–learning environment. In this case, UNESCO is advancing the notion of ICT-based education because many nations are active members of these organizations, and UNESCO is a key tool for advancing the principles of ICT-oriented education in third-world nations as well. The UNESCO framework thus proved to be important and beneficial.

India participates actively in the UN, and UNESCO has incorporated the concepts, tenets, agreements, conferences, etc., in addition to policies such as the NCFTE 2009–2010. The overall professional development of both (in-service and pre-service teachers) was emphasized in this particular policy. The specific policy addressed the technological requirements in the teacher education program. UNESCO attempted to analyze the subsequent implementation of ICT-based learning policies and laws by looking at both their national and international components.

ICT is crucial to the teaching–learning process because it results in high-quality instruction, high-quality student learning, and overall learner professional development, according to UNESCO 2010–2011 endorsed by member states as foreign interventions and programs. To do this, the Indian government created NCFTE 2014 and NCTE 2014–2015 to accelerate the quality and efficacy of teacher education programs. However, in a teacher education program, a paper on ICT and education is taught. ICT has been gradually incorporated into the curriculum and syllabi of several Indian colleges and universities. The most recent policy developed by the Indian government, known as NEP 2020, is a representation of the progressive implementation of those earlier policies. NEP 2020 focuses on additional fields and disciplines and the full implementation of ICT tools and techniques into the teaching-learning environment. Policy's suggested online course demonstrates how the implementation of ICTs in the conducive teaching–learning environment, including teacher preparation, has been increased.

16.1.2 Paradigm Shift in Teaching–Learning Process

The method and content of mass education have undergone a paradigm shift. The Gurukul method, along with the residential and classroom approaches, can be used to chart the transition from older to more modern post-modern eras. ICTs are replacing traditional methods in many important areas of life. The teaching methods have changed with the introduction of ICT. The practices and procedures of commercial and governmental endeavors have experienced a significant shift due to the integration of ICT. In the realm of education, which is inherently a social activity, effective teachers have traditionally been associated with personalized interactions and a deep understanding of their students. The incorporation of ICT in schools has fostered the development of student-centered learning environments, promoting greater engagement and participation. Nevertheless, as the world rapidly embraces digital media and information, the role of ICT becomes increasingly significant in facilitating and enhancing educational experiences.

ICT development that is assertive has refuted long-standing claims about the traditional teaching paradigm by fostering new thinking, transforming the world of education. According to academic opinion, alternative teaching methods are more effective and quicker than conventional teaching styles. Currently, the utilization of ICT in the real classroom process is being endorsed and motivated by both educators and parents (Schwerdt & Wuppermann, 2011). Yet, acquiring technology skills by themselves does not guarantee the capacity to conceptualize and apply effective technology-mediated learning systems. As a result, the relationship between learning and technology must be modeled in teacher education. Implementation of ICT into a conducive learning environment is more pleasant and interesting and researchers have found it in their investigations (Mooji, 2004).

The incorporation of ICT has significantly improved the learning environment by fostering collaboration, creativity, and early evaluation. This has resulted in a paradigm from the conventional teacher-centred approach to a more child-centred approach in the teaching environment with the aid of ICT (Sanchez & Aleman, 2011). The role of the teacher in an environment where technology is dominant has been redefined, and the teaching-learning process now places a strong emphasis on a student-centric approach. Importantly, the use of ICT must be integrated effectively and efficiently for more enjoyable learning (Birch & Irvine 2009; Honan 2008).

16.2 REVIEW OF LITERATURE

A literature review is a prerequisite for any research involving introspection and evaluation of the existing literature on a field or specific topic. Literature refers to scientific research manuscripts containing current knowledge, results, and theoretical and methodological contributions to a particular topic. It provides a theoretical foundation for research studies and aids in determining the nature of the research. Relevant aspects of the literature review must be evaluated critically. There is a plethora of existing research conducted in many countries to review the pivotal role of ICT in the education process (Anderson, 2008; Dede, 2000; Law et al., 2008; Kozma, 2008; Plomp, 2006; Voogt, 2005, 2008). Most of these researches provide the findings intending to study ICT policy implementation.

ICTs have become indispensable tools for accessing information, providing education, and enabling interactive educational activities, regardless of geographical location or time constraints (Mobbs, 2002). The result of the report "World Summit on the Information Society" in 2003 endorsed 175 countries, the importance of supporting developing nations in their efforts to enhance the accessibility of ICTs to open and distance learning opportunities at a reduced cost at remote areas also, thereby enabling all individuals to contribute to the establishment of an information and creative nation (World summit on the information society, 2003). Research findings have shown that high-level integration of ICT positively impacts organizational productivity (Campbell, 2001). Furthermore, ICTs are recognized for their ability to create a more engaging, innovative, and visually appealing learning environment compared to traditional face-to-face instruction. Consequently, educators face the significant challenge of effectively equipped prospective learners with the necessary skills and competencies to effectively and responsibly utilize ICTs, ensuring they're prepared for the demands of the modern workforce (Gibson et al., 2002).

Recent studies by Toledo (2005) and Hennessy et al. (2005) have shed light on teachers' perspectives regarding the integration of ICTs into instruction and the integration of ICTs into classroom practice by non-teachers. The studies suggest that teachers need to develop new strategies to effectively mediate ICT-supported activities. In a study conducted by Jung (2005) that examined approaches to ICT usage in prospective training programs, it was suggested that ICTs have the potential to revolutionize teaching methods. However, if this paradigm shifts toward traditional transformation, the teaching–learning method is not realized, prospective teachers may not have sufficient and not more opportunities to practice and experience effectively implementing ICT for instructional purposes (Ross & Specht, 2008), further emphasizing the importance of positive learning experiences with ICTs as they have a significant impact on successful integration.

Smith and Robinson (2003) presented a novel perspective on the incorporation of technology into the curriculum. They proposed the use of collaborative cohorts as a means to achieve effective implementation and integration, which may be employed in the context of incorporation in the present study. Pompeo's (2004) research findings showed that successful efforts in integrating advanced technologies and ICT also identified various factors that necessitate such effective integration. Qualitative methods were employed to investigate educational institutions with above-average ICT infrastructure. The study revealed that the key factor for successful integration was the presence of adequate resources to support a strong infrastructure. Additionally, it was recommended that the institution should take proactive measures before embarking on integration efforts. These measures encompassed "the creation of administrative policies and delineation of responsibilities, the establishment of a functional infrastructure, effective communication within the organization, and modification of the curriculum to address current needs". Additionally, the continuous professional and technical development of the instructor was found to be crucial.

In the present era, the process of teaching involves the integration of curricular contents with the utilization of digital technologies. ICTs play a crucial role in the day-to-day teaching profession and serve as a means to tackle issues and impediments and find solutions within institutional environments. To effectively engage 21st-century learners, teachers must design learning experiences that use ICT tools. By doing so, these educators could bring creative and innovative personalities into the classroom. Technologies also offer support to educators, enabling them to engage students with the content and enhance the development of skills that extend beyond the mere utilization of these latest technologies.

The usage of digital ICT tools and techniques in the teaching-learning profession has seen a substantial increase in the 21st century, which is an important step toward modern-day education. The fusion of ICT in classrooms can amplify the skills and competencies of education and create a favorable learning environment for learners. The effective use of ICT in education is greatly influenced by organizational factors that impact pre-service teachers. These factors create a supportive environment for both teachers and students, ultimately leading to the successful integration of ICT. One important factor is the professional experience of prospective teachers, as it significantly contributes to their technological growth. Additionally, a supportive environment plays a crucial role in encouraging teachers to experiment with ICT integration. "The successful integration of ICT is also influenced by various other factors, such as the teacher's comfort with computers, reliance on computers as an instructional tool, training, motivation, support, and teaching efficacy". Therefore, to revolutionize the Indian education system and transition it from traditional to digital, it is essential to prioritize the effective integration of ICT in higher education.

16.3 METHODS AND MATERIALS

16.3.1 Objectives of This Study

a. To study the revolution in the Indian education system from traditional to digital transformation into higher education.
b. To describe the transformation in the Indian education system.
c. To study the paradigm shift from the traditional method to the ICT-oriented model.
d. To analyze the attitude of pre-service teachers regarding ICT-oriented teaching.

16.3.2 Research Design

Descriptive research design is a research methodology that is employed to gather information systematically to describe a particular sample effectively. This research design is more focused on providing detailed, factual information about the specific statements of the problem. In this study, the researcher used descriptive research design for an accurate and systematic description of the sample and sampling techniques summarized using statistical measures and presented in an accurate, clear, and organized manner.

16.3.3 Sample and Sampling Technique

The sample size is composed of 100 pre-service teachers registered in two-year B.Ed. (General) program in the Department of Teacher Training and Non-formal Education (IASE), Faculty of Education, Jamia Millia Islamia. The sample was collected through a simple random sampling technique.

16.3.4 Tools Used for Data Collection

A self-constructed questionnaire based on a 5-point Likert scale was used to understand subjects' perception toward the viable scope from traditional method to ICT mode usage in the teaching-learning process. The tool on the perspective of pre-service teachers regarding the possible extent of ICT usage in the teaching-learning process is divided into four sub-dimensions, i.e., ICT-oriented pedagogical skills, strengthening students' learning in techno-oriented classrooms, techno-oriented lectures strengthening students' creativity and involvement in the classroom, and techno-friendly mode of innovative communication.

16.3.5 Procedure of Data Collection

Data collection for the sample was done in one phase. The authorization was taken by the researcher (Head of the Department of Education) to organize data from the trainees. After establishing a proper interaction and instruction with the subject, the questionnaires were distributed to give their responses. The data were collected in a single day and duration of one hour. The researcher collected the filled questionnaires from pre-service teachers and thanked them for participating in the study.

16.3.6 Statistical Analysis

The reliability of the questionnaire's tools, according to Cronbach's alpha, was 0.85. The obtained data were interpreted and analyzed using descriptive statistics, i.e., mean and standard deviation (SD), through SPSS v.23. The mean score was generated to determine the pre-service teachers' insight on the potential transition from traditional methods to ICT mode usage in the classroom. To determine the degree of heterogeneity in the pre-service teachers' responses, SD was determined. To ascertain the perception level of pre-service instructors, frequency and percentage analyses were also computed.

16.4 RESULTS AND DISCUSSION

Results of the statistical analysis are presented in Table 16.1.

The mean and SD values for the four sub-sizes reflecting on four sub-objectives of the research are described as follows:

Mean score for sub-size I, "ICT-related pedagogical skills" was 16.85, with a SD of 1.978. The percentage value suggested a positively high degree of pre-service teachers' perspective about this study sub-dimension, while the mean result shows an above-average level. Therefore, the pre-service teachers' perception of ICT's moderate value in enhancing teachers' pedagogical skills can be confidently inferred.

The mean score of sub-dimension II, i.e., "Strengthening students learning in techno-oriented 21st century classroom" was 15.92 and the value of SD was 1.895. The average value demonstrated a moderate range of attitudes among teachers (pre-service) regarding specific aspect of the research. The result highlights the pre-service

TABLE 16.1
Mean, Standard Deviation (SD) of Pre-service Teachers' Perception toward Possible Scope from Traditional Method to ICT Mode Usage in the Classroom

S. No	Sub-Dimensions	N	Max. Possible Score	Mean	S-D
I	ICT-oriented pedagogical skills	100	20	16.85	1.978
II	Strengthening students' learning in Techno-oriented 21st classroom	100	20	15.92	1.895
III	Techno-oriented lectures strengthen students' creativity and participating involvement in the classroom	100	20	15.96	1.974
IV	Techno-friendly mode of innovative communication	200	20	15.89	1.622

teachers' belief that ICT played a moderately beneficial role in enhancing student learning within the techno-oriented 21st century classroom.

The mean score of sub-dimension III, i.e., "Techno-oriented lecture strengthening students' creativity and participating involvement in the classroom", was 15.97, and SD value was 1.974. Prospective teachers' perception at sub-dimension of research was found to be above average, indicating a favorable attitude regarding the usage of ICT in techno-oriented lectures. Research findings showed that ICT can be moderately useful in enhancing students' creativity and participation in the classroom. Therefore, it can be inferred that pre-service teachers recognize the potential benefits of implementing ICT into their hands-on training and day-to-day practices of teaching professionalism.

Mean score of the sub-dimension IV, i.e., "Techno-friendly mode of innovative communication", was 15.89 and the value of SD was 1.622. The mentioned values signify a higher-than-average degree of awareness among pre-service teachers regarding this particular aspect of the research. Consequently, it can be inferred that prospective teachers hold the perception regarding ICTs that there is a moderate level of usefulness in augmenting the technologically friendly approach to innovative communication within the teaching–learning process. The graphical representation of the mean scores for each sub-size dimension of the research can be observed in Figure 16.1.

The results indicated in Table 16.2, by analysis on item 1, showed that nearly 90% (50%, South Africa (SA), 40% Australia (A)) of prospective teachers agree that ICT is one of the effective sources of teaching-learning process and their environment, whereas 10% were undecided and none of the selected subjects were found to disagree.

The results of the examination of the responses on item no 2, "Use of ICT generates effective teaching skills", showed that approximately 85% (30% SA, 55% A) of pre-service teachers agreed that using technology has advanced their approach to teaching and learning, while 10% were undecided and 5% disagreed.

According to participants on item 3, "ICT-integrated lectures have the potential to convey the topic in the best possible manner" nearly 78% of pre-service teachers

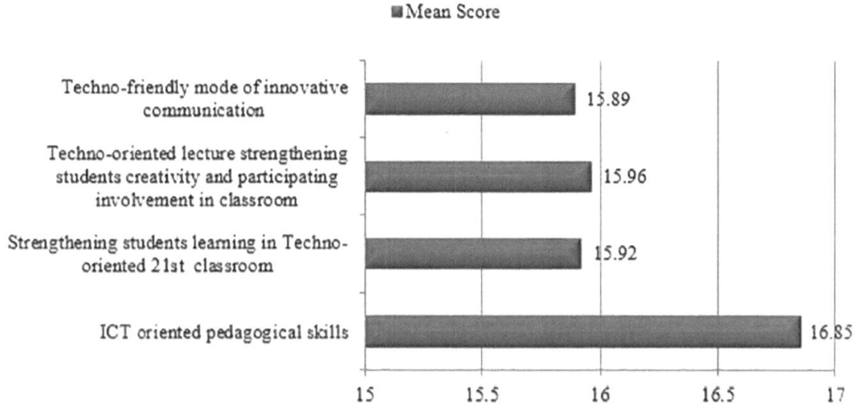

FIGURE 16.1 The study's four sub-dimensions' mean scores are visually represented through a graphical display.

(28% SA, 50% A) agreed with this statement; however, 20% were undecided, and the remaining 2% disagreed. On the analysis item 4, "ICT improves active involvement of learners", research results indicated that 90% of respondents (40 SA and 50% A) agreed with this statement, while 10% were undecided about this item.

The use of ICT promotes learners' creativity. The results of item 5 reveal that approximately 78% (33% SA, 45% A) of pre-service teachers agreed, while 15% were undecided and the remaining 7% disagreed. Item 6 emphasizing that "ICT is a constructive tool for teaching–learning" was analyzed, and it was discovered that almost 85% (40% SA, 45% A) of participants believed it, whereas 10% were unsure and 5% disagreed. Item 7 shows that approximately 92% (38% SA, 54% A) agreed, which says, "Communication becomes effective by using ICT in the classroom", while 8% were unsure and 6.5% disagreed.

Item 8 indicated that "Transfer of information could be more successful by using ICT" and received replies from roughly 87% of respondents (27% SA, 60% A), while 13% were unsure. According to an examination of the replies to item 9, "ICT makes complicated and difficult lessons easy to learn", 34% of respondents from SA and 55% from A agreed with the statement, 8% were unsure, and the remaining 3% disagreed.

Item. 10 highlighted that "ICT enables students to communicate their ideas and views in a better way", which was answered by approximately 98% of respondents (47% SA, 51% A), with 2% disagreeing. Students pay better attention in class when professors use ICT to deliver a lesson, according to an examination of responses to item 11, which found that approximately 88% of respondents (25% SA, 63% A) agreed with this statement, while 10% were unsure and 2% disagreed.

Item 12 showed whether ICT may support the creation of effective lesson plans, indicating that approximately 91% (38% SA, 53% A) of respondents agreed with the statement, although 9% were unsure and 0% disagreed. As per responses to item 13, "Evaluation may be done efficiently by using ICT", approximately 80% (30% SA, 50% A) of respondents agreed with this statement; however, 12% were uncertain and

TABLE 16.2
Responses of Pre-service Teachers (N = 100) on the Possibilities and Scope from Traditional to ICT Mode in Classroom Usage

		Responses in %				
S. No	Items	S-A	A	U-D	D-A	S-D
1	ICT is one of the effective sources of the teaching-learning process	50%	40%	10%	0%	0%
2	The use of ICT creates effective teaching skill	30%	55%	10%	5%	0%
3	ICT-integrated lectures have the scope to cover the content in the best possible manner	28%	50%	20%	2%	0%
4	ICT enhances the active involvement of learners.	40%	50%	10%	0%	0%
5	The use of ICT promotes creativity among learners.	33%	45%	15%	7%	0%
6	ICT is a constructive tool for teaching-learning	40%	45%	10%	5%	0%
7	Communication becomes effective by using ICT in the classroom	38%	54%	8%	0%	0%
8	Transfer of knowledge could be more effective by using ICT.	27%	60%	13%	0%	0%
9	ICT makes complex and difficult lessons easy to understand.	34%	55%	8%	3%	0%
10	ICT enables students to express their ideas and thoughts in a better way.	47%	51%	2%	0%	0%
11	Students pay more attention in class when the teacher teaches the lesson through ICT	25%	63%	10%	2%	0%
12	ICT can help in making effective lesson plans	38%	53%	9%	0%	0%
13	Evaluation can be done effectively by using ICT	30%	50%	12%	8%	0%
14	ICT facilitates collaborative work among students and teachers	35%	55%	10%	0%	0%
15	ICT dynamically enhances teaching-learning by expanding the body of knowledge and covering new domains of quality education	38%	58%	4%	0%	0%
16	ICT provides the facility to conduct learning sessions with a large number of students	28%	55%	12%	5%	0%

8% disagreed. Item 14, "ICT improves collaborative work among students and teachers", received replies from approximately 90% of respondents (35% SA, 55% A), while 10% were undecided.

While responses to item 15 were analyzed, 96% (38% SA, 58% A) of participants agreed with the assertion that "ICT dynamically promotes teaching-learning by growing the body of knowledge and covering new domains of an excellent education", while 4% disagreed. No participants were found to disagree with it. By analyzing the responses to respondent item 16, "ICT provides facility to conduct learning sessions with a big number of students", it was discovered that roughly 83% of respondents (28% SA, 55% A) agreed with the statement, while 12% were unsure and 5% disagreed.

16.5 MAJOR FINDINGS AND DISCUSSION

The research shows that the educational landscape of the country is getting murkier through ICT interventions; the syllabus, curriculum, and policies are monitored by ICT. This study indicated that pre-service teachers effectively adopt ICT skills in their learning process, which enhances the quality of education and makes a degree efficient. The traditional modes of teaching are now upgraded through innovative skills; the teacher education programs play a pivotal role in providing efficient teachers to educational institutions, so the teachers should be techno-friendly and techno-oriented. Therefore, ICT orientation makes pre-service teachers effective and qualitative. Several investigations have shown that ICT creates more content knowledge for learners as well as educators to update their skills and content knowledge and is helpful in the development of complex and rational thinking skills when it is successfully incorporated into a high-quality learning environment (Kozma, 2005; Kulik, 2003; Webb & Cox, 2004).

ICT pedagogical skills play a significant role in shaping the agenda that enhances quality-based education, smart work, effective time management, and participatory involvement of teachers and students. Henceforth, the finding shows that the effective roles of ICT pedagogical skills are vital in the teacher education program and also create possibilities for accessible education. In today's technological environment, traditional teaching methods are dwindling. Traditional and less creative teaching methods have declined as a finding of ICT.

The operation of ICT is hegemonic in today's techno-oriented spaces, including teacher education and academic institutions. The hegemony of ICT-based learning is visible because nowadays, students, parents, and instructors adopt the ICT-oriented teaching-learning process. The upsurge of innovative teaching methods that offer more opportunities and comfortable learning environments, chalk and blackboard instruction is now more prevalent (Siddiqui et al., 2021).

16.6 IMPLICATIONS OF THIS STUDY

The findings of the present research hold significance especially for teachers (pre-service), as they are going to be the teachers of the 21st century who must have enough competence to effectively integrate ICT in their daily teaching practices and to make the effective and efficient teaching–learning process to enrich and be authentic and also to provide first-hand experience of the various concepts and phenomena to the learners.

The study holds significance for teacher educators as they bear the primary responsibility of shaping the perception of pre-service teachers regarding the effectiveness or ineffectiveness of integrating ICT in teaching. They need to introspect on their competence to effectively use ICT resources; only then will they be motivated and encourage their students to use ICT to improve and enhance the skills and competencies in teaching practices.

The study is also significant for the administrators of teacher education institutions because it is their responsibility to abide by the rules and regulations of NCTE regarding the availability of required ICT resources in their institutions and ensure the use of these resources for the purpose they are meant for.

This study is also of significance for policymakers and curriculum planners to re-emphasize the effective integration of ICT in teacher education institutions and send them reminders in this regard from time to time. They should have regular meetings with the administrators of the institutions and teacher educators and have discussions emphasizing the need for effective implementation of ICT to enhance the quality of teacher education programs.

16.7 CONCLUSION

ICT holds a prospect to rejuvenate the mundane teaching–learning practices that have been very prominently used by teachers of the 19th and 20th centuries. Now ICT is recognized as a ground-breaking tool that has significantly influenced nation building. People can now access ICT in third-world countries as well. ICT resources must now be used effectively and efficiently if a teaching–learning process is to be effective in instilling 21st-century skills. ICT use in pre-service teacher education is progressive and improves the quality of the teaching–learning process, according to this study's indication the empirical findings are substantiated by secondary sources. An empirical finding suggests that pre-service teachers are at ease in environments focused on ICT and it is imperative to note that ICT-based learning enhances pedagogical skills, quality improvement, constructive learning, creativity, digital education, etc. Future teachers are moving with techno-fashioned learnings so they can easily implement them in the future teaching–learning process. It is crucial to remember that there are several drawbacks to ICT-based learning, which must be compassionately addressed by key stakeholders. The teachers of the 21st century are clinging to those practices and resist a change in their teaching practices. There is a dire need for teacher educators and pre-service teachers to push their limits out of their comfort zone and welcome the use of ICT resources if they aspire to inculcate 21st-century skills in their learners. Only then we can build a future workforce that is competent to effectively deal with challenges and threats at the work front.

REFERENCES

Anderson, R. E. (2008). Implications of the information and knowledge society for education. In: Voogt, J., Knezek, G. (Eds), International Handbook of Information Technology in Primary and Secondary Education. Springer International Handbook of Information Technology in Primary and Secondary Education, Vol. 20. Boston, MA: Springer. https://doi.org/10.1007/978-0-387-73315-9_1

Bajwa S. B. (2003). ICT Policy in India in the era of liberalization: Its impact and consequences. Global Business and Economic Review GBER, 3(2), 49–61.

Barak, M., & Rafaeli, S. (2004). Online question-posing and peer assessment as means for web-based knowledge sharing in learning. International Journal of Human-Computer Studies, 61(1), 84–103.

Bassi, R. (2011). ICTs in education (ICT4E) policies and plans worldwide. *Global e-School and Communities Initiative (GeSCI) Balanskat, A.(2008). National policies and case studies on the use of portfolios in teacher training*. https://www.eldis.org/document/A64501

Birch, A., & Irvine, V. (2009). Pre-service teachers' acceptance of ICT integration in the classroom: Applying the UTAUT model. Educational Media International, 46(4), 295–315.

Biswas, A. (2021). Recommendations of commissions and committees on LIS education in India: An analytical study. https://doi.org/10.13140/RG.2.2.24807.16800.

Bransford, J. D., Brown, A. L., & Cocking, R. R. (Eds.). (1999). *How people learn: Brain, mind, experience, and school.* Washington, DC: National Research Council/National Academy Press. https://nap.nationalacademies.org/catalog/9853/how-people-learn-brain-mind-experience-and-school-expanded-edition

Campbell, D. (2001). Can the digital divide be contained? The digital divide: Employment and development implications. International Labor Review, 140(2), 119–141.

Darra, M. (2006). Productivity improvements in education: A replay. *European Research StudiesJournal, 9*(3-4), 101-124. https://www.um.edu.mt/library/oar/handle/123456789/31459

Darling-Hammond, L., & Snyder, J. (2015). Meaningful learning in a new paradigm for educational accountability: An introduction. *Education Policy Analysis Archives*, *23*, 7. https://doi.org/10.14507/epaa.v23.1982

Dede, C. (2000). Emerging influences of information technology on school curriculum. Journal of Curriculum Studies, 32(2), 281–303.

Hennessy, S, Ruthven, K, & Brindley, S. (2005). Teacher perspectives on integrating ICT into subject teaching: Commitment, constraints, caution, and change. Journal of Curriculum Studies, 37(2), 155–192.

Honan, E. (2008). Barriers to teachers using digital texts in literacy classrooms. *Literacy*, *42*(1), 36-43. https://doi.org/10.1111/j.1467-9345.2008.00480.x

Ivan S. Gibson, Cathal O'Reilly & Michael Hughes. (2002). Integration of ICT within a project-based learning environment, European Journal of Engineering Education, 271(1): 21–30.

Jung, I. (2005). ICT-pedagogy integration in teacher training: Application cases worldwide. *Educational Technology* & *Society*, 8(2), 94–101.

Jorgenson, D. W., & Vu, K. M. (2016). The ICT revolution, world economic growth, and policy issues. *Telecommunications Policy*, *40*(5), 383-397. https://doi.org/10.1016/j.telpol.2016.01.002

Kozma, R. B.(2005). National Policies that Connect ICT-Based Education Reform to Economic and Social Development. Human Technology: *An Interdisciplinary Journal on Humans in ICT Environments.*1 (2),117-156. Retrieved from http://www.humantechnology.jyu.fi

Kozma, R. B. (2008). Comparative analysis of policies for ICT in education. In *International handbook of information technology in primary and secondary education* (pp. 1083-1096). Springer, Boston, MA. https://doi.org/10.1007/978-0-387-73315-9_68

Kulik, J. A. (2003). Effects of Using Instructional Technology in Elementary and Secondary Schools: What Controlled Evaluation Studies. Arlington, VA: Sri International.

Light, D., & Manso, M. (2006). Educational technology integration in developing countries: Lessons from Seven Latin America School Nets. In *Annual Meeting of the American Educational Research Association. Seattle.* https://www.researchgate.net/publication/211720586_Educational_Technology_Integration_in_Developing_Countries_Lessons_from_Seven_Latin_America_SchoolNets

Law, N., Pelgrum, W. J., & Plomp, T. (Eds.). (2008). *Pedagogy and ICT Use in Schools around the World: Findings from the IEA SITES 2006 Study, 23.* Springer Science & Business Media. Comparative Education Research Center, The University of Hong Kong. https://doi.org/10.1007/978-1-4020-8928-2

Levin, B. (1993). Students and educational productivity. Education Policy Analysis Archives, 1, 5–5.

Mahashevta. (2017). Review of ICT policies in India. International Journal of Education & Research, 7(1), 1–10.

Mathew, A. (2016). Reforms in Higher Education in India: A Review of Recommendations of Commission and Committees on Education. New Delhi: Centre for Policy Research in Higher Education, NUEPA.

Mobbs, P. (2002). Green Net CSIR Toolkit Briefing Glossary and Cross-Reference Index. GreenNet Civil Society Internet Rights Project. Retrieved December 2, 2006, from https://www.fraw.org.uk/library/005/gnirt/gossary.html

Mooij, T. (2004). Optimising ICT effectiveness in instruction and learning: multilevel transformation theory and a pilot project in secondary education. *Computers & Education*, *42*(1), 25-44. https://doi.org/10.1016/S0360-1315(03)00063-0

Pompeo, J. M. (2004). *A study of computer integration on public secondary schools.* Rutgers The State University of New Jersey, School of Graduate Studies. https://www.proquest.com/openview/6a4278992e1eed7bb208a6d6139c709d/1?pq-origsite=gscholar&cbl=18750&diss=y

Plomp, T. (2006). Preparing for Teaching in an Information Society: Various Perspectives. In *Society for Information Technology & Teacher Education International Conference* (pp. 4334-4349). Association for the Advancement of Computing in Education (AACE). Retrieved May 23, 2024 from https://www.learntechlib.org/primary/p/22791/.

Rao, J. P., & Prasad, R. S. (2016). Educational Technology Policies in India and Access. Retrieved from https://oasis.col.org/colserver/api/core/bitstreams/ddb867d5-8709-4c79-9647-8ccf7b451fcf/content

Ross, A. M. (1979). Education and National development. India International Centre, 6(4), 307–315.

Ross, C., & Specht, J. (2008). Identifying discriminating variables between teachers who fully integrate computers and teachers with limited integration. Computers and Education, 51, 1523–1537.

Sarkar, S., Mohapatra, S., Sundarakrishnan, J. (2017). ICT Policies for Primary Education and Different Initiatives Taken in India. In: Information and Communications Technology in Primary School Education, 163-193. Springer, Cham. https://doi.org/10.1007/978-3-319-42441-5_6

Sanchez, J. J. C., & Alemán, E. C. (2011). Teachers' opinion survey on the use of ICT tools to support attendance-based teaching. Computers & Education, 56(3), 911–915.

Schwerdt, G., & Wuppermann, A. C. (2011). Is traditional teaching all that bad? A within-student between-subject approach. Economics of Education Review, 30(2), 365–379.

Shankdher, S. L. (1967). Report of the Education Commission: *COMMENTS AND REACTIONS A STUDY*. SABHA SECRETARIAT NEW DELHI. https://14.139.60.153/bitstream/123456789/1406/1/REPORT%20OF%20THE%20EDUCATION%20COMMISSION-%20COMMENTS%20AND%20REACTIONS-%20A%20STUDY.pdf

Siddiqui, S. A., Alam, M. A., Javed Azmi, M., Ahmad, O., Wiquar, R., & Aziz, A. (2021). Integration of ICT-learning Environments: A paradigm Shift from Traditional Teaching Pattern. *ICIDSSD 2020*, 275. http://dx.doi.org/10.4108/eai.27-2-2020.2303238

Smith, S. J., & Robinson, S. (2003). Technology integration through collaborative cohorts: Preparing future teachers to use technology. Remedial and Special Education, 24(3), 154–160.

Toledo, C. (2005). A five-stage model of computer technology integration into the teacher education curriculum. *Contemporary Issues in Technology and Teacher Education*, 5(2), 177–191.

UNESCO. (2005). Convention on the Protection and Promotion of the Diversity of Cultural Expressions. Retrieved form: https://en.unesco.org/creativity/sites/creativity/files/info-kit_brochure-final-en-web.pdf

UNESCO. (2009). Competency Standards for Teachers: Policy Framework. Published by the United Nations Educational, Scientific and Cultural Organization7, place de Fontenoy, 75352 PARIS 07 SP. Framework. https://unesdoc.unesco.org/ark:/48223/pf0000156210

UNESCO. (2011). ICT Competency Framework. https://iite.unesco.org/pics/publications/en/files/3214694.pdf

Volman, M. (2005). A variety of roles for a new type of teacher Educational technology and the teaching profession. Teaching and Teacher Education, 21(1), 15–31.

Voogt, J. (2008). ICT and curriculum processes: Dilemmas and challenges. In: Voogt, J., Knezek, G. (eds) International Handbook of Information Technology in Primary and Secondary Education. 20, (pp. 117–132). Springer, Boston, MA. https://doi.org/10.1007/978-0-387-73315-9_7

Webb, M., & Cox, M. (2004). A review of pedagogy related to information and communications technology. Technology, Pedagogy, and Education, 13(3), 235–286.

World Summit on the Information Society (WSIS) (2003). https://sustainabledevelopment.un.org/index.php?page=view&type=30022&nr=102&menu=3170

17 Does the Use of Digital Technologies in Education 4.0 Improve Learning Effectiveness in Higher Education in Pakistan?

Rameen Butt, Muskan Butt, and Sana Khalid

17.1 INTRODUCTION

The use of digital technology has brought about a considerable transformation of traditional learning approaches in the current educational setting (Haleem et al., 2022). This paradigm shift, often known as "Education 4.0," heralds a new era in which technology innovations have a significant impact on how people learn (Morris & Rohs, 2023). The investigation of the usefulness of digital technology in improving learning outcomes becomes central to higher education in Pakistan (Khalil et al., 2023). This introduction explores the many aspects of using digital technologies in higher education and looks at how they may improve the quality of instruction in Pakistani institutions.

The integration of digital technologies in higher education has unfolded numerous opportunities for diversifying instructional methods, engaging students, and fostering a more interactive and dynamic learning environment (Frolova et al., 2019). The application of Education 4.0 principles becomes imperative as Pakistan endeavors to stay up to date with international developments in education (Chacon-Diaz & Brown, 2019). The use of digital tools, including online platforms, adaptive learning systems, virtual simulations, and collaborative software, offers a promising pathway to cater to diverse learning styles and facilitate personalized learning experiences (Okoye et al., 2023). This chapter aims to analyze the extent to which these technologies have been adopted and assess their impact on learning effectiveness in higher education institutions across Pakistan.

Evaluating the potential and problems presented by this transformation is another important aspect of investigating the subtleties of integrating digital technology in higher education within the Pakistani setting (Abbas et al., 2019). Although there are a lot of potential advantages, it is important to carefully consider the problems with

DOI: 10.1201/9781003400691-17

digital accessibility, infrastructure, digital literacy, and pedagogical reform (Aslam et al., 2020). It is crucial to comprehend the regional context, cultural ramifications, and infrastructural preparedness to fully use the advantages of Education 4.0 in Pakistan's higher education sector (Farrukh et al., 2023). To offer insights into the efficiency of digital technology adoption and its impact on enhancing learning effectiveness in Pakistani higher education, this chapter aims to investigate and critically assess the existing situation.

17.2 LITERATURE REVIEW

17.2.1 Digital Technologies

The rapid digitalization and integration of technology in education during the 21st century have brought about fundamental changes, revolutionizing the way students learn by promoting critical thinking, enhancing accessibility, and personalizing learning experiences, while also presenting significant challenges within higher education institutions as they adapt to new pedagogical approaches, develop skills, and require faculty professional development to take advantage of the technology (Laurillard, 2008; Haleem et al., 2022). Digital literacy has become a vital ability in the 21st century. The use of technology in the classroom gives students the fundamental digital skills they need to meet the demands of the modern workplace (Haleem et al., 2022). Additionally, integrating technology into education helps students apply their knowledge in real-life situations. Rote memorization and passive learning were frequently emphasized in traditional schooling (Jobirovich, 2021). On the contrary, education now emphasizes developing students' analytical, problem-solving, and critical thinking abilities as a result of digitalization and technology (Srivastava & Dey, 2018). Students are challenged to think creatively and independently via interactive learning activities, preparing them for the complexity of the future. Technology and digitalization are now a crucial part of modern education, changing how information is accessible, shared, and used. Technology has improved learning for both students and teachers by making it more accessible, interesting, and personalized (Nguyen, 2018). The ongoing adoption of technology and digitalization in education will surely pave the way for a more inclusive, creative, and connected future as the 21st century advances (Haleem et al., 2022).

17.2.2 Importance of Digital Technologies

Before the widespread use of technology for education, face-to-face lectures, textbooks, and a traditional classroom setup were the primary forms of teaching and learning strategies. Learning resources were mostly static and restricted to tangible things found in the library of the institution (Aldiab et al., 2019). This made it more difficult to successfully meet each student's learning requirements since there was less access to current knowledge and a variety of learning materials. Students and instructors now have access to a range of knowledge on the internet due to the development of digital technology (Jobirovich, 2021). Through adaptive learning platforms and advanced tutoring systems, digital technologies have made personalized learning pathways possible (Liu et al., 2017). By analyzing student performance and learning preferences, these technologies may personalize material and activities for each student. Learning management systems (LMS) have also made it easier to provide content and monitor student progress, improving both teaching and learning (Aldiab et al., 2019).

17.2.3 Educational Transition to Education 1.0 to Education 3.0

Moving from Education 1.0 through Education 2.0 and toward Education 3.0, there is a gradual shift from passive learning through developing and toward the evolution in education through active learning (Castañeda & Selwyn, 2018).

17.2.3.1 Education 1.0: A Pedagogical and Essentialist Education

According to Gerstein (2014), learners are receptacles of knowledge. The classrooms were teacher centered where the sole purpose of teaching is to transfer knowledge, which involves product-oriented teaching that leads to passive learning (Harkins, 2008). Students were empty vessels, and the only source of knowledge was the teacher. The classroom practices of teachers were mainly based on essentialist philosophy where only the teacher will decide what necessary knowledge must be learnt by students without considering their interest (Martín-Gutiérrez et al., 2017). The standardized system was utilized in teaching, learning, and assessment practices. However, technology was not used in the classrooms. Technology appears to be prohibited in educational institutions (Makrides, 2019).

17.2.3.2 Education 2.0: An Andragogical, Constructivist Approach toward Teaching and Learning

The rise of Web 2.0 and the subsequent focus on social learning and usage of social software like blogs, wikis, podcasts, and virtual worlds caused the move to Education 2.0 (Redecker & Punie, 2010). Traditional methods of learning were improved with the introduction of technology advancement. The conventional method of teaching was changed to problem-based learning, inquiry-based learning, and cooperative learning, but the teacher remained the primary authority in charge of planning and directing the lessons (Williamson, 2016). Through social interaction, knowledge creation predominated over knowledge reproduction. Moreover, students are less passive in the classrooms as compared to Education 1.0 (Keats & Schmidt, 2007).

17.2.3.3 Education 3.0: A Heutagogical Connectivist Approach toward Teaching and Learning

Technology in this period completely changes society, which results in modernization, which gives rise to Education 3.0 (Thomas & Palmer, 2014). As opposed to Education 2.0, as students took an active role in their learning, here, the instructor no longer dominates the classroom (Kannengiesser, 2017). When using technology in the classroom, the teacher serves as a facilitator and helps learners gain new knowledge and skills. Because it educates students about digital abilities, it helps them become autonomous learners in society (Puncreobutr, 2016).

17.2.4 Education 4.0 in Fourth Industrial Revolution

As a result of growing interconnectivity and intelligent automation, the Fourth Industrial Revolution, also known as Industry 4.0, theorizes fast change to technology, industries, and social patterns and processes in the 21st century (Bai et al., 2020). The UK government regarding business, energy, and industrial strategy stated that the blending of technologies like artificial intelligence (AI), gene editing, and advanced robotics that blur the boundaries between the physical, digital, and biological worlds is a component of this period of industrial development (UK, 2021). The digital technologies due to the advent of the Fourth Industrial Revolution, which includes cyber-physical systems (CPS), the Internet of Things (IoT) (Roblek et al., 2016), cloud computing (Hermann et al., 2016), cognitive computing (Kagermann et al., 2013), and AI (Lasi et al., 2014), are essentially a movement toward automation and data sharing in manufacturing technologies and processes. Machines cannot take the place of profound expertise, but they are often more effective than people at doing repeated activities, and the use of machine learning and computer capacity enables machines to complete extremely challenging jobs (Pedota & Piscitello, 2022). The term "Education in the Fourth Industrial Revolution," or "Education 4.0," describes how the educational landscape has changed in response to the opportunities and difficulties presented by Industry 4.0 (Jain & Jain, 2022). Education 4.0 strives to modernize and optimize how we learn, educate, and prepare people for the quickly changing workforce and society in a similar manner to how Industry 4.0 revolutionizes the manufacturing and technology landscape (González-Pérez & Ramírez-Montoya, 2022). The Fourth Industrial Revolution had a significant impact on the worldwide education system, changing the way that students were taught and learned in classrooms. Education 4.0 satisfies the demands of a modern, dynamic world that is evolving quickly from an agrarian to an industrial one (González-Pérez & Ramírez-Montoya, 2022; Jain & Jain, 2022).

17.2.4.1 Adaptation to Technological Advancement

Technology advancement adaptation has been retrieved from the technological adaptation model and is used to improve the learning experiences of students (Palys & Atchison, 2012). Education 4.0 integrates cutting-edge technologies including AI, virtual reality, augmented reality, and IoT (Roblek et al., 2016). Gamification, tests, riddles, puzzles, and some learning applications, such as Duolingo, Byju's, Photomath, Sololearn, and Kahoot, are features of Education 4.0. With the help of these technologies, education may become more interesting and productive by offering personalized learning routes, immersive instructional content, and real-time feedback (Nawaz et al., 2020).

17.2.4.2 Interdisciplinary and Soft Skills

Education 4.0 places a strong emphasis on interdisciplinary learning to better prepare students for a dynamic and interconnected world since Industry 4.0 demands a varied range of abilities (Kundi et al., 2008). Soft skills like creativity, teamwork, and flexibility are also receiving more attention in order to develop well-rounded individuals who can succeed in modern work environments (González-Pérez & Ramírez-Montoya, 2022).

17.2.4.3 Lifelong Learning and Continuous Education

The conventional strategy for gaining learning and quitting at a given point in life becomes ineffective as industries move quickly (Aslam et al., 2021). In order to guarantee that people may upskill and re-skill throughout their careers to remain relevant and competitive, Education 4.0 encourages lifetime learning and ongoing education (Ronchi & Ronchi, 2019).

17.2.4.4 Blended Learning and Online Learning

Blended learning is a method of instruction that combines traditional classroom instruction with online learning environments (Jan et al., 2020). This makes education flexible and accessible, meeting the many requirements of students throughout the world and dissolving temporal and geographic boundaries (Hrastinski, 2019).

17.2.4.5 Teacher's Role as Facilitator

Education 4.0 places a greater emphasis on the role of teachers as mentors and facilitators than merely knowledge suppliers (Rafiq & Ameen, 2012). They assist students in acquiring crucial skills, help them in their learning journeys, and promote critical thinking (Srivastava & Dey, 2018).

17.2.4.6 Data-Driven Decision-Making

Education 4.0 uses data analytics and learning analytics to acquire insights into student performance, preferences, and requirements (Rehman & Khan, 2021). This is similar to Industry 4.0's focus on data and information openness. This information may be used by teachers to personalize education and aid students' learning processes (Liu et al., 2017).

17.2.4.7 Global Collaboration and Mobility

Global communication and cooperation were encouraged by the Fourth Industrial Revolution (Salam et al., 2017). To expose students to other viewpoints and cultures, Education 4.0 promotes worldwide educational alliances, virtual collaborations, and exchange programs (Ehlers & Kellermann, 2019).

17.2.4.8 Preparing Youth for Future Competencies

Education 4.0 aims to match educational initiatives with the demands of the next labor market (Zubairi et al., 2022). In order to provide students with in-demand skills, it concentrates on cutting-edge subjects including AI, robotics, data science, renewable energy, and digital technologies (Nawaz et al., 2020).

17.2.5 Education 4.0-Related Teaching-Learning Practices in Higher Education

It only makes sense that the curriculum of colleges and universities be tailored to the demands of students who must contend with the problems of the real world and "Industry 4.0" given that postsecondary or tertiary education is essential to national success (Ehlers & Kellermann, 2019). What was useful 10 or 20 years ago might not be relevant in the here and now, with more rapid developments in teaching and learning technologies (Doucet et al., 2018); technological developments from different fields must be brought into lecture rooms while taking into account the learning preferences of the current generation (Rachid, 2018). This section reviews the learning practices in higher education as per Education 4.0. The educators in Malaysia are integrating three Education 4.0 learning technologies focused on learning English, namely virtual reality, mobile learning, and chatbot program (Mohd Adnan et al., 2019). A chatbot is an AI program that can simulate a conversation (i.e., a chat) with a human user in natural language through messaging applications, mobile apps, and other electronic platforms (Cahn, 2017), which clearly narrates that all of them understand the necessity to develop and use Education 4.0 technologies.

Moreover, The Slovak University of Technology in Bratislava, according to the findings, implements Education 4.0 tools, which include augmented reality, simulation, radiofrequency identification (RFID), using AI methods in predictive analytics, data transfer, distribution, transformation, adjustment, processing in real time, prediction, sensor designing, creation of complex networks for IoT, and 3D data visualization (Porubčinová et al., 2020). Educational institutions are implementing AI subsets like machine learning, natural language processing, and computer vision to enhance student experience and teaching and learning outcomes through individualized predictive analytics, enabling intelligent digital assistants, and enhancing security. With more than 518,000 schools and 276 million students, the Chinese Ministry of Education reports that their country has the biggest education sector in the world. In order to efficiently manage such massive populations, big data and AI are being studied to provide individualized learning environments (Zhang & Zou, 2019). Individualized learning helps weaker learners, who "will get the opportunity to practice more until they reach the required level" (Halili et al., 2021).

17.2.5.1 Teaching and Learning Practices in Higher Education in Pakistan

Information and communication technology (ICT) and e-learning use have grown in Pakistan during the past 10 years, while the Government of Pakistan's most recent educational strategy has placed a strong emphasis on employing ICTs and e-learning in schools and universities (Mujahid, 2002). The Government of Pakistan adopted a number of policies, but they weren't fully executed because of things that got in the way, such as adequate training, infrastructure, teachers' self-efficacy, their ability to use ICTs, accessibility to resources, lack of internet access, and other things (Butt et al., 2020). According to Asad et al. (2021), the university administration does not encourage the professors to use ICT in their teaching and learning because of their reservations about having insufficient resources and expertise (Ahmed et al., 2020). Moreover, Pakistan's infrastructure is lacking in terms of connection, online accessibility that creates obstacles for the adoption of e-learning, which is another issue the country is dealing with (Adnan et al., 2019). Ultimately, most of the students are not able to utilize ICTs effectively. In contrast, students at private universities have a more positive impression of e-learning practices and their potential than students at public institutions (Ali et al., 2022).

In the case study of the City University of Science and Information Technology (CUSIT), Pakistan's cutting-edge automated infrastructure, management information system (MIS), and LMS were discovered to be present at CUSIT, thereby facilitating faculty, staff, and administration (Habib et al., 2021). Key stakeholders, especially instructors, have access to an integrated and digital platform through the LMS for exchanging course outlines, lesson plans, assignment creation and submission, announcements, and producing assessment reports (Soomro et al., 2020). Similar to that, the student always has access to all course materials, assignments, assessment results, notifications, and other pertinent information (Aldiab et al., 2019).

17.2.6 Impact of Digital Technologies in Education 4.0 on Effectiveness of Learning in Higher Education

A key component of creating skilled human capital in the digital era is Education 4.0. Digital technologies make learning more successful by facilitating individualized, group-based, interactive, and adaptable learning experiences (Qureshi et al., 2021). Digital technologies encourage active learning by allowing students to interact with the subject matter through games, simulations, and interactive media. Students may collaborate on projects, share ideas, and learn from one another with the help of digital tools (Ronchi & Ronchi, 2019). With the aid of digital technologies, kids may learn outside of the classroom because knowledge is readily available. Even the students may learn at their own speed and pace due to digital tools (Jung, 2020). Critical thinking, communication, teamwork, and problem-solving are just a few of the crucial abilities that Education 4.0 helps students develop as they become ready for the workforce of the future (Qureshi et al., 2021).

According to Bujang et al. (2020), Education 4.0, which is being used in Malaysia to adapt to IR 4.0, has a big impact on how well students learn in higher education. Education 4.0 is a brand-new period of the educational system with features that mold pupils for independent study, group learning, and lifetime learning (Ronchi & Ronchi, 2019). In accordance with Education 4.0, the utilization of suitable digital learning platforms and technologies may increase student participation in their studies. Bujan et al. (2020) further elaborated that the impact of Education 4.0 on the effectiveness of learning in higher education is substantial. The goal of Education 4.0 is to provide students with the information and abilities necessary to thrive in the digital era. It emphasizes the use of technology to improve the learning experience and is a student-centered method of teaching (Srivani et al., 2022). Students may study at their own speed, collaborate with their peers, and have access to a multitude of knowledge and resources by using digital learning platforms and tools. Additionally, it is advised to the professors to get ready, adapt, and improve their ICT and digital skills to satisfy the needs of students who desire technology in their education (Miranda et al., 2021). To properly offer their courses, instructors must be adept at utilizing digital learning platforms and technologies (Siddiqui et al., 2021). Additionally, they must be able to modify their instruction to better meet the demands of students who favor online learning. Lecturers may assist students flourish in the digital era and increase the efficacy of learning in higher education by doing this (Salmon, 2019).

17.2.7 Education 4.0-Related Models in Higher Education

17.2.7.1 Education 4.0-Powered Teaching Model for Higher Education

The goal of the Education 4.0-powered teaching model, according to de Amorim Silva and Madeira (2021), is to improve the inclusiveness, sustainability, and resilience of the educational process. This paradigm, which was created for higher education, makes use of pedagogical strategies that are similar to constructivism in order to improve students' knowledge and level of involvement. The model takes into account student characteristics while advising the most effective teaching methods (Auerbach et al., 2018). This indicates that the model develops a customized lesson plan by taking into consideration the unique qualities of each student, such as their goals, interests, and learning styles. By doing this, the model may give each learner a more efficient and interesting learning experience (Hussin, 2018).

The concept uses well-known methods like the ADDIE (Analyze, Design, Develop, Implement, and Evaluate) model for instructional design (Nichols Hess & Greer, 2016) and the BLOOM taxonomy for knowledge-level assessment of students to build this personalized teaching plan (Thompson et al., 2008). A framework called the BLOOM taxonomy divides learning objectives into many tiers of complexity, ranging from easy recall to complicated analysis and assessment. With the use of this framework, teachers may create learning activities that are appropriate for the students' level of understanding (Thompson et al., 2008). The five steps of the ADDIE paradigm for instructional design are analysis, design, development, implementation, and evaluation. This approach assists educators in developing engaging and effective learning opportunities that cater to the requirements of their pupils (Nichols Hess & Greer, 2016).

The teaching paradigm supported by Education 4.0 may be implemented into Agile approaches like SCRUM, XP, Crystal, and PRINCE2, among others. Agile techniques are a collection of rules and procedures that place an emphasis on communication, adaptability, and ongoing development. The approach can adjust to the shifting demands of students by utilizing Agile techniques, which also makes learning more dynamic and responsive (de Amorim Silva & Madeira, 2021).

To sum up, the Education 4.0-powered teaching model is a cutting-edge method of instruction that makes use of pedagogical techniques that are connected to constructivism in order to raise the knowledge and engagement levels of involved students (de Amorim Silva & Madeira, 2021). The methodology may be utilized in both conventional classroom settings and online learning environments and can be implemented into Agile processes (Jamil, 2021). The method offers a more tailored and successful learning experience, making it an essential tool for teachers attempting to raise the standard in the classroom (Hussin, 2018). By employing this method, educators may create a more inclusive and engaging learning environment that caters to the requirements of every single student. This could lead to better academic performance, raised satisfaction among learners, and an all-round more pleasurable learning environment (Wiggins et al., 2017). A promising method of instruction, the Education 4.0-powered teaching model has the potential to revolutionize how pupils are taught in the 21st century (Salas Pilco, 2013).

17.2.7.2 ADDIE Model

Five stages make up the ADDIE model (appeared for the first time in 1975) for instructional design: analysis, design, development, implementation, and evaluation (de Amorim Silva & Madeira, 2021). Widyastuti (2019) explains the following stages.

17.2.7.2.1 Analysis

At this step, the instructional designer determines the goals and demands of the target course in terms of learning. Additionally, the learner characteristics, the learning environment, and the resource pool are all examined by the designer (Aldoobie, 2015).

17.2.7.2.2 Design

During this phase, the instructional designer develops a course plan that includes the learning objectives, underlying assumptions, limitations, and guidelines. The designer also chooses the instructional techniques, approaches, and resources that will be applied to fulfill the learning goals (Reinbold, 2013).

17.2.7.2.3 Development

During this phase, the instructional designer creates the learning resources, including multimedia components, tests, activities, and lesson plans. In order to run the Agile structure, the designer also creates key performance indicators (KPIs) and crucial backlogs (Hsu et al., 2014).

17.2.7.2.4 Implementation

During this phase, the instructional designer puts the lesson plan into action and provides the students with the necessary tools and learning materials. The designer also oversees the learners' development and offers them help and direction (Widyastuti, 2019).

17.2.7.2.5 Evaluation

At this point, the instructional designer assesses the course's efficacy and learning objectives. To enhance the course's design and delivery, the designer gathers input from students and other stakeholders (Muneeb et al., 2020). The course's influence on the students and the learning environment are also evaluated, along with the success of the learning objectives (Widyastuti, 2019).

17.2.8 Challenges Regarding Implementation of Education 4.0 in Higher Education in Pakistan

The Fourth Industrial Revolution has caused a paradigm change in numerous fields, including education policy, teaching and learning practices, and reform. Developed countries routinely adopt new technologies, close the skills gap, and open the doors to new resources (Muneeb et al., 2020). Due to governmental incompetence, long processing times, and lack of sufficient talent, the situation in developing countries is considerably different from that in wealthy countries (Puncreobutr, 2016).

Students in the current world have become habituated to using digital gadgets like smartphones, tablets, and computers, from elementary school students to those in higher education. They often use these gadgets for communication and enjoyment (Sánchez-Cruzado et al., 2021).

The results of the Zeehan et al. (2020) study indicate that instructors' preparation for new technologies is difficult. To improve educational abilities in digital technologies, educators have not yet optimized digital technologies in their instructional practices. Innovative technologies are challenging for both individuals and educational institutions to adapt effectively. Moreover, according to Alda et al. (2020), administrators and faculty members believe they are prepared in terms of their abilities to choose and incorporate digital resources for teaching and learning because they are also given capacity building through seminars and conferences related to technological literacy. But they lack proficiency in using LMS and other online learning tools, as well as in employing virtual reality, robotics, and digital enablers like 3D printing (Nguyen, 2018).

Furthermore, students have a good attitude toward picking up new technology. Personal characteristics including age (Fahmiyah et al., 2023), sex (Setiawan et al., 2018), and professional experience for educational reasons may have an impact on how teachers use technology in the classroom. Moreover, the use of digital technology in the classroom is generally not supported by institutional strategies or models (Mercader, 2019). According to Adnan et al. (2019), instructors have a low opinion of their own digital competence and believe that their digital literacy skills need to be improved. Moreover, according to Ali et al. (2022), Pakistan's infrastructure is lacking in terms of connection, communication and technical gap, and online accessibility, etc., which creates obstacles for the adoption of e-learning, which is another issue the country is dealing with.

The study claims that, after reviewing the issues raised in the literature, ICT and e-learning use have increased significantly in Pakistan over the past 10 years, and that the Government of Pakistan's most recent educational policy has placed a significant emphasis on using ICTs and e-learning in universities as well as adopted a number of digital learning policies (Ali et al., 2022). However, while the rest of the world has adopted Education 4.0, complete with e-learning tools, games, and models, Pakistan has lagged behind in these areas for the past 10 years as well, and it continues to struggle with the same problems that the rest of the world overcame alongside Education 2.0, particularly regarding internet connectivity and accessibility (Zeehan et al., 2020).

17.2.9 Potential of Education 4.0 Approaches in Higher Education in Pakistan

The study recommends that in order to address perceptional, conceptual, and practical shortcomings regarding e-learning practices at the higher education level in Pakistan, e-learning resources, technical training, internet accessibility, and improved communication skills among students and teachers should be made available (Sarkis-Onofre et al., 2021). Research also shows that e-learning practices have a great potential to improve technical skills, critical thinking, positive perception, less time-consuming, cost reduction, result-oriented, and innovative learning to compete with other universities around the world in quality education (Ali et al., 2022).

17.3 METHODOLOGY

A systematic approach was used in this study to collect pertinent data from credible academic sources included in the Scopus database. A thorough and systematic review approach was ensured by adhering to the 2015 Preferred Reporting Items for Systematic Reviews and Meta-Analyses (PRISMA) criteria.

The PRISMA guidelines' limitations (Tricco et al., 2018) were followed in the process of choosing research papers. The extensive coverage of peer-reviewed academic literature across a wide range of fields in the Scopus database led to its selection. Using certain keywords and filters, the search method found academic papers that addressed the application of digital technology in Pakistani higher education. Peer-reviewed papers that primarily examined how Education 4.0 affects learning effectiveness in Pakistani higher education institutions met the inclusion requirements.

A rigorous selection approach (D'Amico et al., 2012) was used to include 95 research papers in the evaluation process. These papers were chosen on the grounds that they met the predetermined standards and were pertinent to the topic under consideration. During the screening procedure, titles and abstracts were first evaluated for conformity with the study goals. Full-text evaluations were then carried out to make sure the chosen articles offered significant insights into the use of digital technologies and their effect on learning outcomes in Pakistani higher education.

Bibliometric analysis of the chosen publications was used to identify keywords related to the research subject. This procedure comprised finding and examining the words, concepts, and phrases that were utilized in the literature, as well as their distribution and frequency (Takkouche & Norman, 2011). Key terminology and concepts were taken from the academic discourse about digital technology integration in Pakistani higher education, allowing for a thorough grasp of the prevalent themes and trends.

PRISMA layout (refer to Figure 17.1) included identifying, screening, determining eligibility for, and research articles in the review process, all in accordance with the PRISMA standards (Rethlefsen et al., 2021). The design guarantees a thorough and methodical approach to the collection and evaluation of literature, improving the validity and reliability of the review's conclusions and results (Tugwell & Tovey, 2021).

17.4 CONCLUSION

The term "Education 4.0" describes the use of digital technology in the classroom to improve learning opportunities. The requirement for teacher professional development, the encouragement of critical thinking, and the customization of learning experiences are some of the difficulties and advantages of this method. The ADDIE model, a five-stage instructional design framework used to create successful learning

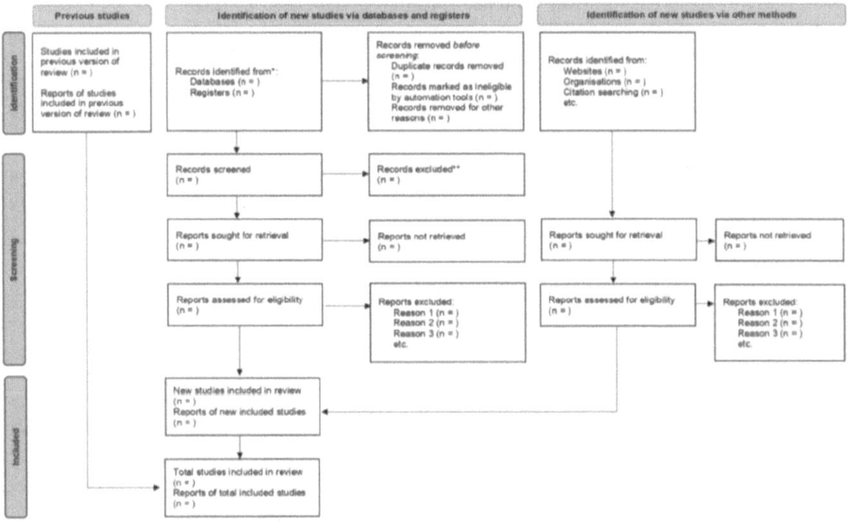

FIGURE 17.1 PRISMA 2020 flow diagram for updated systematic reviews, which included searches of databases, registers, and other sources (Page et al., 2021).

experiences, may be utilized to increase the effectiveness of Education 4.0 practices. The stages of the model are analysis, design, development, implementation, and evaluation. The ADDIE model may be utilized to design personalized learning experiences that take into consideration the goals, interests, and learning preferences of each learner. The study also highlights the Education 4.0-powered teaching and learning model, which is a teaching strategy that employs constructivist-inspired pedagogical techniques to increase students' knowledge and engagement. When recommending the most efficient teaching strategies, the model considers the characteristics of the students. To create a personalized lesson plan, this methodology assesses students' knowledge levels using the ADDIE model and the BLOOM taxonomy. In this chapter, interactive learning techniques that push students to think independently and creatively are highlighted. Case studies, simulations, and problem-based learning are some of these activities. According to this study, these exercises can aid students in honing their analytical, problem-solving, and critical thinking skills. This chapter makes the case that the current teaching methods in higher education in Pakistan are not in line with Education 4.0 because of a number of issues, such as a lack of infrastructure and resources. It argues that implementing Education 4.0 by addressing these problems and utilizing the models defined in this chapter are essential to improving learning outcomes in higher education in Pakistan.

REFERENCES

Abbas, Qaisar, Shafqat Hussain, & Shafqat Rasool. (2019). Digital literacy effect on the academic performance of students at higher education level in Pakistan. *Global Social Sciences Review, 4*(1): 154–165.

Adnan, Malik, Zahid Yousaf, & Aftab Hussain Gilani. (2019). E-learning with digital media and human development: The way forward for Pakistan. *Journal of Educational Research, 22*(2): 137–152.

Ahmed Khuwaja, Hussain Maqbool, Ali Maqbool, Komal Abdul Rahim, Sana Gul, Sanam Hanif, & Shabnoor Karim. (2020). Status of digital learning practices in health sciences education in Pakistan. *Journal of the Pakistan Dental Association*, 19(8): 137–151.

Alda, Rivika, Helen Boholano, & Filomena Dayagbil. (2020). Teacher education institutions in the Philippines towards Education 4.0. *International Journal of Learning, Teaching and Educational Research, 19*(8): 137–154.

Aldiab, Abdulaziz, Harun Chowdhury, Alex Kootsookos, Firoz Alam, & Hamed Allhibi. (2019). Utilization of learning management systems (LMSs) in higher education system: A case review for Saudi Arabia. *Energy Procedia, 160*: 731–737.

Aldoobie, Nada. (2015). ADDIE model. *American International Journal of Contemporary Research, 5*(6): 68–72.

Ali, Romena, Khalid Khurshid, & Vagiha Naz. (2022). Implementing e-learning practices and integration of technology at higher education level in Pakistan: Problems and potentials. *Harf-o-Sukhan, 6*(1): 12–28.

Amorim Silva, Rafael de, & Filipe Madeira. (2021). Toward to an Education 4.0-powered teaching model for higher education. *Revista Da UIIPS, 9*(4), 70–80.

Asad, Muhammad Mujtaba, Nadia Hussain, Maria Wadho, Zahid Hussain Khand, & Prathamesh P. Churi. (2021). Integration of E-learning technologies for interactive teaching and learning process: An empirical study on higher education institutes of Pakistan. *Journal of Applied Research in Higher Education, 13*(3): 649–663.

Aslam, Rabia, Najmonnisa Khan, Muhammad Mujtaba Asad, & Ushaque Ahmed. (2021). Impact of technological pedagogical content knowledge on teachers' digital proficiency at classroom in higher education institution of Pakistan. *Interactive Technology and Smart Education, 18*(1): 119–130.

Aslam, Rabia, Najmonnisa Khan, & Ushaque Ahmed. (2020). Technology integration and teachers' professional knowledge with reference to International Society for Technology in Education (ISTE)-standard: A causal study. *Journal of Education and Educational Development, 7*(2): 307–327.

Auerbach, Anna J., Matt Higgins, Peggy Brickman, & Tessa C. Andrews. (2018). Teacher knowledge for active-learning instruction: Expert-novice comparison reveals differences. *CBE-Life Sciences Education, 17*(1): ar12.

Bai, Chunguang, Patrick Dallasega, Guido Orzes, & Joseph Sarkis. (2020). Industry 4.0 technologies assessment: A sustainability perspective. *International Journal of Production Economics, 229*: 107776.

Bujang, Siti Dianah Abdul, Ali Selamat, Ondrej Krejcar, Petra Maresova, & Ngoc Thanh Nguyen. (2020). Digital learning demand for future education 4.0-case studies at Malaysia education institutions. *Informatics, 7*: 13.

Butt, Rameen, Hammad Siddiqui, Raheel Ahmed Soomro, & Muhammad Mujtaba Asad. (2020). Integration of industrial revolution 4.0 and IOTs in Academia: A state-of-the-art review on the concept of Education 4.0 in Pakistan. *Interactive Technology and Smart Education, 17*(4): 337–354.

Cahn, Jack. (2017). *CHATBOT: Architecture, Design, & Development*. University of Pennsylvania School of Engineering and Applied Science Department of Computer and Information Science.

Castañeda, Linda, & Neil Selwyn. (2018). More than tools? Making sense of the ongoing digitizations of higher education. *International Journal of Educational Technology in Higher Education, 15*(1): 1–10.

Chacon-Diaz, Lucia, & Susan Brown. (2019). *Integrating Digital Technology in Education*, p. 173. School-University-Community Collaboration.

D'Amico, Simone, Per Bodin, Michel Delpech, & Ron Noteborn. (2012). Prisma. In: Marco D'Errico (Ed), *Distributed Space Missions for Earth System Monitoring*, pp. 599–637. New York: Springer.

Doucet, Armand, Jelmer Evers, Elisa Guerra, Nadia Lopez, Michael Soskil, & Koen Timmers. (2018). *Teaching in the Fourth Industrial Revolution: Standing at the Precipice*. Routledge.

Ehlers, Ulf-Daniel, & Sarah A Kellermann. (2019). *Future Skills: The Future of Learning and Higher Education*. Karlsruhe.

Fahmiyah, Indah, Ika Qutsiati Utami, Ratih Ardiati Ningrum, Muhammad Noor Fakhruzzaman, Angga Iryanto Pratama, & Yohanes Manasye Triangga. (2023). Examining the effect of teacher's age difference on learning technology adoption using technology acceptance model. In *AIP Conference Proceedings*, Vol. 2536. AIP Publishing, Surabaya, Indonesia.

Farrukh, Moattar, Tariq Rahim Soomro, Taher M. Ghazal, Haitham M. Alzoubi, & Muhammad Alshurideh. (2023). Perspectives of online education in Pakistan: Post-covid scenario. In: Muhammad Alshurideh, Barween Hikmat Al Kurdi, Ra'ed Masa'deh, Haitham M. Alzoubi, Said Salloum (Eds), *The Effect of Information Technology on Business and Marketing Intelligence Systems*, pp. 519–550. Springer International Publishing.

Frolova, Elena V., Tatyana M. Ryabova, & Olga V. Rogach. (2019). Digital technologies in education: Problems and prospects for Moscow Electronic School" project implementation. *European Journal of Contemporary Education, 8*(4): 779–789.

Gerstein, Jackie. (2014). *Moving from Education 1.0 through Education 2.0 towards Education 3.0*. In L.M. Blaschke, C. Kenyon, and S. Hase (Eds.), *Experiences in Self-Determined Learning*. CreateSpace Independent Publishing Platform.

González-Pérez, Laura Icela, & María Soledad Ramírez-Montoya. (2022). Components of education 4.0 in 21st century skills frameworks: Systematic review. *Sustainability, 14*(3): 1493.

Habib, Muhammad Nauman, Waseef Jamal, Uzma Khalil, & Zunnoorain Khan. (2021). Transforming universities in interactive digital platform: Case of City University of Science and Information Technology. *Education and Information Technologies, 26*: 517–541.

Haleem, Abid, Mohd Javaid, Mohd Asim Qadri, & Rajiv Suman. (2022). Understanding the role of digital technologies in education: A review. *Sustainable Operations and Computers, 3*: 275–285.

Halili, Siti Hajar, Shukri Sulaiman, Hamidah Sulaiman, & Rafiza Razak. (2021). Embracing industrial revolution 4.0 in universities. In *IOP Conference Series: Materials Science and Engineering*, 1088:012111. IOP Publishing.

Harkins, Arthur M. (2008). Leapfrog principles and practices: Core components of Education 3.0 and 4.0. *Futures Research Quarterly, 24*(1): 19–31.

Hermann, Mario, Tobias Pentek, & Boris Otto. (2016). Design principles for Industrie 4.0 scenarios. In *2016 49th Hawaii International Conference on System Sciences (HICSS)*, pp. 3928–3937. IEEE.

Hrastinski, Stefan. (2019). What do we mean by blended learning? *TechTrends, 63*(5): 564–569.

Hsu, Tzu-Chuan, Jane Lee-Hsieh, Michael A Turton, & Su-Fen Cheng. (2014). Using the ADDIE model to develop online continuing education courses on caring for nurses in Taiwan. *The Journal of Continuing Education in Nursing, 45*(3): 124–131.

Hussin, Anealka Aziz. (2018). Education 4.0 made simple: Ideas for teaching. *International Journal of Education and Literacy Studies, 6*(3): 92–98.

Jain, Vishal, and Parul Jain. (2022). From Industry 4.0 to Education 4.0: Acceptance and use of videoconferencing applications in Higher Education of Oman. *Journal of Applied Research in Higher Education, 14*(3): 1079–1098.

Jamil, Sadia. (2021). From digital divide to digital inclusion: Challenges for wide-ranging digitalization in Pakistan. *Telecommunications Policy, 45*(8): 102206.

Jan, A., I. Sultana, & M. Adnan. (2020). Digital media and smart education in Pakistan: Challenges and prospects for the teachers in the age of e-learning. *Global Educational Studies Review, 4*: 51–59.

Jobirovich, Yarashov Mardon. (2021). The role of digital technologies in reform of the education system. *The American Journal of Social Science and Education Innovations, 3*(04): 461–465.

Jung, Jisun. (2020). The fourth industrial revolution, knowledge production and higher education in South Korea. *Journal of Higher Education Policy and Management, 42*(2): 134–156.

Kagermann, Henning, Johannes Helbig, Ariane Hellinger, & Wolfgang Wahlster. (2013). *Recommendations for Implementing the Strategic Initiative INDUSTRIE 4.0: Securing the Future of German Manufacturing Industry; Final Report of the Industrie 4.0 Working Group*. Forschungsunion.

Kannengiesser, Udo. (2017). The future: Obstacles and opportunities. *S-BPM in the Production Industry: A Stakeholder Approach*, pp. 209–230.

Keats, Derek, & J Philipp Schmidt. (2007). The genesis and emergence of Education 3.0 in higher education and its potential for Africa. *First Monday, 12*(3): 3–5.

Khalil, Aysha, Ambreen Haqdad, & Naveed Sultana. (2023). Educational metaverse for teaching and learning in higher education of Pakistan. *Journal of Positive School Psychology, 7*(2): 1183–1197.

Kundi, Ghulam Muhammad, Bahadar Shah, & Allah Nawaz. (2008). Digital Pakistan: Opportunities & challenges. *JISTEM-Journal of Information Systems and Technology Management, 5*: 365–390.

Lasi, Heiner, Peter Fettke, Hans-Georg Kemper, Thomas Feld, & Michael Hoffmann. (2014). Industry 4.0. *Business & Information Systems Engineering, 6*: 239–242.

Laurillard, Diana. (2008). *Digital Technologies and Their Role in Achieving Our Ambitions for Education*. Institute of Education, University of London.

Liu, Min, Emily McKelroy, Stephanie B Corliss, & Jamison Carrigan. (2017). Investigating the effect of an adaptive learning intervention on students' learning. *Educational Technology Research and Development, 65*: 1605–1625.

Makrides, Gregoris A. (2019). *The Evolution of Education from Education 1.0 to Education 4.0: Is It an Evolution or a Revolution*. Beer Sheva, Israel.

Martín-Gutiérrez, Jorge, Carlos Efrén Mora, Beatriz Añorbe-Díaz, & Antonio González-Marrero. (2017). Virtual technologies trends in education. *Eurasia Journal of Mathematics, Science and Technology Education, 13*(2): 469–486.

Mercader, Cristina. (2019). Las Resistencias Del Profesorado Universitario a La Utilización de Las Tecnologías Digitales. *Aula Abierta, 48*(2): 167–174.

Miranda, Jhonattan, Christelle Navarrete, Julieta Noguez, José-Martin Molina-Espinosa, María-Soledad Ramírez-Montoya, Sergio A Navarro-Tuch, Martín-Rogelio Bustamante-Bello, José-Bernardo Rosas-Fernández, & Arturo Molina. (2021). The core components of Education 4.0 in higher education: Three case studies in engineering education. *Computers & Electrical Engineering, 93*: 107278.

Mohd Adnan, Airil Haimi, Rafidah Abd Karim, Mohd Haniff Mohd Tahir, Nurul Nadiah Mustafa Kamal, & Ahmad Muhyiddin Yusof. (2019). Education 4.0 technologies, Industry 4.0 skills and the teaching of English in Malaysian tertiary education. *Arab World English Journal (AWEJ), 10*(4): 330–343.

Morris, Thomas Howard, & Matthias Rohs. (2023). The potential for digital technology to support self-directed learning in formal education of children: A scoping review. *Interactive Learning Environments, 31*(4): 1974–1987.

Mujahid, Yousaf Haroon. (2002). Digital opportunity initiative for Pakistan. *The Electronic Journal of Information Systems in Developing Countries, 8*(1): 1–14.

Muneeb, Faiza, Sana Mehmood, & Saliha Mehboob. (2020). Digital media and technological challenges for media education in Pakistan. *Pakistan Journal of Social Research, 2*: 23–35.

Nawaz, Nishad, Susanne Durst, A Hariharasudan, & Zurabi Shamugia. (2020). Knowledge management practices in higher education institutions – A comparative study. *Polish Journal of Management Studies, 22*.

Nguyen, Dang. (2018). The university in a world of digital technologies: Tensions and challenges. *Australasian Marketing Journal, 26*(2): 79–82.

Nichols Hess, Amanda, & Katie Greer. (2016). Designing for engagement: Using the ADDIE model to integrate high-impact practices into an online information literacy course. *Communications in Information Literacy, 10*(2): 6.

Okoye, Kingsley, Haruna Hussein, Arturo Arrona-Palacios, Héctor Nahún Quintero, Luis Omar Peña Ortega, Angela Lopez Sanchez, Elena Arias Ortiz, Jose Escamilla, & Samira Hosseini. (2023). Impact of digital technologies upon teaching and learning in higher education in Latin America: An outlook on the reach, barriers, and bottlenecks. *Education and Information Technologies, 28*(2): 2291–2360.

Page, Matthew J., Joanne E. McKenzie, Patrick M. Bossuyt, Isabelle Boutron, Tammy C. Hoffmann, Cynthia D. Mulrow, Larissa Shamseer, et al. (2021). The PRISMA 2020 statement: An updated guideline for reporting systematic reviews. *International Journal of Surgery, 88*: 105906.

Palys, Ted, & Chris Atchison. (2012). Qualitative research in the digital era: Obstacles and opportunities. *International Journal of Qualitative Methods, 11*(4): 352–367.

Pedota, Mattia, & Lucia Piscitello. (2022). A new perspective on technology-driven creativity enhancement in the fourth industrial revolution. *Creativity and Innovation Management, 31*(1): 109–122.

Porubčinová, Martina, Ivana Novotná, & Helena Fidlerová. (2020). The use of Education 4.0 tools in tertiary education system in Slovakia. *Information Technologies and Learning Tools, 80*(6): 161–175.

Puncreobutr, Vichian. (2016). Education 4.0: New challenge of learning. *St. Theresa Journal of Humanities and Social Sciences, 2*(2): 92–97.

Qureshi, Muhammad Imran, Nohman Khan, Hamad Raza, Amina Imran, & Fadillah Ismail. (2021). *Digital Technologies in Education 4.0. Does It Enhance the Effectiveness of Learning?*

Rachid, S., Una Cunningham, Kevin Watson, & Jocelyn Howard. (2018). Revisiting the digital divide (s): Technology-enhanced English language practices at a university in Pakistan.

Rafiq, Muhammad, & Kanwal Ameen. (2012). Use of digital media and demand for digitized contents in higher education sector of Pakistan. *The International Information & Library Review, 44*(3): 116–122.

Redecker, Christine, & Yves Punie. (2010). Learning 2.0 promoting innovation in formal education and training in Europe. In *Proceedings on Sustaining TEL: From Innovation to Learning and Practice: 5th European Conference on Technology Enhanced Learning, EC-TEL 2010*, Barcelona, Spain, September 28-October 1, 2010, *5*, pp. 308–323. Springer.

Rehman, Aziz Ur, & Bakhtiar Khan. (2021). Challenges to online education in Pakistan during COVID-19 & the way forward. *Social Science Learning Education Journal, 6*(7): 503–512.

Reinbold, Sarah. (2013). Using the ADDIE Model in Designing Library Instruction. *Medical Reference Services Quarterly, 32*(3): 244–256.

Rethlefsen, Melissa L., Shona Kirtley, Siw Waffenschmidt, Ana Patricia Ayala, David Moher, Matthew J. Page, & Jonathan B. Koffel. (2021). PRISMA-S: An extension to the PRISMA statement for reporting literature searches in systematic reviews. *Systematic Reviews, 10*: 1–19.

Roblek, Vasja, Maja Meško, & Alojz Krapež. (2016). A complex view of Industry 4.0. *Sage Open, 6*(2): 2158244016653987.

Ronchi, Alfredo M, & Alfredo M Ronchi. (2019). E-learning: How teaching and training methods changed in the last 20 years. *E-Services: Toward a New Model of (Inter) Active Community*, 69–113.

Salam, Shafaq, Mengke Yang, Aliya Shaheen, Mahmood Movahedipour, & Jianqiu Zeng. (2017). ICT and students' performance in Pakistan. *Human Systems Management, 36*(4): 277–284.

Salas Pilco, & Sdenka Z. (2013). Evolution of the framework for 21st century competencies. *Knowledge Management & E-Learning: An International Journal, 5*: 10–24.

Salmon, Gilly. (2019). May the fourth be with you: Creating Education 4.0. *Journal of Learning for Development, 6*(2): 95–115.

Sánchez-Cruzado, Cristina, Raúl Santiago Campión, & Ma Teresa Sánchez-Compaña. (2021). Teacher digital literacy: The indisputable challenge after COVID-19. *Sustainability, 13*(4): 1858.

Sarkis-Onofre, Rafael, Ferrán Catalá-López, Edoardo Aromataris, & Craig Lockwood. (2021). How to properly use the PRISMA statement. *Systematic Reviews, 10*(1): 1–3.

Setiawan, A, H Yanti, & Y Z Miraj. (2018). Analysis of ICT literacy competence among vocational high school teachers. In IOP *Conference Series: Materials Science and Engineering*, 306:012097. IOP Publishing. Manado, Indonesia.

Siddiqui, Hammad, Rameen Butt, Raheel Ahmed Soomro, Muhammad Mujtaba Asad, & Fahad Sherwani. (2021). Engineering students' motivation toward Education 4.0: A conceptual framework for Rural Sindh." In *Innovative Education Technologies for 21st Century Teaching and Learning*, pp. 199–214. CRC Press.

Soomro, Kamal Ahmed, Ugur Kale, Reagan Curtis, Mete Akcaoglu, & Malayna Bernstein. (2020). Digital divide among higher education faculty. *International Journal of Educational Technology in Higher Education, 17*: 1–16.

Srivani, V, A Hariharasudan, Nishad Nawaz, & Sabina Ratajczak. (2022). Impact of Education 4.0 among engineering students for learning English language. *PLoS One, 17*(2): e0261717.

Srivastava, Kiran, & Soumen Dey. (2018). Role of digital technology in teaching–learning process. *IOSR Journal of Humanities and Social Science (IOSR-JHSS), 23*(1): 74–79.

Takkouche, Bahi, & Guy Norman. (2011). PRISMA statement. *Epidemiology, 22*(1): 128.

Thomas, Michael OJ, & Joann M. Palmer. (2014). Teaching with digital technology: Obstacles and opportunities. Editors: Alison Clark Wilson, Ornella Robutti, Nathalie Sinclair. In *The Mathematics Teacher in the Digital Era: An International Perspective on Technology Focused Professional Development*, Springer, pp. 71–89.

Thompson, Errol, Andrew Luxton-Reilly, Jacqueline L Whalley, Minjie Hu, & Phil Robbins. (2008). Bloom's taxonomy for CS assessment. In *Proceedings of the Tenth Conference on Australasian Computing Education-Volume 78*, 155–161.

Tricco, Andrea C., Erin Lillie, Wasifa Zarin, Kelly K. O'Brien, Heather Colquhoun, Danielle Levac, David Moher, et al. (2018). PRISMA extension for scoping reviews (PRISMA-ScR): Checklist and explanation. *Annals of Internal Medicine, 169*(7): 467–473.

Tugwell, Peter, & David Tovey. (2021). PRISMA 2020. *Journal of Clinical Epidemiology, 134*: A5–A6.

UK GOV. (2021). *Department for Business, Energy & Industrial Strategy: About Us*. https://www.gov.uk/government/organisations/department-for-business-energy-and-industrial-strategy/about

Widyastuti, E. (2019). Using the ADDIE model to develop learning material for actuarial mathematics. Ed. Martin Lard. In *Journal of Physics: Conference Series*, 1188:012052. IOP Publishing.

Wiggins, Benjamin L, Sarah L Eddy, Leah Wener-Fligner, Karen Freisem, Daniel Z Grunspan, Elli J Theobald, Jerry Timbrook, & Alison J Crowe. (2017). ASPECT: A survey to assess student perspective of engagement in an active-learning classroom. *CBE-Life Sciences Education, 16*(2): ar32.

Williamson, Ben. (2016). Digital education governance: Data visualization, predictive analytics, and 'real-time' policy instruments. *Journal of Education Policy, 31*(2): 123–141.

Zeehan, Farah, Rose Alinda Alias, & Zaidatun Tasir. (2020). Discovering digital technology training challenges for future-ready educator: A preliminary study from trainer perspective. *Universal Journal of Educational Research, 8*(3A): 12–23.

Zhang, Zhihao, & Shuo Zou. (2019). AI-enabled tuition ushers in the 'intelligent age.' *Chinadaily.com.cn*.

Zubairi, Asma, Abdullah Khalayleh, Imdad Baloch, Haani Mazari, Tom Kaye, & Caspar Groeneveld. (2022). *Pakistan Digital Learning Landscape Analysis*. EdTech Hub.

18 A Systematic Review of Multimodal Learning in Higher Education

Raja Nor Safinas Raja Harun and
Amreet Kaur Jageer Singh

18.1 INTRODUCTION: DEFINITION AND CONCEPT

Multimodal learning is an instructional approach which integrates multiple modes of representation, engagement, and expression to foster learning and development (Jewitt, 2008; Jewitt et al., 2016). The term has grown in popularity and usage since it was first coined in the mid-1990s. Since then, there has been a burgeoning of academic discussions centred around multimodality, as evidenced by many conferences organised, monographs and edited volumes published, and other scholarly mediums created to examine multimodality in a detailed manner. Unlike traditional pedagogies, which focus primarily on linguistic and logical representations of knowledge, multimodal pedagogies emphasise the design of learning environments and activities that allow learners to meaningfully engage with ideas and represent their knowledge through diverse semiotic resources such as images, gestures, drama, music, dance, speaking, and writing (Lim & Tan-Chia, 2022; Lin & Tsai, 2021; New London Group, 1996).

As stated by Kress (2010), "learning and cognition rely heavily on learners' interaction with varied representations of meaning" (p. 54). This perspective aligns with multimodal learning philosophies, which build on the views of social semiotics regarding communication and meaning-creation as encompassing much more than language alone (Kress, 2010). Social semiotics posits that meaning is constructed and conveyed through multiple symbolic systems, of which spoken and written language is only one facet. Multimodal theories expand on this idea by proposing that knowledge is obtained and learning happens through social interactions and engagements with multiple modalities and texts. These diverse "modes of representation", as termed by Kress (2010), can be visual, such as diagrams, gestures, illustrations, and graphic organisers. Alternatively, they may be textual, such as written words and language. Other modalities include auditory modes like lectures, music, or discussion and tactile or kinaesthetic modes like hands-on activities, role-play, and simulations. According to multimodal theories, integrating experiences with these various modalities within a social environment allows deep learning and cognition (Chiu & Churchill, 2015; Doherty, 2023; Kao & Luo, 2020; Taxipulati & Lu, 2021).

 DOI: 10.1201/9781003400691-18

Research indicates that multimodal instructional approaches can enhance learning, especially for diverse learners (Sankey et al., 2010; Shanahan, 2013). Multimodal pedagogies align with contemporary literacy views, encompassing proficiency with diverse communicative modes and media, not just print literacy (The New Media Consortium, 2005). By integrating multiple modes of representation into the learning process, multimodal instruction provides opportunities for deeper learning, knowledge application, varied forms of student expression and engagement, and inclusion of diverse learners (Bouchey et al., 2021).

The potential of multimodal learning has led to growing interest in multimodal instructional approaches in higher education contexts (Andrews & Smith, 2011; Camiciottoli & Campoy-Cubillo, 2018). College courses are increasingly recognising the need to prepare students for multimodal literacy practices relevant to academic, professional, and civic life. This chapter discusses essential theoretical foundations, research evidence, and practical strategies for leveraging multimodal pedagogies to promote engaged learning in higher education across disciplines.

18.2 THEORETICAL FOUNDATIONS OF MULTIMODAL LEARNING

Multimodal learning theories derive from social semiotics and multimodal design perspectives, which view communication and representation as inherently multimodal (Bozkurt, 2020; Kress, 2010). These theories posit that meaning-making involves integrating multiple modes, including image, gesture, music, speech, and writing (Jewitt, 2008). Multimodal theories also build on scholarship highlighting the mediated nature of human cognition, which is facilitated by cultural tools and technologies for thinking and communicating (Scollon, 2001; Vygotsky, 1978). According to Mohammed Hassan Al-Ahdal et al. (2022), multimodal learning combines multiple modes of representation, engagement, and expression to foster learning and development.

Several key theoretical frameworks have underpinned the multimodal learning approaches. Social semiotics recognises that signs and symbols have no inherent meanings; instead, meanings are socially constructed through interpretation and representation (Hodge & Kress, 1988). From this view, diverse semiotic resources hold the potential for meaning-making, learning, and expression. Multiliteracies theory has also highlighted the multifaceted, culturally situated nature of literacy and communication (New London Group, 1996). Multimodal pedagogies aim to develop students' repertoires of practice with different modes of meaning-making and texts.

Multimodality theories also build on multimedia learning research on how information delivery via dual channels (visual and verbal) can enhance students' learning (Mayer, 2009). Integrating multiple representations of concepts through different modalities can help build mental connections and deepen understanding (Ainsworth, 2006). Similarly, activity theory has also suggested that working across multiple modalities creates sites for contrastive thinking that can foster learning (Böck & Pachler, 2013).

This theoretical grounding informs multimodal instructional approaches that engage students in learning concepts through diverse modalities. Multimodal activities encourage students to interpret ideas through varied semiotic modes and to develop fluency in negotiating meanings with different symbolic systems. Such opportunities for transformative engagement with concepts can enrich learning and literacy development (Kress, 2010).

Constructivist learning theories have also informed the multimodal instructional approaches (Dalton & Proctor, 2008). Constructivism proposes that learners actively build knowledge and meaning from their experiences. Multimodal activities are in tandem with constructivist views by allowing students to engage with ideas through diverse experiences and representations. Combining multiple modes requires learners to actively integrate information and construct a more profound understanding (Mayer, 2009).

Sociocultural learning theories are also in view of multimodal pedagogies through their emphasis on social interactions and cultural tools which mediate cognition (Cole & Engestrom, 1993). Multimodal instruction leverages various cultural artefacts and technologies as resources for meaning-making. It also involves collaborative learning wherein students pool diverse perspectives and modes of representation (Siegel, 2012). Such social sharing shapes collective knowledge-building.

Integrating multiple modalities in learning activities provides interactive, socially rich environments where learners can grapple with concepts through varied experiences. As Kress (2010) notes, "Multimodality describes the normal state of human communication" (p. 1). Thus, multimodal learning approaches are in alignment with how students naturally leverage different modes to interpret and engage with their world. Allowing learners to work across diverse modalities taps into the multimodal nature of human cognition and communication.

18.3 METHODOLOGY

This systematic review adhered to the fundamental principles and steps for conducting reviews described by Boland et al. (2017) and Shaffril et al. (2020). The review process was adapted to address the current study's specific research questions and scope. The key stages have formulated the research aims, defining inclusion and exclusion criteria, developing and documenting a comprehensive search strategy, selecting relevant papers and extracting data, and analysing the quality of the study. While guided by an established systematic review methodology, each step was customised to ensure a rigorous, unbiased evidence synthesis optimised for the review. The selective review method was intended to give authoritative answers to the review questions based on the highest quality evidence.

18.3.1 Formulating the Review Aims

This systematic review aimed to identify, critically evaluate, and synthesise relevant evidence to provide empirically grounded answers to key questions about using multimodal learning approaches in higher education. Specifically, the review addressed three research questions:

1. What is the significance of integrating diverse media formats in higher education classrooms?
2. What challenges do educators face in implementing multimodal pedagogy?
3. What practical strategies can facilitate effective multimodal instruction?

By conducting an objective, rigorous synthesis of current scholarly literature concerning these questions, the review sought to offer evidence-based insights to inform teaching practices and policies pertaining to multimodal learning approaches in higher education contexts.

18.3.2 Defining Inclusion and Exclusion Criteria

In order to identify the most pertinent research, this review employed rigorous inclusion and exclusion criteria for article selection. Studies were included if they were published in peer-reviewed journals indexed in academic databases, written in English, published between 2003 and 2022, provided full-text access, and used qualitative, quantitative, or mixed methods approaches. Notably, only studies investigating applications of multimodal learning in higher education settings were included. On the other hand, articles were excluded if they were in the form of editorials, conference papers, dissertations, book chapters, notes, surveys, reports, or letters.

18.3.3 Developing and Documenting Search Strategy

A rigorous search strategy was developed to identify relevant literature across four major academic databases: Web of Science, Scopus, Google Scholar, and the Education Resources Information Center (ERIC). Web of Science (WoS) was selected as it is the world's oldest and most authoritative citation index, covering over 34,000 top-tier journals (Birkle et al., 2020). Scopus was chosen for its expansive coverage of over 76 million multidisciplinary records and advanced search capabilities (Baas et al., 2020; Martín-Martín et al., 2021). Google Scholar was included to further widen the search yield across its repository of 389 million documents (Gusenbauer & Haddaway, 2020). Finally, the ERIC database provided access to a more focused coverage, especially on education and social sciences publications. Together, these four databases provided comprehensive coverage of relevant scholarly literature.

A systematic search string was created to identify relevant literature using field codes, phrase searching, and Boolean operators. The following keywords were used: TITLE-ABS-KEY (("multimodal learning" OR "multi-modal learning" OR "multimedia learning" OR "multi-media learning") AND ("higher education" OR "tertiary education" OR "college" OR "university") AND ("teachers" OR "educators" OR "faculty")). This search string was used across the four databases, enabling the identification of 614 relevant articles through this comprehensive search strategy to help answer the review questions.

18.3.4 Selecting Relevant Papers and Extracting Data

The 614 articles obtained from the searches were further filtered based on the predefined inclusion and exclusion criteria. Expressly, articles were excluded if they were published before 2003, appeared in non-indexed journals, were written in the form of literature reviews, book chapters, conference proceedings, theses, and reports, were not written in English, and did not offer open-access publication. This systematic screening process resulted in 153 articles eligible for full-text review. These articles were then imported into Mendeley reference management software for further systematic screening, which led to an additional six duplicate articles being identified and removed. In total, 74 articles within the years 2003–2022 were identified to advance to the eligibility phase.

The full texts of the remaining 74 articles underwent comprehensive screening for relevance to the review objectives. Forty-two articles were excluded because they focused excessively on primary and secondary education contexts, with abstracts deemed insufficiently related to multimodal or multimedia learning topics. Through this rigorous selection process, 32 articles met the criteria to be included in the final qualitative evidence synthesis phase. The Preferred Reporting Items for Systematic Reviews and Meta-Analysis (PRISMA) flow diagram in Figure 18.1 illustrates the systematic process of study identification, screening, eligibility, and final inclusion, detailing the rationale for exclusions at each stage. Only studies meeting the rigorous standards of relevance and quality were incorporated to address the focused research questions.

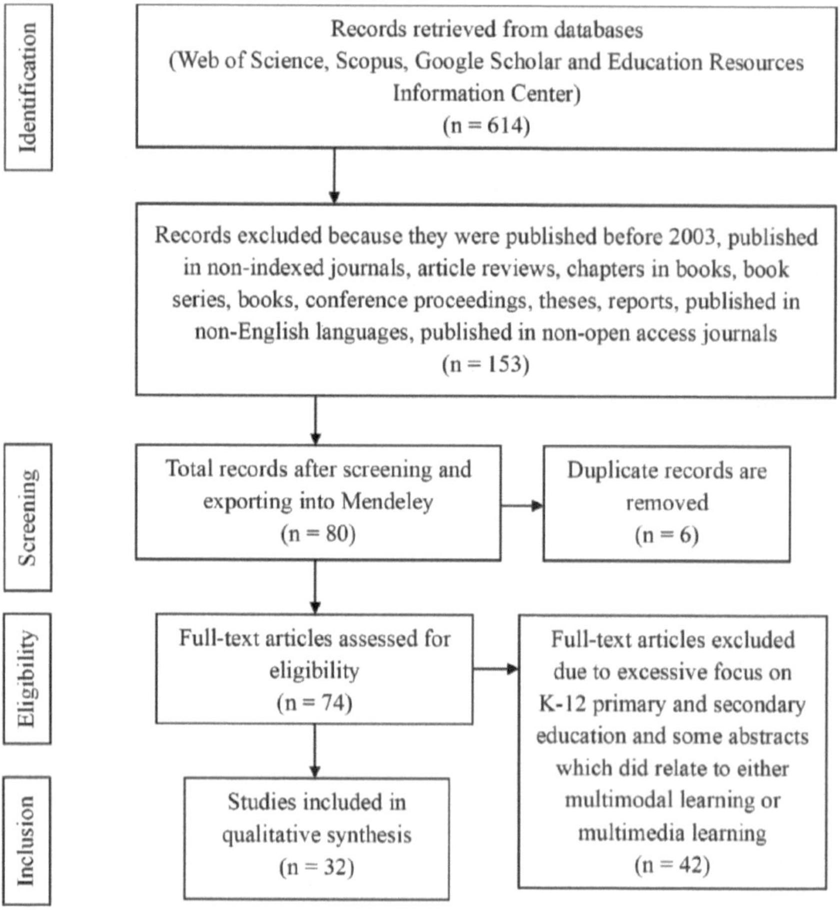

FIGURE 18.1 Review method.

18.3.5 Analysing Quality of Study

The 32 included articles underwent rigorous qualitative content analysis to extract key categories and themes relevant to answering the review questions. The full texts were read thoroughly, and a standardised form was used to systematically code and synthesise findings related to the three research questions. Relevant content was coded into categories based on the research questions. To ensure reliability, the categories were continually refined to eliminate any overlap or redundancy that emerged. Further analysis elicited salient, overarching themes within each defined category.

To validate the emerging categories and themes, member-checking procedures were employed whereby the key findings were verified against the original articles to ensure an accurate representation of content and meaning. The final categories and corresponding themes on the significance, challenges, and practical strategies related to multimodal learning approaches in higher education resulting from this systematic qualitative synthesis are displayed in Table 18.1. This table outlines the key concepts that emerged through the rigorous synthesis process.

The analysis was conducted by thoroughly reading and synthesising the findings from the articles into relevant categories and themes. A qualitative content analysis process was undertaken, where texts were extracted and compared across the articles to identify recurring themes. Through this rigorous synthesis process, the

TABLE 18.1
The Categories Based on the Research Questions

Research Questions	Categories and Themes
What is the significance of integrating diverse media formats in higher education classrooms?	**Category:** The significance of incorporating various forms of media in the classrooms **Themes:** Energising student engagement and motivation; Deepening content mastery and comprehension; Cultivating development of disciplinary literacies; Promoting critical thinking and higher-order thinking skills; Accommodating meeting diverse needs; Enabling authentic disciplinary communication; Promoting interactive and accessible experiences
What challenges do educators face in implementing multimodal pedagogy?	**Category:** Challenges in implementing multimodal learning **Themes:** Requiring careful planning and preparation; Ensuring coherent design; Developing effective methods for assessing multimodal work; Providing sufficient access to technology; Scaffolding required technical skills; Overcoming perceptions of diminished academic quality; Gaining teacher and administrator buy-in
What practical strategies can facilitate effective multimodal instruction?	**Category:** Practical strategies for implementing multimodal learning **Themes:** Employing backward design approaches; Scaffolding instruction to build skills; Allowing flexibility in modes of expression; Adapting materials and activities to diverse needs; Leveraging technology tools and simulations; Drawing on available multimedia creation platforms; Utilising campus resources for support; Establishing norms valuing diverse contributions

researchers were able to aggregate common conceptions related to the research questions that emerged across the diverse sample of studies.

18.4 FINDINGS AND DISCUSSION

Findings from studies on multimodal learning in higher education can be categorised into three sub-sections: the significance of incorporating various forms of media in the classrooms, challenges in implementing multimodal learning, and practical strategies for implementing multimodal learning.

18.4.1 The Significance of Incorporating Various Forms of Media in the Classrooms

A growing body of research demonstrates the pedagogical value of incorporating multimedia and multimodal learning experiences across higher education disciplines. Thoughtfully integrating multimedia instruction that combines visual, audio, and interactive media formats provides diverse pathways for students to engage with course content, develop disciplinary literacies, and demonstrate their knowledge compared to traditional lecturing (Chen et al., 2010; Kay & Kletskin, 2012). For example, studies have revealed the benefits of interactive simulations, virtual reality (VR) environments, podcasts and student-created media for increased comprehension, critical thinking, communication skills, and engagement in disciplines ranging from healthcare to marketing (Aldosari et al., 2018; Ally et al., 2022; Bin Eid et al., 2021). When aligned with course goals, incorporating varied media formats allows educators to meet diverse student needs, cultivate disciplinary literacies, and promote engaged learning across subjects.

Several studies reveal how multimedia instructional approaches enhance student learning and success. Singh (2004) found that developing interactive video tutorial-based problems helps introductory physics students learn problem-solving heuristics. The preliminary evaluation showed that most students who cannot solve the tutorial problems without help can solve similar problems after working through the video tutorial. In other words, the interactive video tutorials helped to reduce cognitive load and provide scaffolding support at every stage of problem-solving, as students could rewatch explanations as needed. Likewise, Kay and Kletskin (2012) implemented problem-based video podcasts in an undergraduate mathematics course. Students who accessed the video tutorials scored significantly higher on problem-solving assessments than peers who only had static textbook resources. The researchers concluded that multimedia learning resources can improve conceptual understanding by providing dynamic explanations of processes.

In nursing education, McCallum et al. (2011) found that nursing students became proactive in their clinical decision-making skills when Second Life simulation modules were added to the curriculum. The results showed that 3D virtual environment-based multimedia scenarios aided in applying theoretical knowledge to care situations. Integrating innovative 3D virtual worlds provided accessible learning experiences. Similarly, Bayram and Çalışkan (2022) designed VR simulations to help nursing students gain knowledge, develop collaboration and critical thinking skills, prepare for clinical practice and communicate effectively with patients. Nursing students with those skills can provide safe care, administer medications correctly, and notice changes in their patients. In other words, students who participate in VR simulations can provide patient safety in actual clinics.

Research also reveals how incorporating discipline-specific media literacy practices strengthens student skills. For example, Cook and Kirchoff (2017) studied English students analysing relations between visual and written rhetoric in graphic novels. Students demonstrated significant growth in their multimodal literacy capabilities to interpret meanings conveyed through both images and texts. Engaging in diverse semiotic materials facilitated critical evaluation skills. Zhang and Wu (2023) found that university-level Chinese English students who analysed image–text relations in online news articles significantly improved their media literacy skills.

Kuchel et al. (2014) and Blacer-Bacolod (2022) found that having biology students produce science documentary videos increased their conceptual understanding, ability to explain concepts and science communication skills, and identify misconceptions. Creating documentaries engaged learners in disciplinary media production practices that deepened their scientific literacies. Additionally, a study by Lavega-Burgués et al. (2020) provided further evidence for the benefits of multimedia instructional approaches on comprehension and critical thinking. The researchers investigated combining physical activity in the form of traditional sporting games with multimedia learning materials. The results demonstrated improved memory and retention of information when traditional games were integrated with multimedia learning content compared to conventional studying. Furthermore, research by Aprianto et al. (2020) demonstrated greater gains in critical thinking and writing proficiency among English as a Foreign Language (EFL) students who engaged with multimedia-enriched online modules versus traditional textbooks. This additional study supports the notion that thoughtfully designed multimedia activities can aid the development of critical thinking skills and deeper comprehension.

Besides science documentary videos, sporting games and multimedia-enriched online modules, music integration enhances content retention in subjects ranging from neuroscience to calculus (Hallam, 2015). Information literacy courses using podcasts, comics, and magazines significantly increased student confidence in finding, evaluating, and ethically using media sources compared to text-only methods (Feekery et al., 2021; Melor et al., 2012). Such studies validate the effectiveness of multimodal media formats for increased engagement, participation, and learning.

Student-generated media also provide impactful learning experiences, as evidenced by Stanley and Zhang (2018) and McCarthy et al. (2021). The outcomes reported in these two studies revealed that students who produced explanatory videos and podcasts on marketing concepts reported higher perceived learning and engagement than peers who wrote in conventional styles. The authentic experience of communicating concepts through self-produced videos promoted multimedia literacy and content mastery.

In summary, numerous studies published in leading education journals provide substantial evidence that intentionally incorporating multimedia and multimodal learning opportunities yields considerable benefits for student engagement, comprehension, disciplinary literacy, critical thinking, and diverse learning needs across higher education contexts.

18.4.2 Challenges in Implementing Multimodal Learning

While research highlights the benefits of multimodal learning, incorporating these pedagogies effectively presents some challenges. Multimodal instruction requires careful planning and preparation to ensure activities align with learning goals and promote meaningful engagement across modes (Jewitt, 2008). Other challenges include assessing multimodal work, providing necessary technology access and skills, and gaining teacher buy-in.

Designing coherent multimodal learning experiences takes concerted effort and time. Instructors must identify appropriate modalities, media formats, and technologies to integrate based on content goals and student needs (Sanders & Albers, 2010). Additionally, crafting assignments that leverage different modalities for deep learning rather than novelty's sake is critical (Rowsell & Walsh, 2011). Without sufficient planning and design, multimodal activities risk being disjointed rather than purposeful.

Teachers can also struggle to develop criteria and methods to assess multimodal projects and activities (Tan et al., 2020). Assessments should evaluate both content mastery and fluency with various communicative modes. This requires balancing, weighing, and synthesising diverse analytic approaches and rubrics (Archer, 2010). Establishing clear expectations around multimodal assessment is, therefore, vital.

Furthermore, the limited access to classroom technology tools and inadequate student technical skills hinder the implementation of multimodal instruction (Hobbs, 2017). Teachers must determine availability and provide guidance on any hardware, software, and platforms needed for activities. Offering technical and design support helps students engage fully with multimodal assignments.

Overcoming perceptions that multimodality diminishes academic rigour or integrity can also be a hurdle (Jewitt, 2005). Gaining teacher and administrator buy-in involves addressing concerns about standards alignment and demonstrating that multimodal pedagogies enhance rather than diminish learning. Making the research case for multimodal instruction is vital. Furthermore, some students may struggle with the lack of necessary technical skills and access to required technology tools (Ramirez et al., 2018), compounding the challenge of limited classroom resources. This aligns with the need to scaffold student skills and ensure the availability of appropriate tools, as noted above.

While not without challenges, research shows that the benefits of multimodal learning warrant concerted efforts to leverage its potential successfully. Attending to the instructional design, assessment, technical, and institutional factors that enable impactful multimodal education remain an evolving endeavour.

18.4.3 Practical Strategies for Implementing Multimodal Learning

Research provides guidance on best practices for implementing multimodal learning across educational contexts. Recommended strategies include employing backward design approaches, scaffolding instruction to build skills, allowing flexibility in modes of expression, adapting materials and activities to diverse needs, leveraging technology tools and simulations, drawing on available multimedia creation platforms, utilising campus resources for support, and establishing norms valuing diverse contributions.

Backward design principles emphasise aligning multimodal choices with learning goals and assessment (Cope & Kalantzis, 2009). Instructors should identify desired outcomes first, then select modalities and activities that support achieving those goals. Multimodality should enrich, not obscure, curriculum objectives. Developing rubrics and models helps clarify expectations (Archer, 2010).

Scaffolding instruction is also advised when introducing multimodal practices and tools to provide needed support (Yawiloeng, 2022). For example, a study by Lee (2019) on using technology-enhanced multimodal activities for EFL presentation skills found that learner training was key to overcoming barriers like video fear and technical challenges. Hobbs (2017) and Lee (2019) suggest that educators provide tutorials, exemplars, and ample work time to build student skills and comfort with new modalities. Establishing class norms valuing diverse contributions can further promote engagement.

Multimodal pedagogies should grant flexibility for learners with diverse needs and backgrounds to engage meaningfully. Allowing choice in modes of expression and topics taps into student strengths and interests (Halbritter, 2006). For instance, a study by Saad Aljebreen and Aseel Alzamil (2022) found that using short films to teach idioms allowed EFL students to engage with content in a more motivating and tailored way. Teachers should also adapt materials and activities to varying linguistic proficiencies and learning styles. Leveraging technology tools can help facilitate this flexibility.

Specifically, virtual simulations, 3D learning environments, and interactive video tutorials have proven effective for multimodal instruction in domains like healthcare, engineering, and physics (Huang, 2023; Aprianto et al., 2020). These types of multimedia tools provide practical options for implementation across disciplines. Further, drawing on available multimodal creation tools and platforms makes implementation more feasible (Bouchey et al., 2021). User-friendly applications like Canva, WeVideo, and SoundCloud allow students to create infographics, videos, podcasts, and more. Utilising campus supports like libraries, media labs, and learning centres can provide additional technology access and guidance.

When intentionally designed and scaffolded, multimodal learning provides dynamic opportunities to meet diverse learner needs, foster engagement, and promote knowledge development. Implementing such pedagogies takes creativity and effort but yields considerable educational benefits.

18.5 RECOMMENDATIONS FOR EDUCATORS

Here are some suggestions for educators when implementing multimodal teaching strategies based on research and best practices. First, it is recommended that educators identify the curriculum objectives and content areas that can benefit from instruction. By reviewing learning goals, course materials, and key concepts, educators can determine which areas could be enhanced through forms of representation. For example, interactive simulations work well to understand interactions and ecosystem dynamics; visual arts can enrich poetry and audio formats to connect to oral traditions (Jewitt, 2008). It is important to identify where multimodality can strengthen comprehension.

Additionally, it is advised to review existing materials and activities to assess the use of different modalities. Educators should take stock of modalities integrated into their teaching practices, such as lectures, discussions, textbooks, videos, and writing assignments. By doing this, they can identify any modes as well as find opportunities to purposefully incorporate new modes that align with students' diverse learning approaches, cultural backgrounds, and interests (Cope & Kalantzis, 2009). Building upon existing modalities provides a foundation before introducing multimodal elements.

Furthermore, it is suggested for educators to implement practices through pilot initiatives. For example, introducing a multimedia text or redesigning a lesson activity using new modes (Halbritter, 2006). Educators should utilise methods such as observation, surveys, interviews, and analysing student work to assess effectiveness and identify support needs. Based on the evidence gathered, educators can then gradually expand the use of practices.

Moreover, the fourth recommendation emphasises the importance of balancing flexibility and clear expectations when implementing multimodal teaching approaches. While allowing students to choose modes and explore topics enhances their engagement, educators must provide guidelines, models, and support structures to facilitate learning in unfamiliar modalities. This includes setting project expectations, co-creating rubrics with students, providing examples for reference, allocating work time, offering technical guidance, and nurturing creativity within adaptable learning environments.

Similarly, the following recommendation encourages educators to analyse the effects of multimodal mentor texts on their subject areas. Educators can develop multimodal literacy skills by guiding students in deconstructing these texts from a different perspective. By making explicit the structural conventions used in texts through analysis exercises, educators can help students understand how meaning is constructed across modes.

Moving forward with recommendation six involves modelling risk-taking behaviours and fostering growth mindsets when using modes of expression. Encouraging students to explore modalities requires embracing their efforts and emphasising the process-oriented nature of learning. Educators have a role in ensuring opportunities for student participation and should respond to their needs by offering support and encouragement instead of setting rigid expectations.

Besides modelling risk-taking behaviours and fostering growth mindsets, educators are encouraged to collaborate with librarians, media specialists, learning technologists, and colleagues. This collaboration would enable them to generate ideas and access the training, resources, and technical assistance (Bouchey et al., 2021). Educators need to tap into the expertise on their campuses.

Lastly, the eighth suggestion highlights the importance of educators being receptive to student feedback. Educators should consistently seek out input from students.

Gather data in order to refine their teaching approaches. They should allow time for integrating evidence-based practices and making adjustments.

To summarise, implementing pedagogies requires educators to invest time in aspects like planning, scaffolding, assessing students' progress, collaborating with others, and incorporating student input on a regular basis. Despite this investment of time and effort, empirical research emphasises the benefits of adopting such an approach.

18.6 CONCLUSION

In conclusion, this systematic review synthesises high-quality empirical evidence demonstrating the pedagogical value of thoughtfully integrating multimodal learning approaches in higher education contexts. The research indicates that incorporating multimedia across disciplines can enhance student engagement, motivation, comprehension, critical analysis, and disciplinary literacy compared to reliance solely on traditional lecturing formats. However, realising these potential benefits requires addressing implementation challenges such as demanding careful planning and preparation, ensuring coherent design, developing effective methods for assessing multimodal work, providing sufficient access to technology, scaffolding required technical skills, overcoming perceptions of diminished academic quality, and gaining teacher and administrator buy-in. Strategies such as employing backward design approaches, scaffolding instruction to build skills, allowing flexibility in modes of expression, adapting materials and activities to diverse needs, leveraging technology tools and simulations, drawing on available multimedia creation platforms, utilising campus resources for support, and establishing norms valuing diverse contributions can help overcome these hurdles. The findings provide authoritative, evidence-based insights and strategies to inform teaching practices and policies regarding multimodal instruction. As emerging multimedia technologies continue to transform communication, multimodal pedagogies will grow increasingly relevant across higher education. Further research can help extend investigations into the long-term impacts of multimodal learning and effective faculty training programmes. Overall, this review highlights the significant benefits of purposefully designed multimodal instruction for fostering participatory, equitable learning environments that can better prepare diverse students for academic and professional success.

REFERENCES

Ainsworth, S. (2006). DeFT: A conceptual framework for considering learning with multiple representations. *Learning and Instruction*, *16*(3), 183–198.

Aldosari, M. A., Aljabaa, A. H., Al-Sehaibany, F. S., & Albarakati, S. F. (2018). Learning style preferences of dental students at a single institution in Riyadh, Saudi Arabia, evaluated using the VARK questionnaire. *Advances in Medical Education and Practice*, *9*, 179–186. https://doi.org/10.2147/AMEP.S157686

Ally, F., Pillay, J. D., & Govender, N. (2022). Teaching and learning considerations during the COVID-19 pandemic: Supporting multimodal student learning preferences. *African Journal of Health Professions Education*, *14*(1), 13–16. https://doi.org/10.7196/AJHPE.2022.v14i1.1468

Andrews, R., & Smith, A. (2011). *Developing Writers: Teaching and Learning in the Digital Age*. McGraw-Hill Education.

Aprianto, E., Purwati, O., & Anam, S. (2020). Multimedia-assisted learning in a flipped classroom: A case study of autonomous learning on EFL university students. *International Journal of Emerging Technologies in Learning*, *15*(24), 114–127.

Archer, A. (2010). Multimodal texts in higher education and the implications for writing pedagogy. *English in Education*, *44*(3), 201–213.

Baas, J., Schotten, M., Plume, A., Côté, G., & Karimi, R. (2020). Scopus as a curated, high quality bibliometric data source for academic research in quantitative science studies. *Quantitative Science Studies*, *1*(1), 377–386.

Bayram, S. B., & Çalışkan, N. (2022). The use of virtual reality simulations in nursing education, and patient safety. In Stawicki, S. P., & Firstenberg, M. S. (Eds.), *Contemporary Topics in Patient Safety* (Vol. 1). IntechOpen. https://doi.org/10.5772/intechopen.94108

Bin Eid, A., Almizani, M., Alzahrani, A., Alomair, F., Albinhamad, A., Albarrak, Y., Alzuaki, M., Alyahya, S., & Abdulrahman, K. B. (2021). Examining learning styles with gender comparison among medical students of a Saudi university. *Advances in Medical Education and Practice*, *12*, 309–318. https://doi.org/10.2147/AMEP.S295058

Birkle, C., Pendlebury, D. A., Schnell, J., & Adams, J. (2020). Web of science as a data source for research on scientific and scholarly activity. *Quantitative Science Studies*, *1*(1), 363–376.

Blacer-Bacolod, D. (2022). Student-generated videos using green screen technology in a biology class. *International Journal of Information and Education Technology*, *12*(4), 339–345.

Böck, M., & Pachler, N. (2013). *Multimodality and Social Semiosis: Communication, Meaning-Making and Learning in the Work of Gunther Kress*. Routledge.

Boland, A., Cherry, M. G., & Dickson., R. (Eds.). (2017). *Doing a Systematic Review: A Student's Guide*. Sage.

Bouchey, B., Castek, J., Thygeson, J. (2021). Multimodal Learning. In: J. Ryoo & K. Winkelmann (Eds), *Innovative Learning Environments in STEM Higher Education*. Springer Briefs in Statistics. Springer, Cham. https://doi.org/10.1007/978-3-030-58948-6_3

Bozkurt, A. (2020). Educational technology research patterns in the realm of the digital knowledge age. *Journal of Interactive Media in Education*, *2020*(1), 18. https://doi.org/10.5334/jime.570

Camiciottoli, B. C., & Campoy-Cubillo, M. C. (2018). Introduction: The nexus of multimodality, multimodal literacy, and English language teaching in research and practice in higher education settings. *System*, *77*, 1–9.

Chen, P., Lambert, A., & Guidry, K. (2010). Engaging online learners: The impact of web-based learning technology on college student engagement. *Computers & Education*, *54*(4), 1222–1232.

Chiu, T. K. F., & Churchill, D. (2015). Exploring the characteristics of an optimal design of digital materials for concept learning in mathematics: Multimedia learning and variation theory. *Computers & Education*, *82*, 280–291.

Cole, M., & Engestrom, Y. (1993). A cultural-historical approach to distributed cognition. In G. Salomon (Ed.), *Distributed Cognitions: Psychological and Educational Considerations* (pp. 1–46). Cambridge University Press.

Cook, M. P., & Kirchoff, J. S. J. (2017). Teaching multimedia literacy through reading and writing graphic novels. *Language and Literacy*, *19*(4), 76–95.

Cope, B., & Kalantzis, M. (2009). Multiliteracies: New literacies, new learning. *Pedagogies: An International Journal*, *4*(3), 164–195.

Dalton, B., & Proctor, C. P. (2008). The changing landscape of text and comprehension in the age of new literacies. In J. Coiro, M. Knobel, C. Lankshear, & D. J. Leu (Eds.), *Handbook of Research on New Literacies* (pp. 297–324). Lawrence Erlbaum Associates.

Doherty, C. (2023). Using web log analysis to evaluate healthcare students' engagement behaviours with multimedia lectures on YouTube. *PLoS ONE*, *18*(4), 1–15. https://doi.org/10.1371/journal.pone.0284133

Feekery, A., Chisholm, K., Jeffrey, C., & Diesch, F. (2021). Enhancing students' professional information literacy: Collaboratively designing an online learning module and reflective assessments. *Journal of Information Literacy*, *15*(2), 150–165.

Gusenbauer, M., & Haddaway, N. R. (2020). Which academic search systems are suitable for systematic reviews or meta-analyses? Evaluating retrieval qualities of Google Scholar, PubMed and 26 other resources. *Research Synthesis Methods*, *11*, 181–217.

Halbritter, B. (2006). Musical rhetoric in integrated-media composition. *Computers and Composition*, *23*, 317–334. https://doi.org/10.1016/j.compcom.2006.05.005

Hallam, S. (2015). *The Power of Music*. Music Education Council.

Hobbs, R. (2017). *Create to Learn: Introduction to Digital Literacy*. Wiley Blackwell.

Hodge, R. & Kress, G. R. (1988). *Social Semiotics*. Polity Press.

Huang, Y. (2023). Open learning environment for multimodal learning based on knowledge base technology. *International Journal of Emerging Technologies in Learning*, *18*(11), 38–51.

Jewitt, C. (2005). Multimodality, 'reading', and 'writing' for the 21st century. *Discourse: Studies in the Cultural Politics of Education*, *26*(3), 315–331.

Jewitt, C. (2008). Multimodality and literacy in school classrooms. *Review of Research in Education*, *32*(1), 241–267.

Jewitt, C., Bezemer, J., & O'Halloran, K. L. (2016). *Introducing Multimodality*. Routledge.

Kao, C.-C., & Luo, Y.-J. (2020). Effects of multimedia-assisted learning on learning behavior and student knowledge in physical education lessons: Using basketball game recording as an example. *International Journal of Emerging Technologies in Learning*, *15*(1), 119–139. https://doi.org/10.3991/ijet.v15i01.11393

Kay, R., & Kletskin, I. (2012). Evaluating the use of problem-based video podcasts to teach mathematics in higher education. *Computers & Education*, *59*(2), 619–627.

Kuchel, L., Cokley, J. D., & Stevens, S. (2014). A documentary video assignment to enhance learning in large first-year science classes. *International Journal of Innovation in Science and Mathematics Education*, *22*(4), 48–64.

Kress, G. (2010). *Multimodality: A Social Semiotic Approach to Contemporary* Communication. Routledge.

Lavega-Burgués, P., Luchoro-Parrilla, R. A., Serna, J., Salas-Santandreu, C., Aires-Araujo, P., Rodríguez-Arregi, R., Muñoz-Arroyave, V., Ensenyat, A., Damian-Silva, S., Machado, L., Prat, Q., Sáez de Ocáriz, U., Rillo-Albert, A., Martín-Martínez, D., & Pic, M. (2020). Enhancing multimodal learning through traditional sporting games: Marro360°. *Frontiers in Psychology*, *11*, 1384. https://doi.org/10.3389/fpsyg.2020.01384

Lee, Y.-J.J. (2019). Integrating multimodal technologies with VARK strategies for learning and teaching EFL presentation: An investigation into learners' achievements and perceptions of the learning process. *Australian Journal of Applied Linguistics*, *2*(1), 17–31. https://doi.org/10.29140/ajal.v2n1.118

Lim, F. V. K., & Tan-Chia, L. H. (2022). *Designing Learning for Multimodal Literacy: Teaching Viewing and Representing*. Routledge. https://doi.org/10.4324/9781003258513

Lin, H.-Y., & Tsai, S.-C. (2021). Student perceptions towards the usage of AR-supported STEMUP application in mobile courses development and its implementation into English learning. *Australasian Journal of Educational Technology*, *37*(3), 88–103.

Martín-Martín, A., Thelwall, M., Orduna-Malea, E., & Delgado López-Cózar, E. (2021). Google Scholar, Microsoft Academic, Scopus, Dimensions, Web of Science, and Open Citations' COCI: A multidisciplinary comparison of coverage via citations. *Scientometrics*, *126*, 871–906.

Mayer, R. E. (2009). *Multimedia Learning* (2nd ed.). Cambridge University Press.

McCallum, J., Ness, V., & Price, T. (2011). Exploring nursing students' decision-making skills whilst in a Second Life clinical simulation laboratory. *Nurse Education Today*, *31*(7), 699–704. https://doi.org/10.1016/j.nedt.2010.03.010

McCarthy, S., Pelletier, M., & McCoy, A. R. (2021). Talking together: Using intercollegiate podcasts for increased engagement in marketing education. *Marketing Education Review*, *31*(2), 125–130.

Melor, M. Y., Siraj, S., & Hamat, A. (2012). Effects of using digital comics to improve ESL writing. *Research Journal of Applied Sciences, Engineering and Technology*, *4*(18), 3462–3469.

Mohammed Hassan Al-Ahdal, A. A., Alqasham, F. H., & Mohammed Qarabesh, M. A. (2022). Lateral multimodal learning in a Saudi EFL context: Investigating the perceptions of teachers and M2E female learners. *F1000Research*, *11*(369), 1–19.

New London Group. (1996). A pedagogy of multiliteracies: Designing social futures. *Harvard Educational Review*, *66*(1), 60–93.

Ramirez, J. D., Hernandez-Leon, E., Figueroa-Sandoval, B. E., & Aillon-Newman, M. (2018). Learning and experience: Aesthetics of multimodal texts in higher education. *Digital Education Review*, *33*, 170–184.

Rowsell, J., & Walsh, M. (2011). Rethinking literacy education in new times: Multimodality, multiliteracies, and new literacies. *Brock Education Journal*, *21*(1), 53–62.

Saad Aljebreen, & Aseel Alzamil. (2022). The impact of using short films on learning idioms in EFL classes. *World Journal of English Language*, *12*(7), 250–264.

Sanders, J., & Albers, P. (2010). Multimodal literacies: An introduction. In P. Albers & J. Sanders (Eds.), *Literacies, the Arts and Multimodality* (pp. 1–25). National Council of Teachers of English.

Sankey, M., Birch, D., & Gardiner, M. (2010). Engaging students through multimodal learning environments: The journey continues. In C. H. Steel, M. J. Keppell, P. Gerbic & S. Housego (Eds.), *Curriculum, technology & transformation for an unknown future. Proceedings ascilite Sydney 2010* (pp. 852–863).

Scollon, R. (2001). *Mediated Discourse: The Nexus of Practice*. Routledge.

Shaffril, H. A. M., Samsuddin, S. F., & Samah, A. A. (2020). The ABC of systematic literature review: The basic methodological guidance for beginners. *Quality & Quantity*, *55*, 1313–1346.

Shanahan, L. E. (2013). Multimodal representations: A fifth-grade teacher influences students' design and production. *Pedagogies: An International Journal*, *8*(2), 85–102.

Siegel, M. (2012). New times for multimodality? Confronting the accountability culture. *Journal of Adolescent & Adult Literacy*, *55*(8), 671–680.

Singh, C. (2004). Interactive video tutorials for enhancing problem-solving, reasoning, and meta-cognitive skills of introductory physics students. *AIP Conference Proceedings*, *720*, 177–180. https://doi.org/10.1063/1.1807283

Stanley, D. E., & Zhang, Y. (2018). Student-produced videos can enhance engagement and learning in the online environment. *Online Learning*, *22*(2), 5–26. https://doi.org/10.24059/olj.v22i2.1367

Tan, L. H., Zammit, K. E., D'warte, J., & Gearside, A. (2020). Assessing multimodal literacies in practice: A critical review of its implementations in educational settings. *Language and Education*, *34*(2), 97–114. https://doi.org/10.1080/09500782.2019.1708926

Taxipulati, S., & Lu, H.-D. (2021). The influence of feedback content and feedback time on multimedia learning achievement of college students and its mechanism. *Frontiers in Psychology*, *12*(706821), 1–14. https://doi.org/10.3389/fpsyg.2021.706821

The New Media Consortium. (2005). *A Global Imperative: The Report of the 21st Century Literacy Summit*.

Vygotsky, L. S. (1978). *Mind in Society: The Development of Higher Psychological Processes*. Harvard University Press.

Yawiloeng, R. (2022). Using instructional scaffolding and multimodal texts to enhance reading comprehension: Perceptions and attitudes of EFL students. *Journal of Language and Linguistic Studies*, *18*(2), 877–894.

Zhang, D., & Wu, Y. (2023). Becoming smart "digital natives": Cultivating Chinese English majors' new media literacy via Journalism English reading and listening. *Journal of China Computer-Assisted Language Learning*, *3*(2), 263–286. https://doi.org/10.1515/jccall-2022-0029

19 Perspective on the Evolution of Teaching Competencies for Integration in Industry 4.0 Technologies with Education 4.0

A. Sanjeev and R. Sharma

19.1 INTRODUCTION: BACKGROUND AND DRIVING FORCES

From time to time, rapid technological advancements have revolutionized the manufacturing of mechanized goods and paved the way for varied industrial revolutions. In the 1760s, the First Industrial Revolution, aka the age of mechanical production, was marked by mechanized manufacturing, while the Second Industrial Revolution ushered in a new world with the massive use of electrical energy as a medium of power. Beginning in the 1950s, the Third Industrial Revolution brought about a paradigm shift in manufacturing with the advent of digitalization. The development of semiconductors, mainframe computers, personal computing, and the internet contributed to its emergence, among other things. Industry 4.0, which provides digital goods and services, involves running connected virtual and physical assets, optimizing customer-facing operations, and digitizing the whole value chain.

The term "Industry 4.0" refers to a new breed of the preceding industrial revolution. It emphasizes smart factories as its main topic (Kagermann et al., 2013) and even changes how people work (Stock et al., 2018). In addition to professional sectors, Industry 4.0 has also inevitably impacted higher education institutions (HEIs), as they must acquire the competencies necessary for Industry 4.0 acceptance. Today's industrial transformation requires expertise in a variety of fields, including real-time data collection and analysis, the Internet of Things (IoT), cloud services, big data, and knowledge of the most recent technical developments (Lee et al., 2015). HEIs must thus adapt to the changes brought about by Industry 4.0 and be the frontrunners in honing the competencies of their students to meet the standards of Industry 4.0. To be relevant, the educators delivering Education 4.0 need to pivot and reposition to fulfill the expectations of Industry 4.0.

DOI: 10.1201/9781003400691-19

The use of artificial intelligence (AI) was accelerated by the Fourth Industrial Revolution, which maximized the immense potential of data and put a thrust on machine learning, robots, the IoT, and automation. Consequently, these transformations in Industry 4.0 due to technological advancements have been widely documented and researched by eminent scholars (Lasi et al., 2014).

One of the most crucial instruments for supplying the human abilities required in response to the economic, social, and technical changes of this century is education. A new paradigm that re-examines concepts of education may be described as "Education 4.0."It redefines areas of education, students, teachers, and schools considering the demands of Industry 4.0. A precondition for bringing innovation to education and the development of innovative skills is the delivery of Education 4.0, which places its focus on an education that will create the producers and users of Industry 4.0. Individual skills, interpersonal skills, life skills, practical skills, labor skills, non-cognitive skills, creativity and learning capabilities, expertise, media and technology skills, and life and career skills are a few examples of 21st-century skills (McComas, 2014).

In the Education 4.0 era, the student is the active agent of the learning process, and the instructor acts as a tutor in this process (Fisk, 2017). Students study according to their rhythm, timetable, and needs. Additionally, more time is given to work on difficulties on one's own or with others, even when not in a formal setting (Zainuddin et al., 2019). Agarwal et al. (2021) linked Industry 4.0 and Education 4.0 and found that the parameters imposed by Industry 4.0 are already having an impact on the educational landscape through the use of cognitive and cloud technologies, computing, the IoT, cyber-physical systems (CPS), and other elements. Qureshi et al. (2021) stated that the employment of digital technologies in Education 4.0 not only improves learning but also thoroughly investigates the relationship between digital technologies and education. Studies depict the increased standards for professional training imposed by the Fourth Industrial Revolution (Industry 4.0).

Education 4.0, the present educational period, is one in which institutions of higher learning use new instructional and management techniques, smart and environmentally friendly buildings, and up-and-coming information and communication technologies(ICTs), increasing the methods of knowledge creation and information transfer. The amalgamation of such materials will aid in the instruction and growth of desirable key competencies in today's pupils. The model of flipped classrooms is one such illustration of an innovative teaching and learning strategy that is a component of Education 4.0. Flipped classes allow students to learn what they need outside of conventional educational settings by researching digital resources for lessons, such as films, presentations, and e-materials. As a result, students can make use of class time for discussions, analyses, and problem-solving (Youngkin, 2014).

Himmetoğlu et al. (2020) asserted that the teachers of Education 4.0 should possess the following personal qualities: curiosity, patience, openness to change, adaptability, and investigative skills. The need to acquire lifetime learning abilities is a prerequisite for Education 4.0 teachers. For this reason, Education 4.0's lifetime learning skills for teachers were differentiated as adopting cutting-edge learning strategies and supporting ongoing professional development. It was found that guiding abilities are the ones that teachers in Education 4.0 need the most. In this context, emphasis should be placed on directing students technically and academically, building emotional connections with students, organizing, motivating, leading, establishing a participative learning environment, assisting students in finding the appropriate material and in setting their objectives (Ersoy, 2021).

19.2 INDUSTRY 4.0: DEFINITION AND CHARACTERISTICS

Lopez et al. (2017), Kagermann et al. (2013), and Gaj et al. (2012) referred to Industry 4.0 as the digital transformation of manufacturing and production-based businesses, which is fueled by networked technologies like cloud computing, cognitive computing, CPS, and IoT. With the primary objective of optimizing resource utilization and meeting consumer demand, Industry 4.0 makes it possible to communicate, observe, coordinate, and control value streams from physical processes, such as machinery and equipment, in real time and without centralized control. As a result, there are four ways to look at Industry 4.0's distinctiveness: interconnection is the first perspective, followed by disclosure of information and technical support in the second and third perspectives, and decentralized decision-making in the fourth.

Hernandez-de-Menendez et al. (2020) stated that Industry 4.0's primary objective is to increase the productivity and adaptability of factories so they can meet future needs. Different technologies enable the use of this strategy. There are several of these, including autonomous robots, simulation, horizontal and vertical system integration, the industrial IoT, cyber security, the cloud, additive manufacturing, augmented reality (AR), and big data and analytics. All of these make it possible for businesses to connect and make better use of the massive volume of information created.

IT-driven changes in manufacturing are the hallmark of Industry 4.0, which is defined by the incorporation of technology and intelligent manufacturing. These changes have had widespread organizational ramifications. Consequently, a change from product to service orientation is inevitable. In the industrial industry, it has also resulted in the formation of new forms of organizations with fresh roles. The latest advancements in Industry 4.0 include smart factories, CPS, new methods for distribution and procurement, product and service creation, and the advent of new production systems that are focused on meeting human needs. Industry 4.0's creative and innovative production methods are designed with sustainability and resource efficiency in mind. According to Moraes et al. (2023), Industry 4.0 will increasingly be included in higher education using simulation, AR, the IoT, and virtual reality (VR). Together, these tools improve student engagement, interactivity, cost and risk reduction, simulation of real-world work situations, development of soft skills, and learning about technologies, their applications, and working principles. The entire learning process is essentially supported by Industry 4.0 technology.

Li (2022) reviewed workforce development in Latin America, Europe, Asia, and Africa; new technology and skills have been adopted considerably more quickly than they did 10years ago. Lifelong learning is therefore a reality and must be considered while setting strategic objectives for a company. Finally, the educational ecosystem of the Industry 4.0 age offers guidance to individuals, academic institutions, and companies that commit to reskilling and upskilling. The importance of aligning higher education and industry skills is emphasized in the study. The workforce of the future should be conscious of diversity and accept multiple cultures and governance structures since Industry 4.0 places a strong focus on establishing a connected world through CPS, IoT, and industrial IoT. To provide prosperity to the world's workers is the aim of Industry 4.0.

19.3 WORKFORCE SKILL REQUIREMENTS TO MEET THE CURRENT INDUSTRIAL REVOLUTION INDUSTRY 4.0

The job market is shifting to match the skill requirements to cater to the rapid technological advancements in the manufacturing and services sectors. Alhloul and Kiss (2023) categorized personal skills as relating to an individual's ability to learn the dynamic environment of Industry 4.0, social and interpersonal skills depicting leadership and communication competence, and technical skills for working in the face of rapid technological advancements. Pena et al. (2021), who adopted a worker-oriented approach, found that four distinct but related skill clusters—cognitive, functional business, managing people, and strategic—must be developed to meet Industry 4.0's demands. These skill clusters can differ depending on the differences between leaders and their subordinates in the workplace.

Pena et al. (2021) established in their research that from an Industry 4.0 perspective, cognitive, functional business, strategic, and managing people abilities may be regarded as important worker abilities. Peña-Jimenez et al. (2021) concluded that four distinct but connected skill clusters (cognitive, functional business, people management, and strategic) are significant assets to be cultivated to tackle the difficulties of Industry 4.0, which can further vary depending on work position distinctions, particularly between leaders and their subordinates. Based on these studies, the skills imperative for bridging the skill gap in the era of Industry 4.0 can be classified as:

- **Cognitive Skills**

  Cognitive capacities are brain-based abilities used to recognize, evaluate, and apply various types of information for problem-solving. These may further be classified as logical reasoning, creativity, active learning, sensitivity to problems, cognitive flexibility, oral expression, integrative analysis, and complex problem-solving. Active learning is the process of understanding the consequences of new information on problem-solving while logical reasoning involves incorporating facts to produce an overall picture that allows for successful conclusion-drawing. Cognitive flexibility is the capacity to switch between ideas and conceive of several ideas at once. Clear and effective communication of information is required for spoken expression. Integrative analysis employs reasoning to point out the benefits and drawbacks of event-planning strategies. Even though these settings are frequently unknown, complex issue-solving defines new job difficulties in complicated contexts.

- **Functional Business Skills**
  These include business report analysis, report formulation, resource management, and service orientation. These skills will help one understand the functioning of the business and identify and improve work-related processes.
- **Strategic Skills**
  These comprise goal-setting and evaluating various aspects of different organizational systems for overall system effectiveness. These abilities include imagining a system in its optimum state, evaluating systems, identifying consequences and causes, identifying problems, evaluating solutions, and making decisions.
- **Managing People Skills**
  Social perception, coordination with others, negotiation, time and people management, persuasion, coordination with others, training and teaching, and financial management are the main skills that the workforce needs to develop in response to the technological advancements happening in the current times.

19.4 EDUCATION 4.0: DEFINITION AND CHARACTERISTICS

Ouhrir et al. (2019), Daud et al. (2019), and Prieto et al. (2019) said that education has a substantial impact on the growth of the economy and the development of human capital. The Fourth Industry Revolution ensured that the educational ecosystem continued to be dynamic and current. The use of various technologies in conjunction with a teaching approach backed by academic research results in active learning. They further described that with the help of growing technology and sophisticated infrastructure, instructors and students have been able to enhance the educational practices used in higher education. The paradigms of educational procedures are also being adjusted and reoriented toward innovation in the teaching pedagogy to satisfy the needs of a technological world that is always evolving. Because of this, it is believed that modern knowledge generation transcends pedagogy and andragogy and instead favors an approach that blends heutagogy, peeragogy, and cybergogy. Humanistic and constructivist principles that are student-centered for both learning and teaching via pedagogy are the foundation of Education 4.0, which promotes self-learning. Metacognition, or awareness of one's learning process, and self-reflection are promoted. Peeragogy is the framework for group learning. It has to do with the variety of educational strategies that support peer learning. Technology development and the proliferation of the internet have encouraged educational offers and contributed to the development of cybergogy. It is a learning methodology that uses ICTs to enhance learning experiences that are time and place-independent.

Himmetoğlu et al. (2020) considered open access, individualized instruction, mental transformation, digital technology integration into the classroom, seamless learning environments, lifelong learning, exploratory learning, and transdisciplinary learning as some of the major characteristics of Education 4.0. Open educational materials and massive open online courses can be considered apt examples in the domain of open-access learning in Education 4.0. In the context of customized education, faculty members are expected to talk about the creation of tailored learning materials, the creation of adaptive learning environments, and the use of learning analytics. The importance of the philosophical reform of human capital within the context of mental change must be emphasized more. Smart campuses, AR, cloud computing, and augmented education are a few examples of the technological advancements used to enhance Education 4.0. Among the most prominent concepts in the framework of Education 4.0, lifelong learning has been linked to continual improvement, sustainability, and figuring out how to learn.

Bonfeld et al. (2020) emphasized that to improve learning experiences and prepare students for the workforce of the future, Education 4.0 calls for integrating cutting-edge technology like AI, VR, and AR into the educational system. The recently proposed concept of Education 4.0 envisions higher education, which aspires to train highly competitive specialists capable of delivering innovative solutions with the aid of the right physical and digital resources. As a result, Education 4.0 re-imagines education as an all-encompassing, lifelong experience in which it is up to the student to develop their skills, with instructors and mentors acting as enablers and facilitators.

Further, it seeks to enable instructors to be more effective at their jobs, leading to enhanced student learning. Additionally, it aims to assist teachers in increasing their ability to guide students on cutting-edge technologies, their ramifications, and how to protect themselves from cyber-attacks. Overall, the position of the teacher becomes increasingly important in Education 4.0 as teachers become more involved in mentoring and advising pupils.

With the rapid expansion of technology, the skills required of students are changing. Students must be adaptive and open to change. Students should develop self-study skills because learning is a lifelong process that requires them to always learn on their own. Education 4.0 provides a solid basis for students to build self-learning skills and improve their learning outcomes. Education in the current times encourages students to use the internet's resources, such as online courses and other learning materials, and to learn at their leisure. It allows students to connect to a larger network and improve communication platforms. Education 4.0 develops desirable abilities in students to make them imaginative and creative entrepreneurs and employees with the ability to adapt to the usage of technology. Ramrez-Montoya et al. (2022) examined the components of Education 4.0. They primarily focus on researching complex thinking as a main competence in educational settings, together with the sub-competencies of critical, systemic, scientific, and creative thinking.

Moraes et al. (2023) emphasized that Education 4.0 is prevalent in higher education. They work together to enhance study opportunities without being constrained by time or geography, boost content immersion, student engagement, and interpersonal contact, decrease costs and risks, replicate real-world work situations, cultivate soft skills, and learn about technologies and their applications.

Conclusion: Industry 4.0 technologies support the entire learning process, but their full potential is not yet realized. They are still mostly limited to academic institutions and manufacturing-related courses. It is envisaged that this effort can help advance Education 4.0 at all teaching levels.

The two major characteristics of Education 4.0 are support for a student-centered model and the transformation of teaching methods.

- ***Support a Student-Centered Model***
  According to Education 4.0, students will have the freedom to select the subjects they wish to study. Furthermore, each student will have a personalized learning experience. In addition, they will need to demonstrate a specific level of expertise before progressing to the next level. Additionally, students will have a choice of learning resources. In addition, the learning process will be more project-based for students to help them build time management and interpersonal skills that will prepare them for employment.
- ***Transformation of Teaching Methods***
  Education 4.0 will improve instructors' abilities to deliver instruction in novel ways. Improving teacher training is cited in research as one way to improve teachers' performance. Teachers are also involved in the development of a comprehensive teacher capacity-building program that combines training with career development.

19.5 RESEARCH DESIGN

The research study employs a cutting-edge literature review procedure. The nature of the chosen technique enables to meet the broad goal of the study with sufficient scientific rigor to support an understanding of each issue.

19.6 EDUCATION 1.0 TO EDUCATION 4.0: THE TRANSITION

The educational system has always adapted to the evolving requirements and beliefs of the times (Aliyu and Talib, 2020). In response to the numerous industrial revolutions brought forth by technological advancements, the education sector has also evolved to remain relevant to the industry during its various transitions. Figure 19.1 summarizes higher education's evolution from Education 1.0 through to the current version of Education 4.0.

Education 1.0 began with the growth of the education industry following the first industrialization at the end of the 18th century. It was characterized by mechanized systems, which led to the development of educationally useful technologies including the typewriter, printing, ballpoint pen, and papermaking machine. This time was notable for its essentialism, behaviorism, and instructive educational ideas. The pupil, on average, had a passive role, while the teacher was revered as a wise man. As a result, the teacher was the center of education during the Education 1.0 period and was in charge of choosing and imparting critical information to their disciples.

Early in the 20th century, which was characterized by mass manufacturing, industrialization, and electricity, Education 2.0 emerged; this was considered to be the second revolution in the education sector. Open-source library resources served as the main information sources throughout this time. The first electronic learning tools, including printers, calculators, and computers, were developed because of the technical breakthrough in Education 2.0. The predominant educational ideas at this time were constructivism and andragogy. The student's function remained passive, while the teacher's role evolved from that of a wise old man to that of a source of information and references to assist in creating the instruments for professional

	Education 1.0	Education 2.0	Education 3.0	Education 4.0
Period	Late 18th Century	Early 20th Century	Late 20th Century	Present
Philosophy	Essentialism, behaviorism, and instructivism	Andragogical, constructivist	Heutagogical, connectivist	Heutagogical, peeragogical and cybergogical
Educator role	Sage	Guide, information source	Orchestrator, curator, and collaborator	Mentor, coach, collaborator, reference
Student role	Largely passive	Emerging active "owning of the knowledge"	Active, "Knowledge ownership", initial independence	Active, high independence, trajectory designer
Approach	Teacher-centered	Peer assessment encouraged, high teacher importance	Co-constructed, first student-centered	Mostly student-centered
Learning outcome	Grades, graduation degree	License to professional practicing	Prepared for practice and scenario analysis	Training of key competencies both soft and hard
Enablers	Mechanical printing, graphite pencil, ballpoint pen, typewriter	First computers, electronic devices and calculators	Computers and widespread use of the internet	ICTs tools and platforms powered by IoT
Information source	Standard texts	Adopted texts and open-source material (physical)	Texts, case studies, second hand experience	Based on online sources
Facilities	Universities / classrooms	Blended laboratories and classrooms	Blended and flexible physical shared spaces	Cyber and physical spaces both shared and individual
Industrial technology	Mechanical systems, steam powered	Mass production, industrialization and electricity	Internet access, automatization and control	Connectivity, digitalization and virtualization

FIGURE 19.1 Transition in education. *Source:* https://ars.els-cdn.com/content/image/1-s2.0-S0045790621002603-gr1_lrg.jpg.

implementation. However, students started to play a more active role and took on the title of "owners of the knowledge."

The close of the 20th century can be considered the era of Education 3.0, which focused mostly on computerization, automation, and control. The student and instructor started adjusting to the idea that learning would no longer require them to be present during a synchronous session because of the new communication era. The teaching–learning processes were reinforced by a variety of materials, including multimedia, internet tools, and virtual laboratories. More heutagogical and connectivist principles were used in this teaching and learning approach. Every teacher was viewed as a partner, curator, and organizer.

What is being referred to at the moment as "Education 4.0" includes the technology, cutting-edge pedagogical approaches, and best practices that characterize the Fourth Industrial Revolution. The works of numerous writers, whose research components were incorporated in this part, served as the infographic sources for defining its structure (Miranda et al., 2021); Salmon, 2019) (Figure 19.2).

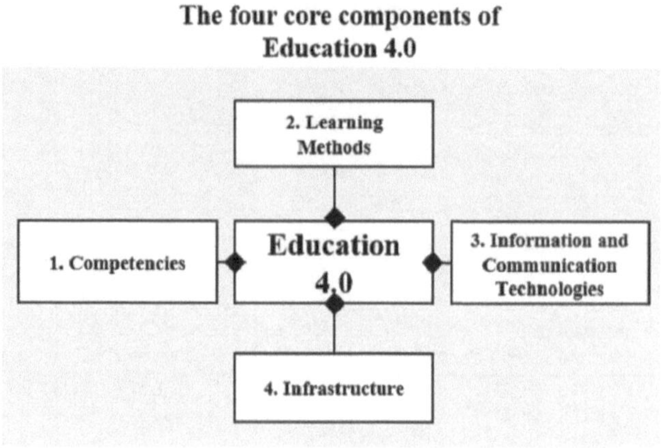

FIGURE 19.2 The four components of education. *Source:* https://ars.els-cdn.com/content/image/1-s2.0-S0045790621002603-fx1_lrg.jpg.

19.7 THE FOUR COMPONENTS OF EDUCATION

i. *Competencies* (the development and training of desired essential capabilities),
ii. *Learning Methods* (adoption of cutting-edge teaching techniques),
iii. *ICTs* (application of latest ICTs), and
iv. *Infrastructure* (utilization of facilities, services, and delivery methods for education)

Furthermore, Figure 19.3 depicts the implementation of the components of education in the Fourth Industrial Revolution of this sector:

1. **Competencies**

   Transversal and disciplinary competencies are the two critical competencies that need to be developed in the higher education sector.

   - **Transversal competencies:** These can be categorized as critical thinking, teamwork, collaboration, communication, creativity, and innovation. The development of these abilities promotes the presentation of ideas for improving any kind of service, process, system, or practice to maximize the utilization of resources. The application of diverse analytical techniques and critical thinking encourages students to get immersed in real-world issues. Activities that foster the individual participation of group members by dividing up work so that each person oversees taking care of a particular aspect of the issue boost cooperation.

     Collaboration is demonstrated through group activities; students must exhibit their ability to interact and work in a team on collaborative projects. Communication competency should encourage students to successfully express their thoughts orally, graphically, or in writing,

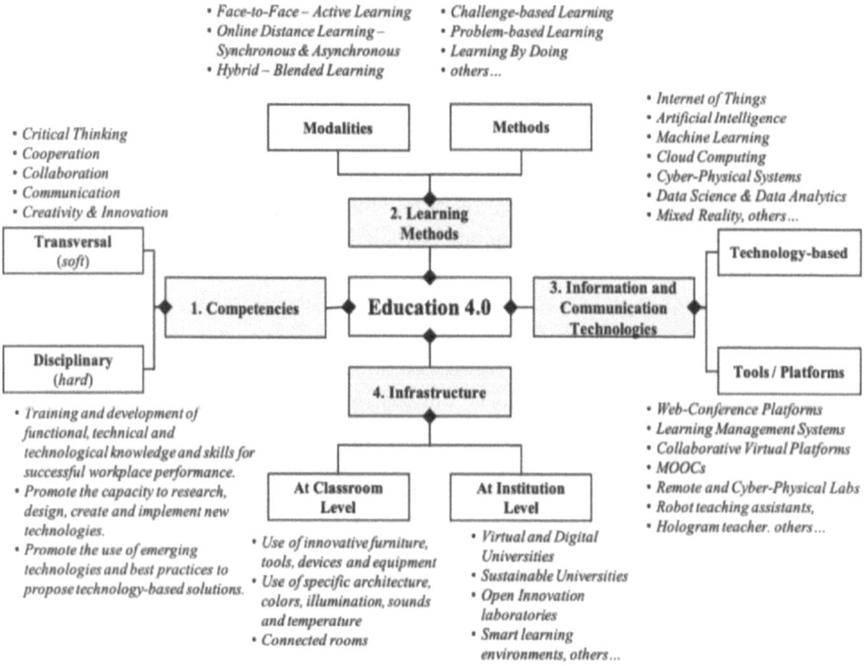

FIGURE 19.3 The framework of Education 4.0. *Source:* https://ars.els-cdn.com/content/image/1-s2.0-S0045790621002603-gr2_lrg.jpg.

as well as via media or any technology tools. Hence, students can apply their developed communication skills in more difficult communication situations involving discussions, proposals, and justifications for projects. Activities that encourage students to build original and creative problem-solving techniques help nurture creativity and innovation.

- **Disciplinary Competencies:** Among them are (i) the instruction and advancement of practical, scientific, and technological knowledge as well as effective workplace performance abilities; (ii) the capacity for research and design; the capacity to develop and use new technologies; (iii) the ability to propose technology-based solutions using emerging technologies.

2. Learning Methods:

The frequently employed learning-delivery modalities are: (i) The foundation of in-person learning is primarily active learning. (ii) Through both asynchronous and synchronous tasks that immerse students in digital models, online distance education makes use of the most up-to-date technology platforms to carry out remote procedures. (iii) Hybrid learning optimizes learning processes and resources through strategies like blended learning or the flipped classroom.

Along with pedagogical strategies like learning through challenges, solving problems, doing, and through games (simulative) learning, creative learning techniques built on teamwork and cooperation are also being increasingly adopted as the learning methods to align teaching pedagogy with the requirements of Industry 4.0.

To fulfill the current requirement for producing a generation of highly competent professionals, new learning methodologies are emerging in Education 4.0. Thus, new programs incorporate learner-centered models that encourage active participation from students and make use of current and evolving technologies to augment their learning. As a result, cooperative and collaborative learning techniques as well as pedagogical strategies including challenge-based, problem-based, gamification-based, and learning by doing are developed.

3. **Information and Communication Technologies**

Kagermann et al. (2013), Lasi et al. (2014), and Nunes et al. (2017) propounded that ICTs have led to new industry opportunities, often referred to as Industry 4.0. As educational institutions transition from a conventional to a student-centered model, the use of ICTs has become crucial, enabling students to learn the skills necessary to adapt to the aforementioned changes (Valentín et al., 2013).

There are two ways to think about the ICT elements in Education 4.0:

- *Technology-based*, which incorporates the basic principles of various methods and technologies to create solutions; and
- *Tools and platforms* are state-of-the-art technological solutions that merge several technologies for management and instructional objectives.

ICT is a group of technical tools that make it easier to access, gather, and distribute information. It lets students and instructors work in collaboration with each other and adds value to teaching pedagogy by incorporating solutions that merge different technologies.

AI, data analytics, virtual image processing, and cloud computing are examples of some of the technological advancements that facilitate providing technology-based solutions. Subsequently, the tools and platforms incorporating these advances, like web conference technologies (Zoom, Webex, and Meets) and learning management systems (Blackboard, Moodle, Canvas, and Google Classroom), have immensely enhanced student-centric learning.

4. **Infrastructure Innovations**

The learning infrastructure at the classroom and institutional levels is designed to meet learning needs and counter challenges in Education 4.0. At this level, it is considered that the classroom should be furnished with cutting-edge furniture, interconnected equipment, and other didactic and educational materials that exhibit certain design traits to positively impact students' learning. The services, facilities, and systems in an educational institution are designed to provide comfort, recreation, sustainability, and accessibility. Post-COVID-19, the adoption of virtual and digital infrastructure was considerably hastened, and online libraries, learning management systems, web conferencing platforms, open innovation laboratories, etc., are now being widely used.

19.8 TEACHER'S COMPETENCIES

Kamaruzzaman (2018), Bermúdez and Juárez (2017), Dasgupta (2012), and Mace et al. (2005) asserted competency as a set of behaviors, capabilities, analytical thinking, data-driven decision capabilities, and information-sharing capabilities. Additionally, competencies are personal talents that demonstrate the capability to regularly execute at a sustained average or above-average level in a particular work function. The three categories of technical, behavioral, and contextual competencies might be used to further categorize these competencies. Contrarily, organizational definitions of competencies include a learning environment, the examination of prior experience, ongoing learning, the process of improving one's skills, the continual generation of new knowledge, the sharing of knowledge and expertise, and quick learning.

HEIs require teachers to have both innovative and practical skills. This is done so that children with various learning preferences will benefit. Furthermore, institutions ought to guarantee that educators are always improving their methods of instruction and meeting the capabilities required by present educational needs. These are techniques for ensuring that the standard of instruction at institutions of higher learning satisfies the qualifications and standards for appointment or advancement, as well as serving as a mechanism for determining the award of exceptional instructors and expert teachers. To reach an active learning outcome, Teaching 4.0 uses a variety of technologies in combination with academically validated instructional approaches (Prieto et al., 2019). There are five teaching competencies that teachers are required to have according to Education 4.0, viz., Functional and Research Competency, Pedagogical Psychology and Assessment, Learning and Mentoring, Digital Competency, and Self-Management and Interactive Competency (Tai et al.,).

Tan et al. (2021), Buasuwan and Studies (2018), and Bermúdez and Juárez (2017) emphasized that the development of the capabilities required for the deployment of Industry 4.0 is taking place in HEIs. To be able to equip learners with an Industry 4.0 mindset or set them up for the future working environment, HEIs must compete with creative, accessible, and adaptable learning environments. As a result, lecturers need to acquire new skills and be able to adapt to the new learning environment, which includes online communities, global connectedness, technological convergence, and digital innovation. Additionally, instructors must be innovative when it comes to the curriculum by offering tasks that are problem-based, integrated, and dependent on others. The four major components of Industry 4.0 suggested by prior research are organizational learning, innovative management, and environment, according to findings from a survey of the literature on ICT.

Teaching 4.0 significantly affects the teaching approach, which allows educators greater room for creativity and innovation. As a result, instructors must know how to implement diverse technologies based on shifting trends and their applicability. The teachers should not completely rely on technology as a teaching tool but rather modify their methods by utilizing differentiated strategies like self-learning techniques, hybrid learning, or online learning, all of which must be adapted to meet student learning requirements.

The teaching and learning processes have benefited from the inclusion of various technologies in Teaching 4.0. To ensure that education is revolutionized, these technologies must be implemented in HEIs. The institutions imparting higher education must make sure that professors and lecturers can utilize a 4.0-based teaching approach and creatively diverse educational innovations. Teachers must possess the following competencies: (i) ability to learn by utilizing various digital media platforms; (ii) critical thinking; (iii) analytical decision-making; (iv) skill competencies; (v) ability to self-evaluate; (vi) individual capacity to attain specific performance; (vii) responsibility; (viii) teaching autonomy; and (ix) capacity to instruct in a productive learning environment.

Brown-Martin (2018) emphasized that in the industrial era of 4.0, teachers must be capable of not only imparting information but should also emphasize the development of skills applicable to the current times. For the improvement of teacher ability during the Industrial Revolution, instructors need to have: (1) educational competency; (2) technical commercialization proficiency; (3) globalization—a world without insulation; speaking about other cultures without stuttering; (4) problem-solving skills; (5) competence in future strategies; and (6) an expert counselor. Indira et al. (2020) suggested that instructors should prepare themselves to complete their duties as creators of knowledge that correlate with Technology 4.0, in addition to acting as knowledge communicators to students. To use 4.0 technologies effectively, instructors must be proficient in technology and understand the issues—including psychological ones—that students face.

Indira et al. (2020) summarized the competencies of teachers in Education 4.0 as follows: (1) Educational competency; (2) Technical commercialization proficiency; (3) Globalization: a world without insulation; speaking about other cultures without stuttering; hybrid; and problem-solving skills; (4) Competence in future strategies because the world is dynamic and quick to change, it can be predicted precisely what will transpire as a result of a plan; (5) An expert counselor.

Buasuwan and Studies (2018) suggested that the development of the capabilities required for the deployment of Industry 4.0 is seen as taking place in HEIs. To equip students with an Industry 4.0 mindset or set them up for the future working environment, HEIs must compete with creative, accessible, and adaptable learning environments. As a result, lecturers need to acquire new skills and be able to adapt to the new learning environment, which includes online communities, global connectedness, technological convergence, and digital innovation (Karakas). Additionally, instructors must be innovative when it comes to the curriculum by offering tasks that are problem-based, integrated, and depend on others.

19.9 CONCLUSION

The chapter elaborates on the distinct characteristics of Industry 4.0, brought on by the rapid technological advancements in the 21stcentury. AI, IoT, robotics, automation, etc., are some of the discoveries that have acted as a spur to this industrial revolution. Consequently, the workforce skill requirements have also undergone a sea change to meet these industry requirements. The education sector is also undergoing a major transformation to train employees who meet the expectations of Industry 4.0.

The changed worker skill set comprises cognitive and functional business skills. The changed Education 4.0 paradigm consists of the incorporation of ICT in education delivery, usage of a newer basket of learning methods, emphasis on developing transversal and disciplinary competencies among the HEI students, and incorporating changes in the infrastructure to provide adequate facilities and support to the delivery of Education 4.0 to train the potential employees of Industry 4.0. Furthermore, HEIs must impart Teaching 4.0 for which they should ensure that the educators equip themselves with innovative and creative skills in designing their teaching pedagogies, be able to instruct with the aid of the latest instructional mediums, possess a global outlook, and be lifelong learners.

REFERENCES

Agrawal, S., Sharma, N., & Bhatnagar, S. (2021). Visão da educação 4.0 Para a indústria 4.0: Tendênciasatuais e visãogeral. *Em Recent Advances in Smart Manufacturing and Materials*, Springer, Singapura, pp. 475–485.

Alhloul, A., & Kiss, E. (2023). Industry 4.0 as a challenge for the skills and competencies of the labor force: A bibliometric review and a survey. *Science*, *4*(3), 34.

Aliyu, F., & Talib, C.A. (2020). Integration of augmented reality in learning chemistry: A pathway for the realization of industrial revolution 4.0 goals. *Journal of Critical Reviews*, *7*(7), 854–859.

Bermúdez, M. D., & Juárez, B. (2017). Competencies to adopt Industry 4.0 for operations management personnel at automotive parts suppliers in Nuevo Leon, In *Proceedings of the International Conference on Industrial Engineering and Operations Management Bogota*, Colombia, pp. 736–747.

Bonfeld, C. A., Salter, M., Longmuir, A., Benson, M., & Adachi, C. (2020). Transformation or evolution? Education 4.0, teaching and learning in the digital age. *Higher Education Pedagogies*, *5*(1), 223–246.

Brown-Martin, G. (2018). Education and the fourth industrial revolution (learning to thrive in a transforming world). In *11th Annual International Conference of Education, Research and Innovation* (p. 7270). Seville, Spain.

Buasuwan, P. J. A. E., & Studies, D. (2018). Rethinking Thai higher education for Thailand 4.0. *Asian Education and Development Studies*, *7*(2), 157–173.

Dasgupta, M. (2012). *Organizational Learning and its Practices: A Review*. SAGE Open.

Daud, W. A. A. W., Teck, W. K., Ghani, M. T. A., & Ramli, S. (2019). The needs analysis of developing a mobile learning application for Cybergogical teaching and learning of Arabic language proficiency. *International Journal of Academic Research in Business and Social Sciences*, *9*(8), 33–46.

Ersoy, M. (2021). An idea for design pedagogy: Devising instructional design in higher education 4.0. *Design and Technology Education: An International Journal*, *26*(1), 118–136.

Fisk, P. (2017). *Education 4.0 the Future of Learning will be Dramatically Different, in School and Throughout Life* [article], available at: www.peterfisk.com/2017/01/future-education-young everyone-taught-together

Gaj, P., Jasperneite, J., & Felser, M. (2012). Computer communication within the industrial distributed environment– A survey. *IEEE Transactions on Industrial Informatics*, *9*(1), 182–189.

Giang, N. T. H., Hai, P. T. T., Tu, N. T. T., & Tan, P. X. (2021). Exploring the readiness for digital transformation in a higher education institution towards industrial revolution 4.0. *International Journal of Engineering Pedagogy*, *11*(2), 4–24.

Hernandez-de-Menendez, M., Morales-Menendez, R., Escobar, C. A., & McGovern, M. (2020). Competencies for industry 4.0. *International Journal on Interactive Design and Manufacturing (IJIDeM)*, *14*, 1511–1524.

Himmetoğlu, B., Ayduğ, D., & Bayrak, C.(2020). Education 4.0: Defining the teacher, the student, and the school manager aspects of the revolution. *Turkish Online Journal of Distance Education*, *21*, 12–28.

Indira, E. W. M., Hermanto, A., & Pramono, S. E. (2020). Improvement of teacher competence in the Industrial Revolution era 4.0. In *International Conference on Science and Education and Technology (ISET 2019)* (pp.350–352). Atlantis Press.

Kagermann, H.,et al. (2013). *Recommendations for Implementing the Strategic Initiative INDUSTRIE 4.0: Securing the Future of the German Manufacturing Industry, Final Report of the Industries4.0 Working Group*, 4(0), 82.

Kagermann, H., Wahlster, W., & Helbig, J. (2013). *Securing the Future of German Manufacturing Industry: Recommendations for Implementing the Strategic Initiative INDUSTRIE 4.0; Final Report of the Industrie 4.0 Working Group.* Forschungsunion: Berlin, Germany, pp. 1–84.

Kamaruzzaman, S. N. (2018). Developing facilities management (FM) competencies for Malaysia: Reference from International Practice. *Journal of Facilities Management*, *16*(2), 157–174.

Lasi, H., Fettke, P., Kemper, H. G., Feld, T., & Hoffmann, M. (2014). Industry 4.0. *Business & Information Systems Engineering*, *6*, 239–242.

LeeBagheri, B., & Kao, H. (2015). A cyber-physical systems architecture for industry 4.0-based manufacturing systems. *Manufacturing Letter*, *3*, 18–23.

Li, L. (2022). Reskilling and upskilling the future-ready workforce for industry 4.0 and beyond. *Information Systems Frontiers*, 1–16. https://doi.org/10.1007/s10796-022-10308-y

Lopez, M., Carro, B., Esguevillas, A. S., & Lloret, J. (2017). Conditional variational autoencoder for prediction and feature recovery applied to intrusion detection in IoT. *Sensor Networks for Collaborative and Secure Internet of Things*, *17*(9), 1967.

Mace, N. L., Coons, D. H., & Weaverdyck, S. E. (2005). *Teaching Dementia Care Skill and Understanding*. Johns Hopkins University Press.

McComas, W. F. (Ed.). (2014). *The Language of Science Education: An Expanded Glossary of Key Terms and Concepts in Science Teaching and Learning*. Rotterdam, The Netherlands: Sense Publisher.

Miranda, J., Navarrete, C., Noguez, J., Molina-Espinosa, J. M., Ramírez-Montoya, M. S., Navarro-Tuch, S. A., & Molina, A. (2021). The core components of education 4.0 in higher education: Three case studies in engineering education. *Computers & Electrical Engineering*, *93*, 107278.

Moraes, E. B., Kipper, L. M., Hackenhaar Kellermann, A. C., Austria, L., Leivas, P., Moraes, J. A. R., & Witczak, M. (2023). Integration of Industry 4.0 technologies with Education 4.0: Advantages for improvements in learning. *Interactive Technology and Smart Education*, *20*(2), 271–287.

Nunes, M. L., Pereira, A. C., & Alves, A. C. (2017). Smart products development approaches for Industry 4.0. *Procedia Manufacturing*, *13*, 1215–1222.

Ouhrir, S., Lotfi, S., & Talbi, M. (2019). Online Peeragogy: Effects of videos developed by students on peer learning and their impact on academic results. *International Journal of Emerging Trends in Engineering Research*, *7*(11), 576–583.

Peña-Jimenez, M., Battistelli, A., Odoardi, C., & Antino, M. (2021). Exploring skill requirements for the Industry 4.0: A worker-oriented approach, *Analesde Psicologia*, *37*(3), 577–588.

Prieto, M. D., Sobrino, Á. F., Soto, L. R., Romero, D., Biosca, P. F., & Martínez, L. R. (2019). Active learning-based laboratory towards engineering Education 4.0. In *Proceedings of the 24th IEEE International Conference on Emerging Technologies and Factory Automation (ETFA)*, Zaragoza, Spain, pp. 776–783.

Qureshi, M. I., Khan, N., Raza, H., Imran, A., & Ismail, F. (2021). Digital technologies in education 4.0. Does it enhance the effectiveness of learning? A systematic literature review. *International Journal of Interactive Mobile Technologies*, *15*(4), 31–47.

Ramírez-Montoya, M. S., Castillo-Martínez, I. M., Sanabria-Z, J., & Miranda, J. (2022). Complex thinking in the framework of Education 4.0 and Open Innovation-a systematic literature review. *Journal of Open Innovation: Technology, Market and Complexity*, *8*(1), 4.

Salmon, G. (2019). May the fourth be with you: Creating Education 4.0. *Journal of Learning Development*, *6*(2), 95–115.

Stock, T., Obenaus, M., Kunz, S., & Kohl, H.(2018). Industry4.0 as enabler for a sustainable development: A qualitative assessment of its ecological and social potential. *Process Safety and Environmental Protection*, *118*, 254–267.

Tai, M. K., Omar, A. K., & Khalip, M. (2021). Development of Teacher Competency Model to Meet the Demands and Challenges of Education 4.0 (Unpublished research report which was funded by Fundamental Research Grant Scheme [Code: FRGS/1/2019/SSI09/UPSI/02/4], Ministry of Higher Education Malaysia).Perak DarulRidzuan, Malaysia: Sultan Idris Education University.

Valentín, A., Mateos, P. M., Gonzalez-Tablas, M. M., Perez, L., Lopez, E., & Garca, I.(2013). Motivation and learning strategies in the use of ICTs among university students. *Computers and Education*, *61*(1), 52–58.

Youngkin, C. A. (2014). The flipped classroom: Practices and opportunities for Health Sciences librarians. *Medical Reference Services Quarterly*, *33*(4), 367–374.

Zainuddin, Z., Zhang, Y., Li, X., Chu, S.K.W., Idris, S., & Keumala, C.M. (2019). Research trends in flipped classroom empirical evidence from 2017 to 2018: A content analysis. *Interactive Technology and Smart Education*, *16*(3), 255–277.

20 Critical Digital Pedagogy
A Collaborative Teaching Approach

J. Koh

20.1 INTRODUCTION

Although critical pedagogy (CP) has been well explored in educational literature, critical digital pedagogy (CDP) is an emerging field (Masood & Haque, 2021). CP was developed by Brazilian educator Paulo Freire in 1968 with the aim to enable individuals to critically analyze and transform the social, political, and economic structures that shape their lives (Freire, 1970). CDP is an extension of that work as a consequence of the influence of digital technologies on education and can be viewed as having four main characteristics (Morris & Stommel, 2018):

1. Its core focus revolves around fostering community and promoting collaboration.
2. To stay inclusive of diverse international perspectives, it demands innovative approaches to rethink how communication and collaboration can transcend cultural and political barriers.
3. It shouldn't be restricted to a solitary perspective; instead, it should embrace a multitude of voices.
4. Its utility and relevance should extend beyond conventional educational institutions.

This chapter delves deeper into the realm of CDP, inviting educators, administrators, and scholars to explore the dynamic intersection between CP and digital technologies. Envisioned as a compass, this chapter sets out to steer readers through the background and evolution of CDP, understanding the theoretical underpinnings, looking at the role of digital Artificial Intelligence (AI) literacy in CDP, followed by exploring how to design learner-centered digital environments, before addressing ethical dilemmas in CDP, and promoting social justice through digital activism before looking at the challenges and future directions.

20.2 BACKGROUND

This section will be split into three sections, the first looking at the underlying theory (CP), the second looking at the context of technology in education, and finally how these two are combined to form an emerging field of CDP.

 DOI: 10.1201/9781003400691-20

20.2.1 Background: CP

To understand CDP, we have to start from CP, which was borne out of Freire's seminary work, *Pedagogy of the Oppressed* (Freire, 1970). The basic premise of the book was an argument against what Freire terms a 'banking model' of education, where the teacher is the depositor (of knowledge) and the students are the depositories. This is also known in modern contexts as the traditional 'sage on a stage' teaching model, where the teacher is seen as the subject matter expert and the students as powerless recipients of the knowledge. The traditional model typically serves a few functions, mainly to impart standardized information packages to students, maintaining order, and is bureaucratically neat. Freire even goes so far as to say that the traditional 'banking model' dehumanizes the students as it takes away any agency of the student. Later researchers such as Henry Giroux have also acknowledged the dehumanizing impact of the traditional model of education and has since suggested that introducing democracy into the classroom is an antidote to this power imbalance (Giroux, 2004).

As such, CP can be viewed as social and educational inequalities from the traditional model of teaching, drawing inspiration from critical theory. CP delves into the interactive dynamics between the act of teaching and the act of learning. Advocates of this approach contend that it constitutes an ongoing cycle of unlearning, learning, and relearning, reflection, evaluation, and the effect that these actions have on the learner, in particular students identified as experiencing oppression (Black & Kirkwood, 2022).

20.2.2 Background: The Historical Context of Technology in Education

The students of today can't imagine a world without technology. They use technology for educational, social, and entertainment purposes, to state a few. The history of technology in education is a rich and evolving one. With the invention of the radio, education leveraged on these mass broadcast mediums for instructional purposes, such as broadcasting educational programs and films to schools. From the time of the radio's invention in 1897, by the end of the 1920s, 176 educational institutions had radio broadcasting licenses (Kentnor, 2015).

The introduction of television and film for education started off as a visual complement to the radio counterparts. However, the evolution of technology in education started to move away from distance education to bring these technologies into the classroom. Television was used as an in-classroom teaching tool for demonstrating and explaining concepts, which resulted in a decline in distance education (Moore et al., 1992).

As personal computers and the internet started becoming more accessible and popular in the latter half of the 21st century, the advent of the digital age introduced a paradigm shift in education. With the proliferation of computers, the internet, and an array of digital tools, classrooms became dynamic spaces where information could be accessed instantaneously, transcending geographical barriers (Singh, 2021). As educators began integrating technology into their pedagogical practices, the promise of enhanced engagement and enriched learning experiences seemed imminent.

However, amidst the excitement surrounding the potential of digital technology, a growing concern emerged (Stanistreet, 2021). The rapid integration of digital tools brought forth a digital divide, further accentuating existing inequalities in education. Students from marginalized backgrounds faced barriers to access, and the digital world inadvertently replicated and sometimes exacerbated the same oppressive structures that CP sought to dismantle.

20.2.3 Background: The Genesis of CDP

The historical context of CP and the integration of technology in education have paved the way for the emergence and development of CDP, which seeks to leverage digital technologies to create inclusive, collaborative, and learner-centered learning experiences. Building on the foundational principles of CP, scholars and educators began to critically examine the impact of digital technologies on teaching and learning. As the first two decades of the 21st century unfolded, and further exacerbated by the outbreak of the COVID-19 pandemic, the CDP movement gained momentum, driven by the urgency to address these issues and transform education in a way that fosters equity, inclusivity, and active citizenship (Boczar & Jordan, 2022).

Despite the perspective in some quarters that technology is the 'silver bullet' for education, CDP challenges this notion that technology alone can be a panacea for educational challenges and instead asserts that pedagogy, informed by critical thought, should guide the integration of digital tools. CDP aims to bridge the digital divide and democratize access to knowledge while questioning the power dynamics inherent in technology use. Hence, educators should cultivate a reflective and questioning stance toward technology, its purposes, and its implications. By doing so, educators can harness the potential of technology to foster meaningful connections, facilitate knowledge creation, and promote social justice.

In the digital age, media literacy and digital literacy have become essential skills, enabling learners to navigate the vast sea of information critically. CDP recognizes the importance of nurturing these literacies, empowering students to become discerning consumers and active contributors to the digital realm. As education starts to delve into the age of AI-enabled education, the importance of critical thinking and genuine use of technologies becomes all the more relevant.

As CDP continues to evolve, it responds dynamically to the changing educational landscape and the emergence of new technologies. It adapts its principles to accommodate innovative teaching practices, recognizing that effective integration of digital tools is contingent upon an ongoing dialogue between theory and praxis.

20.3 THEORETICAL UNDERPINNINGS OF CDP

As earlier described in the introduction, CDP can be viewed through the lens of community and collaboration, openness to diverse voices, the gathering of a multitude of perspectives, and applicability outside the traditional institutions (Morris & Stommel, 2018). Taking into further consideration from the works of Giroux and Freire, and contextualizing it in the context of AI, equitable distribution of power and criticality are core underpinnings of CDP. As such, a brief exploration of each theoretical underpinning will be explored and contextualized for the modern age in light of AI.

Community and Collaboration: CDP centers its practice on community and collaboration, recognizing that learning is a social process. This approach fosters collaboration among learners, promotes active engagement, and encourages the exchange of diverse perspectives and ideas. Although collaborative learning has been well explored, CDP goes one step further to incorporate community into the collaborative learning. CDP centers its practice on community, recognizing that learning is a social process that thrives in environments that allow learners to engage with one another, share ideas, and learn from diverse perspectives in the active participation, peer learning, and the co-construction of knowledge (Morris & Stommel, 2018).

Openness to Diverse Voices: Understandably, for community and collaboration to function, an attitude and commitment to remaining open to diverse, international voices, valuing the contributions of individuals from different cultural and political backgrounds is paramount. It encourages learners to critically examine their own assumptions, biases, and preconceived notions (Morris & Stommel, 2018). By engaging with diverse voices, learners gain a deeper appreciation for different cultural, social, and intellectual perspectives, fostering empathy and promoting a more comprehensive understanding of the world. This is especially relevant in today's AI-enabled education age. Most of the AI systems and technologies are developed within specific cultural, social, and educational contexts (Holmes & Tuomi, 2022). Openness to diverse voices also encourages educators to reimagine the ways in which communication and collaboration can happen across cultural and political boundaries and to carefully consider the cultural and contextual implications of AI systems, ensuring that they are inclusive and equitable for all learners.

Gathering a Multitude of Perspectives: CDP rejects the notion of being defined by a single voice and values the multitude of perspectives. This approach encourages critical thinking, challenges dominant narratives, and promotes a more nuanced understanding of complex issues. In the age of AI, and especially that of generative AI, CDP will encourage a questioning of the responses that AI provides, pushing learners and educators to look beyond just the outputs from AI, but questioning the biases of the algorithms and if there are underlying oppressive structures that cause the AI to generate outputs discriminating against certain learners and individuals based on various natural factors.

Application Outside Traditional Institutions: CDP extends beyond traditional institutions of education, acknowledging that learning can happen in a variety of contexts and settings, both formal and informal. Digital technologies provide opportunities for learning that transcend the boundaries of physical classrooms. CDP in non-traditional learning environments emphasizes the importance of creating open and liberating digital spaces for learning. This is evidenced by the increasing use of open digital spaces such as asynchronous forums and synchronous chat channels, although the use of these open digital spaces by educators remains largely a 'banking model'. By assigning the roles of critical co-investigators of knowledge to both learners and educators, a journey of a co-construction of knowledge is embarked on, allowing for collaborative and transformative learning experiences (Masood & Haque, 2021). In the context of AI, CDP recognizes the dehumanizing effect of technology and the need for learners to be empowered to use technology in ways that promote social justice and equity, resulting in a more humane and humanistic use of AI technology (Koseoglu et al., 2023).

Equitable Distribution of Power: Equitable distribution of power emphasizes the importance of challenging and transforming existing power dynamics in educational settings to promote social justice and create a more inclusive learning environment. It empowers learners to critically analyze and transform the world around them, preparing them for active participation in a global society. CDP argues that because learning in the modern age is a process of discovery, educators should act as guides for students and travel with them on this journey of learning. In this learning journey, learners have agency, voice their perspectives, and contribute to the co-construction of knowledge. This shift in power dynamics fosters a sense of ownership and engagement, allowing learners to take control of their learning experiences (Abraham, 2014). The equitable distribution of power achieves another objective of CDP, which is furthering social justice. CDP seeks to address and dismantle oppressive structures that marginalize certain individuals or groups based on factors such as race, gender, socioeconomic status, or ability. By redistributing power, CDP aims to create a more equitable and inclusive educational experience for all learners (Aronson & Laughter, 2016).

Criticality: Criticality is a core theoretical underpinning of CDP, encouraging learners to question, analyze, and challenge existing knowledge and power structures (Bontly et al., 2017), engaging in reflective and nuanced thinking about the subject matter, promoting a critical consciousness and an awareness of the social, political, and ethical implications of digital technology. Criticality prompts educators to consider how digital tools can be leveraged to promote social justice, equity, and meaningful learning experiences and brings learners and educators together to critically examine the ways in which digital technologies shape communication, collaboration, and knowledge production. Especially with the advent of AI, criticality allows learners to have a transformative and emancipatory educational experience, empowering learners to become active participants in their own education and challenging dominant narratives.

20.4 THE ROLE OF DIGITAL AI LITERACY IN CDP

As its name suggests, CDP has a strong digital component in the pedagogy, which also underscores the role that digital literacy plays in CDP, as it empowers learners to navigate the complexities of the digital realm with critical consciousness. Digital literacy is "ability to access, manage, understand, integrate, communicate, evaluate and create information safely and appropriately through digital technologies for employment, decent jobs and entrepreneurship" (Law et al., 2018). However, with AI's ubiquitous influence in education, the examination of the role of digital literacy has to evolve accordingly. Hence the role of digital AI literacy should be explored in the context of CDP. That said, given the newness of this field, there are no clear consensus on the definition of what constitutes digital AI literacy (Ng et al., 2021). Looking through the literature, digital AI literacy can be classified into three elements, namely Critical Consumption of Information, AI Ethics, and Authentic Use of AI. However, framed in the context of CDP, the role that digital AI literacy plays is even more critical, promoting social justice and the co-construction of knowledge within the elements of digital AI literacy.

Critical Consumption of Information: In a world saturated with information, digital literacy equips learners with the skills to critically evaluate and assess the credibility, reliability, and bias of online sources. This also includes critically challenging the notion that technology is inherently neutral or deterministic. Learners critically examine the role of technology in education, with the appreciation that technology is shaped by human values, biases, and interests. Only then can students become active participants in shaping the direction and purpose of digital tools. By encouraging students to question the information they encounter, CDP nurtures a generation of discerning individuals who are less susceptible to misinformation and manipulation.

AI Ethics looks at not only the understanding and discussion of the data ethics and data privacy, but also the discourse around inclusivity and accessibility. In the age of big data and constant digital tracking, digital AI literacy empowers learners to understand their rights and responsibilities regarding data privacy. CDP questions the underlying oppressive structures that determine the digital ethics, emphasizing the importance of protecting personal information and understanding the potential consequences of digital footprints. But going beyond digital ethics, CDP also tries to promote social justice in addressing the digital divide and ensuring inclusivity in education. By equipping learners with the necessary digital skills, educators can bridge the gap between students from different socioeconomic backgrounds and provide equal opportunities for all to participate meaningfully in digital learning experiences.

Authentic Use of AI looks at empowering learners to use AI in a responsible manner that can solve real-world problems. This includes "designing, developing, and deploying AI systems and solutions with good intentions to empower employees and businesses, and fairly impact society" (Accenture, 2021). In the context of CDP, it also means empowering students to be co-constructors of knowledge. Students express their ideas, perspectives, and creativity through various digital media forms, aided by AI to speak their voice. This empowerment fosters a sense of agency, ownership, and self-efficacy, as learners become active participants and contributors to their own learning journey in a digital landscape.

20.5 DESIGNING LEARNER-CENTERED DIGITAL ENVIRONMENTS

Creating inclusive and learner-centered digital learning spaces is essential for promoting equitable and effective learning experiences. Although CDP as a design philosophy is still relatively new, there are some strategies that can be considered, especially in light of the upcoming advent of AI into the classroom.

Inclusive Safe Learning Spaces: To facilitate the active participation and discourse that CDP requires, fit-for-purpose learning spaces have to be set up. These learning spaces should be designed based on a tripartite framework addressing the diverse physiological, cognitive, and cultural needs of learners (Holeston, 2020). What this means in practice is a learning space that learners can feel safe to share their views, have respectful and healthy discourse, and facilitate collaborative knowledge co-construction and exploration. Although many online spaces do have such features embedded in them, such as forums, these are not the main online learning space where learning and knowledge construction occur. With AI now available to be implemented in the classroom, peer collaboration can be further enhanced. For instance, a (human) teacher could initiate conversations around a particular topic by providing students with opportunities to engage in meaningful conversations with their peers. An AI-powered chatbot could then also act as an initial knowledge repository for referencing and the provision of information. With that knowledge and the appreciation of possible biases of the AI system, teachers and students can then engage in a robust debate of ideas, and in the process co-construct the relevant knowledge.

Re-designing Learning Spaces also includes rethinking the design of the learning platform itself. This includes the materials, the user interface, and the accessibility of the platform itself. For instance, the learning platform could use larger fonts and increased font spaces as these have been shown to be more accessible for dyslexic learners (Duranovic et al., 2018). Other methods of promoting inclusivity and accessibility could also mean integrating AI-powered chatbots. Students feel less self-conscious when dealing with chatbots (Drouin et al., 2022), lowering barriers to participating in the discussions, and empowering them to contribute critically.

Teaching Practice Redesign is also important in this re-designing process. Although there is recognition that current online learning spaces need a rethink of the relevant teaching practices (Koh & Hulbert, 2022), these have not fully translated into the adoption of CDP in the teaching practices. For instance, the Community of Inquiry (CoI) model (Garrison et al., 1999) recognizes the importance of social learning influences, but does not sufficiently take into account the need for critical discourse and dialogic approaches. In the coming AI age, a CDP approach will involve providing students with a sense of belonging, offering an open and welcoming environment, promoting active listening, and creating a psychologically safe environment where students feel comfortable sharing their thoughts and ideas. To integrate AI into the new teaching process, frameworks such as the AI-Teacher spectrum (Koh et al., 2023) can be used as a guide to facilitate the dialogic process among AI, the learner, and the educator. For instance, in group discussions, AI chatbots probe students by asking more open-ended questions to encourage further dialogue and conversation. Human teachers can then move up the AI-Teacher spectrum by encouraging the critical thinking process via relevant contextual questions. AI chatbots can then further support the teacher's prompts by providing personalized feedback based on the students' responses, demonstrating a conversation between themselves, the AI and the human teacher working in tandem to guide them toward deeper thinking and the development of higher-order thinking skills.

20.6 ADDRESSING ETHICAL DILEMMAS IN CDP

Despite the strong emphasis that CDP places on ethics and social justice, the approach is not without its ethical dilemmas. Two of the more prominent ethical dilemmas, especially in the age of AI, are the inclusive learner-centered practices vs facilitating robust discourse and prescriptive instruction vs facilitating discourse.

20.6.1 Inclusive Learner-Centered Practices vs Facilitating Robust Discourse

The ethical dilemma between freedom of speech and safe inclusive spaces is a complex issue. On the one hand, freedom of speech is a fundamental aspect of fostering open dialogue, robust debate, and the free exchange of ideas (Bonotti & Seglow, 2022). On the other hand, creating safe and inclusive spaces is essential for promoting a learning environment where all individuals feel respected, valued, and free from harm or discrimination (Aronson & Laughter, 2016). The tension between these two principles poses an ethical dilemma for CDP practitioners. CDP practitioners have to balance the encouragement of debates and discourse, which may result in certain speech that may perpetuate harm, marginalize certain groups, or create an unsafe environment with the need for space and inclusive spaces, while some argue that restricting speech can infringe on individuals' rights and limit intellectual diversity (Behrent, 2020). The challenge becomes even more complex with online learning. Studies have shown that online education may cause some students to behave more poorly online (Hanno et al., 2022). With AI in the mix, this complexity is further exacerbated. With the prevalence of AI-powered applications, students may use AI-powered text generation or chatbots to impersonate teachers or classmates in digital communications, causing confusion, spreading false information, or damaging relationships within the educational community.

Addressing this ethical dilemma requires careful consideration and a nuanced approach. It involves finding a balance between protecting individuals' right to express their opinions and ensuring that the learning environment is inclusive, respectful, and free from harm. This may involve implementing policies and practices that promote respectful discourse, providing clear guidelines for acceptable behavior, and fostering a culture of open-mindedness and empathy (Aronson & Laughter, 2016).

20.6.2 Prescriptive Instruction vs Facilitating Discourse

The other ethical dilemma in CDP is the tension between prescriptive instruction and facilitating discourse. In the case of academically weaker students who need structure and guidance to understand concepts and complete tasks, prescriptive instruction can help ensure that learners acquire foundational knowledge and skills efficiently (Maniates, 2010). The step-by-step approach is familiar to many students but may cause an over-dependency on blindly following instructions and limiting learners' autonomy, critical thinking, and creativity. A critical discussion may prove as a barrier to access for these students in terms of participation and contribution. On the other hand, although facilitating discourse promotes active learning, collaborative problem-solving, and the development of critical thinking skills, it also faces resource challenges. However, facilitating discourse may require more time and effort, and it can be challenging to manage diverse perspectives and ensure productive discussions (Harbour & Denham, 2021). In the age of generative AI, CDP practitioners will face the dilemma of being more prescriptive to manage academic misconduct. In collaborative group work, students may use AI-generated content, such as text or graphics, to contribute less to group projects or manipulate results, putting an unequal workload on their peers and undermining collaborative skills. And because it can be so difficult to differentiate between AI-generated content and authentic AI use, teachers also face the dilemma of having to be more prescriptive in their instruction, robbing learners of their self-efficacy in their learning and redistributing power back to the teachers.

To address this dilemma, educators can incorporate a variety of instructional strategies, such as blended learning and flipped classroom approaches (Weinhandl et al., 2020), or incorporate learning frameworks like the AI-Teacher spectrum (Koh et al., 2023) that combine prescriptive instruction with opportunities for discourse. CDP practitioners can also draw upon ethical decision-making models, such as the American Counselling Association's (ACA) ETHICS model (Ling & Hauck, 2017) to navigate ethical dilemmas and make informed choices in their instructional practices.

20.7 PROMOTING SOCIAL JUSTICE THROUGH DIGITAL ACTIVISM

Promoting social justice through digital activism is a key aspect of CDP (Morris & Stommel, 2018). Digital activism utilizes social media and online platforms to raise awareness, build communities, unite ideas, and share narratives related to social justice issues. CDP can encourage learners to engage with digital activism by amplifying marginalized voices and promoting narratives that challenge social injustices. Students are encouraged to take in other perspectives and experiences, questioning the power structures that oppress certain group(s).

CDP can facilitate digital activism via the creation of inclusive and learner-centered digital learning spaces where learners can engage in discussions, share resources, and collaborate on projects related to social justice. These spaces are also fertile grounds for learners to use their digital skills to create campaigns, raise funds, and engage in advocacy efforts that address social injustices.

Although CDP can help promote and facilitate digital activism, it is important to note that digital activism also comes with challenges and risks, such as the spread of disinformation and the potential for governments to suppress activists (Powell, 2022). To some extent, CDP can minimize these challenges by promoting media literacy, critical evaluation of information, and responsible digital citizenship, but the risks remain. By integrating digital activism into CDP, educators can empower learners to become active participants in promoting social justice, fostering empathy, critical thinking, and civic engagement. This approach aligns with the principles of CDP, which aim to challenge oppressive systems and create inclusive and equitable learning environments.

20.8 CHALLENGES AND FUTURE DIRECTIONS

As with any new pedagogy, CDP faces significant challenges as well. The first challenge is that of a resistance to change. CDP requires a shift in the traditional approach to teaching and learning, which can be challenging for some educators and learners. Overcoming resistance to change requires effective communication, collaboration, and support from stakeholders (Young, 2018). Despite the swath of literature on managing change, however, moving from traditional well-established pedagogy to CDP could prove difficult due to the newness of the pedagogy, lack of training and infrastructure resources allocated, and management enthusiasm (Mincu, 2022). Student reception to these new ways of learning may also prove to be mixed, given the amount of responsibility that learners have to take for their own learning (Boronski, 2021).

As CDP is underpinned by digital technologies, the digital divide will remain a significant challenge. Not all learners have equal access to technology and internet connectivity, which can limit their ability to participate in digital learning environments (Stanistreet, 2021). As the world becomes more interconnected digitally, addressing the digital divide requires implementing policies and practices that promote equitable access to technology and internet connectivity.

Looking forward to the near future, the next frontier of CDP would be how AI now factors into this. Frameworks like AI-Teacher spectrum (Koh et al., 2023) can inform how learning tasks and material can be redesigned to reflect CDP principles in the integration of AI. Discussions around the design of AI technologies such as chatbots (Koh et al., 2022) could also reflect CDP principles and its subsequent implementation. However, the challenge of integrating AI into CDP remains, given that frameworks regarding AI in education have yet to be fully formed.

As AI plays a greater role in education, the need for digital AI literacy also becomes the direction for the near future. CDP can continue to foster digital AI literacy by promoting critical thinking, responsible digital citizenship, and an understanding of the social and ethical implications of digital technologies (Morris & Stommel, 2018). To this end, an increasing knowledge of AI technologies needs to be incorporated in the classrooms of today (Ng et al., 2021), developing the knowledge and skills needed to understand, use, evaluate, and ethically navigate AI. Educators can draw upon the AI literacy model, which focuses on four core areas of AI literacy: knowing and understanding AI, using and applying AI, evaluating and creating AI, and AI ethics.

20.9 CONCLUSION

In conclusion, CDP emerges as a transformative force, bridging CP with digital technologies to reshape education in the digital era. By nurturing digital AI literacy, CDP empowers learners to critically consume information and create digital content responsibly. It goes beyond technical proficiency, fostering inclusive and equitable learning environments, where students actively co-create knowledge. Despite the various challenges, CDP envisions a future where technology serves as a tool for liberation, promoting lifelong learning and compassionate global citizenship. As educators, we must embrace the horizon of CDP, guiding us toward a transformative educational landscape that prioritizes critical thinking, digital literacy, and the limitless potential of human creativity. Together, we can shape an inclusive and empowered generation of learners, redefining education for generations to come.

REFERENCES

Abraham, Getahun Yacob. (2014). *Critical Pedagogy: Origin, Vision, Action and Consequences. KAPET, 10*(1), 90–98. ISSN 1653-4743

Accenture. (2021). *AI Ethics & Governance*. Accenture.

Aronson, B., & Laughter, J. (2016). The theory and practice of culturally relevant education. *Review of Educational Research, 86*(1), 163–206. https://doi.org/10.3102/0034654315582066

Behrent, Michael. (2020). *"A Tale of Two Arguments about Free Speech on Campus."* AAUP. https://www.aaup.org/article/tale-two-arguments-about-free-speech-campus

Black, Ernest, & Kirkwood, Kirk. (2022). *Twice as Good to Get Half: Content and Context of Black Male Teachers and Administrators*, pp. 17–27. https://doi.org/10.4018/978-1-7998-9000-3.ch002.

Boczar, Amanda, & Jordan, Sydney. (2022). Continuity during COVID: Critical digital pedagogy and special collections virtual instruction. *IFLA Journal, 48*(1): 99–111. https://doi.org/10.1177/03400352211023795.

Bonotti, Matteo, & Seglow, Jonathan. (2022). Freedom of speech: A relational defence. *Philosophy & Social Criticism, 48*(4): 515–29. https://doi.org/10.1177/01914537211073782.

Bontly, Susan, Khalil, Samar, Mansour, Tahani, & Parra, Julia Lynn. (2017). Starting the conversation: A working definition of critical digital pedagogy. In *Society for Information Technology & Teacher Education International Conference (SITE)*. Austin, TX, 383–388.

Boronski, Tomas. (2021). *Critical Pedagogy*. London: Routledge. https://doi.org/10.4324/9781315101811.

Drouin, Michelle, Sprecher, Susan, Nicola, Robert, & Perkins, Taylor. (2022). Is chatting with a sophisticated Chatbot as good as chatting online or FTF with a stranger? *Computers in Human Behavior, 128*(March): 107100. https://doi.org/10.1016/j.chb.2021.107100.

Duranovic, Mirela., Senka, Smajlagic., Babic-Gavric Branka. (2018) Influence of increased letter spacing and font type on the reading ability of dyslexic children. *Ann Dyslexia, 68*(3): 218–228. https://doi.org/10.1007/s11881-018-0164-z.

Freire, Paulo. (1970). *Pedagogy of the Oppressed*. London: Penguin Classics.

Garrison, D. Randy, Anderson, Terry, & Archer, Walter. (1999). Critical inquiry in a text-based environment: Computer conferencing in higher education. *Internet and Higher Education, 2*(2–3): 87–105. https://doi.org/10.1016/S1096-7516(00)00016-6.

Giroux, Henry A. (2004). Critical pedagogy and the postmodern/modern divide: Towards a pedagogy of democratization. *Teacher Education Quarterly, 31*(1): 31–47.

Hanno, Emily C., Fritz, Lily S., Jones, Stephanie M., & Lesaux, Nonie K. (2022). School learning format and children's behavioral health during the COVID-19 pandemic. *JAMA Pediatrics, 176*(4): 410. https://doi.org/10.1001/jamapediatrics.2021.5698.

Harbour, Kristin E., & Denham, André R. (2021). Supporting all students: Productive mathematical discourse in online environments. *Intervention in School and Clinic, 57*(2): 87–94. https://doi.org/10.1177/10534512211001849.

Holeston, Richard. (2020). Toward inclusive learning spaces: physiological, cognitive, and cultural inclusion and the Learning Space Rating system. *EDUCAUSE Review*. https://er.educause.edu/articles/2020/2/toward-inclusive-learning-spaces

Holmes, Wayne, & Tuomi, Ilkka. (2022). State of the art and practice in AI in education. *European Journal of Education, 57*(4): 542–70. https://doi.org/10.1111/ejed.12533.

Kentnor, Hope. (2015). Distance education and the evolution of online learning in the United States. *Curriculum and Teaching Dialogue, 17*(1 & 2), 21–34.

Koh, Josiah, M. Cowling, M. Jha, & K. N. Sim. (2022). Work-in-progress-exploring model based automated response systems to enhance student outcomes. In *8th International Conference of the Immersive Learning Research Network (ILRN)*, pp. 131–33. https://doi.org/10.23919/iLRN55037.2022.9815987.

Koh, J., Cowling, M., Jha, M., & Sim, K. N. (2023). *The human teacher, the AI teacher and the AIed-teacher relationship. Journal of Higher Education Theory and Practice, 23*(17). https://doi.org/10.33423/jhetp.v23i17.6543

Koh, Josiah, & Tara Hulbert. (2022). The role of nonverbal communication in asynchronous talk channels. *Journal of Open, Flexible and Distance Learning, 26*(2): 29–50. https://www.jofdl.nz/index.php/JOFDL/article/view/517/353.

Koseoglu, Suzan, Veletsianos, George, & Rowell, Chris. (2023). *Critical Digital Pedagogy in Higher Education*. Athabasca University Press.

Law, Nancy, Woo, David, de la Torre, Jimmy, & Wong, Gary. (2018). *A Global Framework of Reference on Digital Literacy Skills for Indicator 4.4.2.* UNESCO Institute for Statistics. https://unesdoc.unesco.org/ark:/48223/pf0000265403.locale=en.

Ling, Thomson J., & Hauck, Jessica M. (2017) The ETHICS model: Comprehensive, ethical decision making. *Vistas Online*. Article 18, 1–12.

Maniates, Helen. (2010) *When Highly Qualified Teachers Use Prescriptive Curriculum: Tensions Between Fidelity and Adaptation to Local Contexts*. PhD in Education, University of California, Berkeley. https://escholarship.org/content/qt85w2j5fb/qt85w2j5fb.pdf?t=mtg9dp.

Masood, Marwa Mohammad, & Haque, Md. Mahmudul. (2021). From critical pedagogy to critical digital pedagogy: A prospective model for the EFL classrooms. *Saudi Journal of Language Studies, 1*(1): 67–80. https://doi.org/10.1108/SJLS-03-2021-0005.

Mincu, Monica. (2022). Why is school leadership key to transforming education? Structural and cultural assumptions for quality education in diverse contexts. *Prospects, 52*(3–4): 231–42. https://doi.org/10.1007/s11125-022-09625-6.

Moore, Michael G., Verduin, John R., & Clark, Thomas A. (1992). Distance education: The foundations of effective practice. *The Journal of Higher Education, 63*(4): 468. https://doi.org/10.2307/1982124.

Morris, Sean Michael, & Stommel, Jesse. (Ed.) (2018). An urgency of teachers: The work of critical digital pedagogy. In *Hybrid Pedagogy*. PressBook.

Ng, Davy Tsz Kit, Leung, Jac Ka Lok, Chu, Samuel Kai Wah, & Shen, Qiao Maggie. (2021). Conceptualizing AI literacy: An exploratory review. *Computers and Education: Artificial Intelligence,* 2: 100041. https://doi.org/10.1016/j.caeai.2021.100041.

Powell, Catherine. (March 2022). *The Promise of Digital Activism-and Its Dangers*. Council on Foreign Relations.

Singh, Mayengbam Nandakishwor. (2021). Inroad of digital technology in education: Age of digital classroom. *Higher Education for the Future, 8*(1): 20–30. https://doi.org/10.1177/2347631120980272.

Stanistreet, Paul. (September 2021). Closing the digital divide: It's about more than access. *Only Connect: The Lifelong Learning Blog from UIL*. UNESCO Institute for Lifelong Learning.

Weinhandl, Robert, Lavicza, Zsolt, and Houghton, Tony. (2020). Designing online learning environments for flipped approaches in professional mathematics teacher development. *Journal of Information Technology Education: Research, 19*: 315–37. https://doi.org/10.28945/4573.

Young, Cherelle. (September 2018). *Critical Digital Pedagogy in the Modern Classroom: Expectations vs Reality*. Teaching Hub.

Index